I0868899

Advances in Electronic Government, Digital Divide, and Regional Development (AEGDDRD) Book Series

ISSN:2326-9103
EISSN:2326-9111

Editor-in-Chief: Zaigham Mahmood, University of Derby, UK & North West University, South Africa

MISSION

The successful use of digital technologies (including social media and mobile technologies) to provide public services and foster economic development has become an objective for governments around the world. The development towards electronic government (or e-government) not only affects the efficiency and effectiveness of public services, but also has the potential to transform the nature of government interactions with its citizens. Current research and practice on the adoption of electronic/digital government and the implementation in organizations around the world aims to emphasize the extensiveness of this growing field.

The Advances in Electronic Government, Digital Divide & Regional Development (AEGDDRD) book series aims to publish authored, edited and case books encompassing the current and innovative research and practice discussing all aspects of electronic government development, implementation and adoption as well the effective use of the emerging technologies (including social media and mobile technologies) for a more effective electronic governance (or e-governance).

COVERAGE

- Public Information Management, Regional Planning, Rural Development
- E-Citizenship, Inclusive Government, Connected Government
- ICT Infrastructure and Adoption for E-Government Provision
- Case Studies and Practical Approaches to E-Government and E-Governance
- Citizens Participation and Adoption of E-Government Provision
- ICT within Government and Public Sectors
- Knowledge Divide, Digital Divide
- Electronic Government, Digital Democracy, Digital Government
- Social Media, Web 2.0, and Mobile Technologies in E-Government
- E-Governance and Use of Technology for Effective Government

IGI Global is currently accepting manuscripts for publication within this series. To submit a proposal for a volume in this series, please contact our Acquisition Editors at Acquisitions@igi-global.com or visit: http://www.igi-global.com/publish/.

The Advances in Electronic Government, Digital Divide, and Regional Development (AEGDDRD) Book Series (ISSN 2326-9103) is published by IGI Global, 701 E. Chocolate Avenue, Hershey, PA 17033-1240, USA, www.igi-global.com. This series is composed of titles available for purchase individually; each title is edited to be contextually exclusive from any other title within the series. For pricing and ordering information please visit http://www.igi-global.com/book-series/advances-electronic-government-digital-divide/37153. Postmaster: Send all address changes to above address. Copyright © 2023 IGI Global. All rights, including translation in other languages reserved by the publisher. No part of this series may be reproduced or used in any form or by any means – graphics, electronic, or mechanical, including photocopying, recording, taping, or information and retrieval systems – without written permission from the publisher, except for non commercial, educational use, including classroom teaching purposes. The views expressed in this series are those of the authors, but not necessarily of IGI Global.

Titles in this Series

For a list of additional titles in this series, please visit:
http://www.igi-global.com/book-series/advances-electronic-government-digital-divide/37153

The COVID-19 Pandemic and the Digitalization of Diplomacy
Floribert Patrick Calvain Endong (University of Dschang, Cameroon)
Information Science Reference • © 2023 • 263pp • H/C (ISBN: 9781799883944) • US $235.00

Handbook of Research on Current Advances and Challenges of Borderlands, Migration, and Geopolitics
Rui Alexandre Castanho (WSB University, Poland)
Information Science Reference • © 2023 • 444pp • H/C (ISBN: 9781668470206) • US $270.00

Smart Village Infrastructure and Sustainable Rural Communities
Mohammad Ayoub Khan (University of Bisha, Saudi Arabia) Bhumika Gupta (Govind Ballabh Pant Institute of Engineering & Technology, Pauri Garhwal, India) Agya Ram Verma (Govind Ballabh Pant Institute of Engineering & Technology, Pauri Garhwal, India) Pushkar Praveen (Govind Ballabh Pant Institute of Engineering and Technology, India) and Cathryn J. Peoples (Ulster University, UK)
Engineering Science Reference • © 2023 • 315pp • H/C (ISBN: 9781668464182) • US $250.00

Handbook of Research on Cyber Approaches to Public Administration and Social Policy
Fahri Özsungur (Mersin University, Turkey)
Information Science Reference • © 2022 • 692pp • H/C (ISBN: 9781668433805) • US $315.00

Advances in Deep Learning Applications for Smart Cities
Rajeev Kumar (Teerthanker Mahaveer University, India) and Rakesh Kumar Dwivedi (Teerthanker Mahaveer University, India)
Engineering Science Reference • © 2022 • 335pp • H/C (ISBN: 9781799897101) • US $250.00

For an entire list of titles in this series, please visit:
http://www.igi-global.com/book-series/advances-electronic-government-digital-divide/37153

IGI Global
PUBLISHER of TIMELY KNOWLEDGE

701 East Chocolate Avenue, Hershey, PA 17033, USA
Tel: 717-533-8845 x100 • Fax: 717-533-8661
E-Mail: cust@igi-global.com • www.igi-global.com

Editorial Advisory Board

Ansary Ahmed, *Asia e University, Malaysia*
Afshar Alam, *Jamia Hamdard University, India*
Narend Baijnath, *Commonwealth of Learning, Vancouver, Canada & Council on Higher Education, South Africa*
Taner Dursun, *TÜBITAK, Turkey*
Salil Kanhere, *University of New South Wales, Australia*
Kinshuk, *University of North Texas, USA*
Liew Teik Kooi, *Help University, Malaysia*
Srinivas Mahankali, *Master Mentors Advisory Pvt. Ltd., India*
Muhsin Menekse, *Purdue University, USA*
Alexander Mikroyannidis, *Knowledge Media Institute, The Open University, UK*
Vis Naidoo, *Southern Alberta Institute of Technology, Canada*
Isa Sertkaya, *BILGEM Blockchain Research Lab, Turkey*
Ahmed Tlili, *Beijing Normal University, China*
John Traxler, *University of Wolverhampton, UK*
Junhong Xiao, *Shantou Radio and Television University, China*

Table of Contents

Detailed Table of Contents

This chapter looks at blockchain technology from a legal perspective and discusses the opportunities, risks, and challenges that surround the use of this technology. It adopts a three-pronged approach to discussing blockchain technology, how it can be leveraged, and the challenges that must be addressed before it can be adopted. The three areas of discussion are 1) legal and regulatory compliance issues; 2) the various business applications, possible risks, and how they are addressed, and 3) the possibilities for harnessing this technology for well-being and welfare of the general public.

Blockchain technology is highly regarded for its potential to revolutionize various economic sectors and meet consumer needs. In the tourism industry, blockchain has attracted investor attention due to its characteristics, such as decentralisation, disintermediation, security, transparency, and immutability. Blockchain applications like smart contracts, tokens, cryptocurrencies, decentralised applications (Dapps), and decentralised autonomous organisations (DAOs) have the potential to transform the

period of using online web pages in supplier selection. In this research, purchasing processes in industrial markets and new area digital transformation processes are examined.

Blockchain technology, which is one of the powerful tools that can be used in the development of education systems, is among the technologies that are abstained from for environmentally friendly activities due to its high level of energy consumption and gas emissions. However, there is not enough data that blockchain technology harms the environment or consumes more energy than other technologies. It is thought that this technology, which is more flexible than many technologies and will strengthen education systems, can be used in the design of environmentally friendly learning environments. Based on this idea, the authors drew a general framework about eco-friendly learning, green technologies, and blockchain in the chapter and presented suggestions for the design of environmentally friendly learning environments in the context of use cases of blockchain technology in education.

This chapter provides an exploration to the enormous potential of blockchain and its strategic benefits. The application of blockchain has the potential to revolutionize existing business models. Blockchain is a peer-to-peer model that can speed up processes resulting in robust tracking and reducing costs of transactions. It is essential for business professionals to understand the significance of the implications of this far-reaching technology on business activities. Consequently, it is imperative that business leaders comprehend aspects of the technology to be integrated in strategic initiatives. The purpose of this chapter is to inform readers about the strategic benefits of blockchain technology. The chapter explains blockchain in the context of business strategies while highlighting its applications. This chapter is structured to explain the idea of blockchain technology, expound on the importance of the technology's business applications, and illustrate how professionals can integrate this technology in business.

Section 3
Education Implications

The potential of implementation of blockchain technology in education to manage different internal processes seems quite obvious. The chapter discusses different areas of the open and distance learning system in general and learner support services in particular in which use of BCT can be a promising pay-off. The chapter focuses on the introduction of blockchain: how it started and recent developments, open and distance learning system in India, learner support services in open and distance learning, BCT in India, strategies of the government of India to augment blockchain, application of blockchain in open and distance learning in general and learner support services in particular, challenges in implementation of blockchain in open and distance learning and learner support in the country, and its implications.

The fourth industrial revolution is peeping through blockchain technology since the invention of the steam engine, electricity, and information technology. It has come up with a lot of expectations and is now considered a 'disruptive technology' stirring the fabric of society to its roots. This desk research concentrates on evaluating its potential for Indian schools. How could blockchain contribute to effective schooling in India? What could be the possible framework to implement this technology for progressive changes in schools? A thorough review of the existing literature and the Indian context reveals that blockchain is promising for Indian schools providing enhanced digital accuracy, openness, cost-saving operations, and a heightened constructivist learning paradigm. Based on these assumptions, the author has proposed a framework for blockchain implementation (FBI) in school operations which could be a substantial contribution to the 'glocal' management of blockchain technology in schools.

Foreword

Education could not possibly be what it is today if it weren't for technological advances. Looking back at the history of education, each new generation of technology is a driving force behind educational development. For example, the use of parchment for writing was the first step to perpetuating knowledge, greatly facilitating education for later generations. At the same time, papermaking and printing could be acclaimed as the first revolutionary catalyst for the massification of education. No other education sector is more reliant on technology than distance education, an umbrella term for all modes of learning other than campus-based and/or face-to-face study. This is why distance education is often classified according to the dominant type of technology used, be it a three-generation theory or a five-generation typology. Technological development has been a prerequisite for the success of distance education at different historical stages since the 19th century.

Today, distance education, in this sense, has become part of the (new) normal of education in general. This convergence means that education, both campus-based and at a distance, needs to leverage technology, particularly technology of the contemporary times, to accomplish its mission.

We are now embracing the Fourth Industrial Revolution, a concept formulated by Schwab (2016), the founder of the World Economic Forum, in 2016. This revolution "will fundamentally alter how we live, work, and relate to one another". Like its predecessors, the Fourth Industrial Revolution is already and will continue impacting all aspects of human society, including education. Since education is the foundation of social development, the significance of how education should make the most and best use of contemporary technologies is self-evident.

What are the representative technologies of the Fourth Industrial Revolution? According to Marr (2020), ten top technology trends are driving the Fourth Industrial Revolution and will define the next decade: (1) artificial intelligence and machine learning, (2) the internet of things, (3) big data, (4) blockchains, (5) cloud and edge computing, (6) robots and cobots, (7) autonomous vehicles, (8) the 5G network, (9) genomics and gene editing, and (10) quantum computing. Augmented reality and virtual reality and 3D and 4D printing, are also included on the list elsewhere. Given this revolution's unanticipated scale, scope, and complexity, it is most likely that there are technologies that we do not know yet today but are to emerge in the days to come. Some of these key technologies are already being applied in education, including learning and teaching as well as administration and management. Artificial intelligence and big data are two cases in point. 5G technology has also begun to be utilized for educational purposes. For example, many top universities in China aim to build 5G-enhanced campuses in the next few years. Similarly, another key technology, blockchain, is attracting interest from the educational community worldwide.

According to Wikipedia (n.d.):

A blockchain is a decentralized, distributed and public digital ledger that is used to record transactions across many computers so that the record cannot be altered retroactively without the alteration of all subsequent blocks and the consensus of the network.

It has its primary features: immutable, distributed, decentralized, secure, consensus, unanimous, and faster settlement. These affordances have implications for education and will disrupt education when effectively and appropriately applied. Nevertheless, the applications of blockchain in education are currently more in the form of generic rhetoric than in action. For example, Purohit (2022) discusses the possible impact of blockchain on education in five areas: smart contracts for courses and assignments; degrees, report cards, and paperwork; incentivization of education; streamlining fee payments; and universal access and lower cost. Tom Vander Ark, CEO of Getting Smart, came up with a more specific list of 20 blockchain applications in education: transcripts, badges, student records, identity, infrastructure security, ridesharing, cloud storage, energy management, prepaid cards, smart contracts, learning marketplace, records management, retail, charity, human resources, governance, libraries, publishing, public assistance, and bonds (Ark, 2018). Four chapters included in this volume are about applications of blockchain in education, with three of them exploring the possibilities and/or pilot attempts in India and the remaining one proposing a framework for eco-friendly learning based on use cases of blockchain technology in education. This is a welcome shift from generic discourse towards more concrete operation levels.

Given the primary features mentioned above, the impact of blockchain is not limited to education. As a matter of fact, Tom Vander Ark's arguments for blockchain in education are based on its applications in and/or affordances for other sectors of society. Other chapters in this book elaborate on the use of blockchain in such areas as climate crisis, accounting, auditing, assurance, retail supply chain, cyber security governance in banking, and purchasing processes in industrial markets, usually with a particular context in mind. One chapter looks at the opportunities, risks, and challenges about the use of blockchain from a legal perspective. At the same time,another compares the impacts of development policies and institutional changes on blockchain-enabled ecosystems between China and India.

I particularly appreciate the glocal perspective that the authors adopt in their argumentation. A glocal thinking of technological affordances is essential to effective application, which will not be possible unless technology is used by considering its concrete socio-cultural and socio-economic circumstances. To this end, this book is a good start.

Junhong Xiao
Open University of Shantou, China

REFERENCES

Ark, T. V. (2018, August 27). *20 Ways Blockchain Will Transform (OK, May Improve) Education.* Retrieved from https://www.gettingsmart.com/2018/08/27/20-ways-blockchain-will-transform-ok-may-improve-education/

Blockchain. (n.d.). In *Wikipedia.* https://en.wikipedia.org/wiki/Blockchain

Marr, B. (2020, May 4). *The Top 10 Technology Trends of the 4th Industrial Revolution.* https://www.forbes.com/sites/bernardmarr/2020/05/04/here-are-the-top-10-technology-trends-of-the-4th-industrial-revolution/

Purohit, A. (2022, August 18). *5 Ways Blockchain Impacts The Education Industry In 2022 And Beyond.* Retrieved from https://elearningindustry.com/ways-blockchain-impacts-education-industry-in-2022-and-beyond

Schwab, K. (2016, January 14). *The Fourth Industrial Revolution: What it means, how to respond.* Retrieved from https://www.weforum.org/agenda/2016/01/the-fourth-industrial-revolution-what-it-means-and-how-to-respond/

Preface

Blockchain is a disruptive technology that allows multiple parties to securely and transparently record, verify, and transfer assets without the need for intermediaries (Casino et al., 2019). At its core, blockchain is a decentralized database that maintains a continuously global growing list of records, called blocks, that are linked and secured using cryptography. The first and most well-known application of blockchain technology is Bitcoin (Nakamoto, 2008), a digital currency that allows for peer-to-peer transactions without the need for financial institutions. However, the potential applications of blockchain technology go far beyond just currency and financial transactions. Blockchain can be used to track and manage digital assets, verify the authenticity of documents and data, and even enable new forms of decentralized governance and decision-making (Teles & Azevedo, 2019).

At its most basic level, a blockchain is a series of blocks that contain data. Each block contains a unique code, called a hash, which is generated based on the data it contains. Each block also contains a hash of the previous block, creating a chain of blocks that are linked together. This makes it virtually impossible to alter or tamper with the data contained in any one block without affecting the entire chain. In addition, being secure and tamper-proof, blockchain technology is also highly transparent. Because each block contains a hash of the previous block, it's possible to trace the entire history of a transaction or data record from its first date to the present. This creates a high level of transparency and accountability, which is especially important for applications that need third-party trust.

One of the most major features of blockchain technology is its decentralized nature (Sharma, Yildirim, & Kurubacak, 2020). Unlike traditional databases, data is not stored on a single server or set of servers but the blockchain stores the data on a distributed network of computers, known as nodes. This means that there is no central point of failure or control, and no single entity has the power to alter or manipulate the data on the blockchain.

To maintain the integrity of the blockchain and prevent fraudulent or malicious transactions, the network uses a consensus mechanism to validate new blocks of data. In the case of Bitcoin, this consensus mechanism is known as proof of work

(PoW), which involves miners competing to solve complex mathematical problems in order to add new blocks to the chain. Other consensus mechanisms, such as proof of stake or delegated proof of stake, rely on different methods to validate new blocks and maintain the security of the network.

Blockchain technology is a powerful tool that has the potential to transform many industries and create new opportunities for economic growth and development (Sahani, Priyadarshini, & Chinara, 2021). By providing a secure, transparent, and decentralized way to record and transfer data, blockchain can help build more sustainable and inclusive ecosystems that benefit both local and global communities. Blockchain technology has the potential to transform many industries by improving efficiency, security, and transparency. Here are some examples of the global impact of blockchain technology on different areas:

- **Finance:** Blockchain technology has many well-known use cases, including in the financial sector, where it has the potential to revolutionize the way we make and verify financial transactions (Thomason, 2022). Blockchain-based cryptocurrencies such as Bitcoin and Ethereum provide a decentralized alternative to traditional banking systems (Bansal, Malyadri, Singh, & Pervez, 2021).
- **Supply Chain Management:** Blockchain technology can help improve transparency and accountability in supply chain management by providing a secure and tamper-proof way to track the movement of goods from origin to destination (Wang et al., 2019). This can help prevent fraud, reduce waste, and improve sustainability (Mahmood, 2021).
- **Healthcare:** Blockchain technology has the potential to improve the security and privacy of patient data by providing a secure and transparent way to store and share medical records (Skiba, 2017). This can help reduce administrative costs and improve patient outcomes by providing healthcare professionals with faster and more accurate access to patient data (Varadharajan, Bansal, Nair & Leevinson 2021).
- **Energy:** Blockchain technology can help facilitate the transition to renewable energy by creating a decentralized energy market that allows consumers to buy and sell energy directly from one another (Polas, Saha, & Ahamed, 2022). This can help reduce the reliance on fossil fuels and promote the development of sustainable energy sources.
- **Education:** Blockchain technology can help store and verify educational credentials, making it easier for students and employers to prove qualifications and reduce the risk of fraud (Grech & Camilleri, 2017). This can also help improve access to education by providing a more secure and transparent way to transfer credits between institutions.

- **Government:** Blockchain technology has the potential to improve transparency and accountability in government by providing a secure and tamper-proof way to record and verify government transactions (Alketbi et al., 2018; Ølnes et al., 2017). This can help reduce corruption and improve public trust in government institutions.
- **Real Estate:** Blockchain technology can help simplify and streamline the process of buying and selling property by providing a secure and transparent way to record and verify property transactions (Joshi & Choudhury, 2022). This can help reduce administrative costs and improve the efficiency of the real estate market (Alisawi, Al-Dawoodi, Mohammed Wahab, Hammood, Yaseen Nawaf & Ghazi, 2022).

Blockchain technology has the potential to transform many different sectors by providing a secure, transparent, and decentralized way to record and transfer data. By improving efficiency, security, and transparency, blockchain technology can help create more sustainable and inclusive ecosystems that benefit both local and global communities (Thomason, Bernhardt, Kansara, & Cooper, 2019). While the technology is still in its early stages, its potential impact on a wide range of industries is already being felt around the world.

As blockchain technology continues to grow and mature, governments and regulatory bodies around the world are beginning to develop policies and regulations to govern its use. Policies related to blockchain have their impact in varied ways:

- **Digital Identity:** Many governments are exploring the use of blockchain technology for digital identity management, which can help improve security and reduce fraud. As digital identity becomes increasingly important for accessing online services and participating in the digital economy, blockchain technology could be used to create more secure and decentralized identity systems (Patel & Thakur, 2021). However, this also raises policy considerations around data privacy, security, and governance.
- **Anti-Money Laundering (AML) and Know Your Customer (KYC) Regulations:** Due to the potential for blockchain to be used for illicit activities, many governments are implementing AML and KYC regulations for blockchain-based businesses (Üçoğlu, 2022). These regulations help ensure that blockchain-based businesses are operating within legal frameworks and not facilitating money laundering or terrorist financing.
- **Data Protection:** With the rise of blockchain-based data storage and sharing, many governments are developing policies to protect personal data and ensure compliance with privacy regulations (Manzoor, 2023). The European

Union's General Data Protection Regulation (GDPR) is one such example, which aims to ensure that personal data is protected and not misused.

- **Smart Contracts:** Many governments are exploring the use of smart contracts, which are self-executing contracts with the terms of the agreement directly written into code (Boily, 2022). These contracts can help reduce administrative costs and improve efficiency (Madakam & Harshita 2021), but also require policies and regulations to ensure legal compliance and dispute resolution (Asharaf & Adarsh, 2017).
- **Intellectual Property:** With the potential for blockchain to be used for intellectual property management, many governments are developing policies and regulations to protect intellectual property rights (Pina, 2022). This includes the development of blockchain-based patent registries and copyright management systems.
- **Taxation:** As blockchain-based transactions and assets become more prevalent, governments are developing policies and regulations to ensure that these transactions are properly taxed (Önkan & Arikan, 2022). This includes the development of regulations for cryptocurrencies and other blockchain-based assets, as well as guidelines for taxation of blockchain-based businesses (Bhattacharjee, Saha, & Chatterjee, 2022).

POLICY IMPLICATIONS

The future of blockchain technology is constantly evolving, and as the technology matures, new trends are emerging that will have policy implications for governments and regulatory bodies. Let's examine some of the potential future trends on blockchain and their policy implications:

- **Central Bank Digital Currencies (CBDCs):** Many countries are exploring the use of CBDCs, which are digital versions of their fiat currencies. CBDCs could have implications for monetary policy, financial stability, and privacy, and require careful policy considerations around issuance, distribution, and governance (Ahiabenu, 2023).
- **Decentralized Finance (DeFi):** DeFi refers to financial applications built on blockchain technology, such as lending and trading platforms (Thomason, 2022). DeFi has the potential to disrupt traditional finance by providing more accessible and transparent financial services, but also raises policy questions around consumer protection, market integrity, and regulatory oversight (Popescu, 2022).

- **Internet of Things (IoT):** As the number of IoT devices continues to grow, blockchain technology could be used to secure and manage the vast amounts of data generated by these devices. However, this also raises policy concerns around data privacy, security, and ownership (Mahmood, 2021).
- **Interoperability:** As different blockchain networks and platforms emerge, there is a growing need for interoperability between them. This could involve the development of standards and protocols to ensure seamless communication between different networks, as well as policy considerations around data privacy and security (Williams, 2020).
- **Sustainability:** As the environmental impact of blockchain technology comes under scrutiny, there is a growing focus on developing sustainable blockchain solutions. This could involve the development of more energy-efficient blockchain protocols, as well as policy considerations around the carbon footprint of blockchain-based activities (Zakaria, Zainuddin & Mohamad, 2022).

RELEVANCE OF BLOCKCHAIN IN THE WORLD TODAY

The future of blockchain technology is full of potential, but also raises important policy questions and considerations. As the technology continues to evolve and mature, governments and regulatory bodies will need to carefully consider the implications of new trends and applications on areas such as financial stability, consumer protection, data privacy, and sustainability (Al-khateeb, 2021). By taking a proactive and collaborative approach to policy development, policymakers can help ensure that blockchain technology is developed and used in a way that benefits society as a whole.

Blockchain technology is one of the most transformative innovations of our time. Its decentralized and transparent nature has the potential to create a more equitable and sustainable world. However, realizing this potential requires more than just technical expertise. It requires policies and strategies that are grounded in the needs and values of local communities, while also addressing global challenges such as climate change, economic inequality, and digital inclusion.

Blockchain technology is transforming the way we conduct business, store data, exchange value, and interact with each other. Its decentralized and transparent nature has the potential to revolutionize many industries and create new opportunities for economic growth and development (Mason, Halgamuge & Aiyar, 2021).

In this book, we explore the concept of "glocal" policy and how it can help us build sustainable blockchain ecosystems that benefit both local and global communities. We start by introducing the basics of blockchain technology and its potential applications in various industries. We then dive into the concept of glocal policy, examining how it can be applied to blockchain ecosystems to promote sustainability, inclusivity, and innovation.

Throughout the book, we feature case studies and examples of successful blockchain initiatives from around the world. From digital identity systems in Africa to supply chain tracking in Southeast Asia, these examples demonstrate how blockchain can be used to solve real-world problems and create new opportunities for economic growth and development.

We also examine the challenges and opportunities of building sustainable blockchain ecosystems, including issues of governance, regulation, and social impact. We explore different models of governance and decision-making in blockchain ecosystems, as well as the role of regulation in promoting innovation while protecting users' rights and interests.

By the end of the book, readers will have a comprehensive understanding of how to build sustainable and inclusive blockchain ecosystems that benefit both local and global communities. We hope that this book will serve as a practical guide for policymakers, entrepreneurs, and innovators who are working to build a better future with blockchain.

We believe that blockchain technology has the potential to create a more equitable and sustainable world, but it will take all of us working together to make this a reality. We invite readers to join the conversation and contribute to the development of sustainable blockchain ecosystems that benefit both local and global communities. Together, we can build a better future with blockchain.

As the world becomes increasingly interconnected, we need policies and strategies that address both global and local needs. This is especially true for blockchain technology, which has the potential to create a truly global ecosystem. In this book, we explore the concept of 'glocal' policy and how it can help us build sustainable blockchain ecosystems that benefit both local and global communities.

We wrote this book because we believe that blockchain technology has the potential to create a more equitable and sustainable world. But to achieve this, we need policies and strategies that are grounded in the needs and values of local communities, while also addressing global challenges such as climate change, economic inequality, and digital inclusion. This is the reason why we choose the "glocal" setting for this book. Our hope is that this book will serve as a practical guide for policymakers, entrepreneurs, and innovators who are working to build a better future with blockchain.

TARGET AUDIENCE

The target audience for this book includes policymakers, government officials, business executives, entrepreneurs, blockchain developers, researchers, and academics who are interested in understanding the potential and policy implications of blockchain technology in revolutionizing traditional financial processes, such as accounting and auditing, and its impact on global markets and industries. It will also be useful for professionals from various fields such as computer science, management, law, healthcare, and social sciences, who want to implement blockchain-based solutions in their respective areas. Additionally, students pursuing degrees in these areas of study will find this book informative and helpful for their academic research. This book aims to provide valuable insights into the policy implications of blockchain technology, and how it can be effectively applied across various industries and sectors.

This book will also interest professionals, scholars, and learners from diverse fields -such as law, computer sciences, education, information and communication technologies, psychology, sociology, etc.- who would like to concentrate on how Blockchain merges the gap between research needs and priorities as well as their transformative models and egalitarian practices. In addition, the book will be highly valuable to those researchers in the field interested in keeping abreast of activist developments in the digital information age. It would be an important resource for e-law, e-health, e-decision makers, e-commerce, eLearning environment designers, librarians as well. Finally, this book will provide current, strategic, and creative information to those interested in omnipresent Blockchain with its transformative models.

Policymakers, Computer scientists, programmers, communication designers, the learners themselves, support staff, managers, and stakeholders need to keep abreast of the latest research in the blockchain policy. Furthermore, key workers should incorporate such research into their decision-making processes and focus on future trends. Identifying future research needs and priorities for Blockchain is necessary to foster these improvements. Future research needs must be identified and priority given to the social and political impact of new management, communication, pedagogy, technology, and evaluation-based developments on society and the relation of those developments to the idea of a free society. This book will have a scholarly value that all of those concerned with and involved in Blockchain must consider the impact of the digital world not just on themselves but at a global and international level.

ORGANISATION OF CHAPTERS

The chapters of the book have been organised under four sections.

Section 1 deals with policy matters.

In the first chapter, Kalpana Muthireddi focuses on exploring Blockchain technology through a legal lens, with a particular emphasis on identifying the opportunities, risks, and challenges that come with utilizing this cutting-edge technology. To provide a comprehensive view, the chapter takes a three-pronged approach to the discussion, examining how Blockchain can be leveraged to create positive change, while also recognizing the obstacles that must be overcome before it can be widely adopted. The first area of exploration involves delving into the legal and regulatory compliance issues associated with Blockchain. Next, the chapter considers various business applications for the technology and examines the possible risks and how they can be mitigated. Finally, the chapter explores the potential for Blockchain to be harnessed for the betterment of society, examining how it can be used to improve the well-being and welfare of the general public.

Second chapter looks into Blockchain-based tourism industry, in which Stelios Ioannidis, provides an overview of the potential impact of blockchain technology on the tourism industry. The decentralized and transparent nature of blockchain, along with its other characteristics, such as disintermediation and immutability, have attracted investors to explore its applications in tourism. Smart contracts, tokens, cryptocurrencies, and DAOs hold the potential to enhance trust, automate processes, and reduce costs in the tourism industry. However, the broad-scale adoption of blockchain in tourism is limited due to several challenges. This chapter explains these obstacles, such as a lack of awareness and expertise, high energy consumption, and regulatory gaps in data management and income taxation. Despite these challenges, the chapter concludes that blockchain technology has enormous potential benefits and highlights the need for further exploration and development of this area in the future.

In the chapter on strategic planning and policy framework for implementation of Blockchain technology in education in India, Anjana and Siran Mukerji, report that the Blockchain technology is widely regarded as one of the most ground-breaking innovations in information technology of the 21st century. Its applications span a wide range of sectors, including education. Given the growing popularity, increased acceptance, and the numerous benefits of this technology, the Indian government has taken steps towards its adoption. Blockchain technology is a critical area of research, with the potential for deployment in various domains such as governance, banking and finance, education, cybersecurity, and more. This chapter aims to achieve several objectives, such as highlighting the strategic planning and policy framework for the implementation of blockchain technology in India. Additionally, it provides insights into the international landscape and the current status of the implementation of blockchain technology across various sectors in India. The chapter also examines the multifaceted use of blockchain, analyzes the challenges involved in its implementation, and outlines a way forward for the application of this technology in the education sector in India.

Section 2 has chapters having industry implications.

The fourth chapter is by Ayla Avci who makes an investigation of purchasing applications after digital change in industrial markets. This chapter deals with the dynamic and ever-changing global markets, where consumer demands and preferences undergo continuous transformations. The industrial markets have also undergone changes, as a multitude of products and services are exchanged. The internet has played a crucial role in bringing world countries together as a common market, which has led to diverse consumer needs and wishes shaped by cultural interactions. With the widespread adoption of online shopping, e-commerce has become increasingly popular, thanks to the fast and timely delivery options for products. The complex industrial markets have seen a rise in demand for speedy transactions and diverse product offerings, as more people make purchases. In recent years, many companies have shifted towards using online web pages to select suppliers. This research examines the purchasing processes in industrial markets and explores the latest digital transformation trends.

The next chapter is by Selin Göçen, and Gülsün Kurubacak, who explains that blockchain technology is a powerful tool that can aid in the development of education systems. However, due to its high level of energy consumption and gas emissions, it is often avoided in environmentally friendly activities. Despite this, there is insufficient evidence to suggest that blockchain technology is environmentally harmful or consumes more energy than other technologies. As a flexible technology, it has the potential to strengthen education systems and can be utilized in designing eco-friendly learning environments. To this end, the authors have presented a general framework that includes eco-friendly learning, green technologies, and blockchain, offering suggestions for designing environmentally friendly learning environments through the use of blockchain technology in education.

The chapter by Mohan Menon and Ashraf Mady discusses the vast potential of blockchain technology and its strategic benefits. They report that that applications of blockchain have the potential to transform existing business models by speeding up processes, creating a robust tracking system, and reducing transaction costs. It is crucial for business professionals to understand the implications of this groundbreaking technology on business activities. Therefore, it is imperative for business leaders to comprehend the aspects of this technology in order to integrate it into strategic initiatives. The purpose of this chapter is to inform readers about the strategic advantages of blockchain technology. The chapter provides an overview of blockchain in the context of business strategies while highlighting its various applications. The structure of the chapter involves explaining the concept of blockchain technology, emphasizing the importance of the technology's business applications, and illustrating how professionals can integrate it into their business practices.

Section 3 has two chapters that deal with educational implications for open and distance learning for higher education sector and school sector.

The first chapter in this section deals with augmenting learner support services with use of Blockchain technology where Pulist explains the potential of implementing Blockchain technology in education for managing various internal processes. It discusses various aspects of the Open and Distance Learning (ODL) system, particularly the learner support services, where the use of Blockchain technology can yield promising results. The chapter provides insights into the introduction of Blockchain technology, its evolution, the ODL system in India, the learner support services offered by ODL, and the government's strategies to promote Blockchain technology in the country. It also delves into the application of Blockchain technology in ODL, specifically in the learner support services, while highlighting the challenges of implementing this technology in the Indian context. The chapter concludes by examining the implications of Blockchain technology for ODL and its learner support services.

The chapter by Arnab Kundu deals with the evaluation of blockchain technology in school education, and its potential to contribute to effective schooling. The study explores how this technology could be utilized in developing countries such as India, and the possible framework to implement it for positive change in schools. A comprehensive literature review was conducted, and the findings indicate that blockchain could be particularly useful for underprepared schools that are willing to provide quality education for their students. It offers digital accuracy, greater transparency, cost-saving operations, and a heightened constructivist learning experience. Blockchain can also enhance transparency, reduce manual practices for record keeping, prevent fraud by eliminating manual verification of transcripts and other documents, increase transparency in assessments, facilitate curriculum personalization, and provide a platform for collaboration between students and opportunities through smart contracts.

Section 4 has three chapters that deal with implications for retail supply chain and auditing processes.

Xinyu Zheng and Syed Imran Ali explore the potentials of internet of things (IoT) and blockchain technology in the collaboration and integration of retail supply chain. This chapter explains the growing importance of integrating Industry 4.0 technologies into the supply chain of organizations. These disruptive technologies have contributed to improving efficiency, performance, and sustainability in organizational ecosystems. Despite these advancements, the globalized retail industry faces significant challenges in terms of collaboration and integration. Therefore, it is imperative to conduct a comprehensive review of Industry 4.0 technologies, such as the Internet of Things (IoT) and blockchain technology, to explore their potential in addressing the challenge of achieving collaboration and integration across retail supply chains. To this end, a systematic literature review (SLR) methodology is adopted to gather current academic publications from the SCOPUS database spanning 2018-2021. Descriptive and thematic analyses are conducted to reveal the characteristics, themes, potentials, and relationships necessary for achieving collaboration and integration across retail supply chains. The findings suggest potential solutions for small and medium-sized enterprises and highlight future directions for research in this field.

Seval Kardes Selimoglu, Gülsün Kurubacak, and Mustafa Hakan Saldi present a case study of the Turkey's Bank Sector by examining the internal audit functions in cyber security governance. This chapter looks into the critical risk category that cyber-attacks have attained with the rapid rise in technology usage and emerging digital transformation trends, particularly in service sectors like banking. It is imperative to ensure the validity of activities coordinated within the cyber security governance mechanisms of banks to maintain financial transactions within acceptable risk levels. Risk management operations for cyber security processes should be controlled with the assistance of audit units within banks. In this chapter, based on a doctoral thesis research study, the responsibility fields of internal audit teams in cyber security management processes are thoroughly examined using the Delphi technique. To propose a set of solutions for continuous audit methods in cyber security governance, blockchain technology is explored in terms of its technical concepts and policy framework.

The last chapter is by Seval Selimoglu, Gul Yesilcelebi, and Mehtap Altunel who have examined the impact of blockchain technology on accounting, auditing, and assurance practices. To achieve this objective, the pertinent national and international literature on the accounting, auditing, and assurance practices of blockchain technology is being reviewed. The study assesses the advantages and disadvantages of blockchain technology and categorizes them accordingly. As a result of the analysis, it is concluded that blockchain technology has a significant impact on accounting, auditing, and assurance practices.

CONCLUSION

This book makes a significant contribution to the field of Blockchain technology and its applications by adopting a policy-based approach. By presenting the features, advantages, and disadvantages of Blockchain, the book provides a solid foundation for experts, academics, students, and researchers to understand the technology and its potential uses. Additionally, by discussing the applications and platforms of

Blockchain and how they can ensure quality for diverse environments, the book offers practical insights into selecting the most suitable Blockchain applications.

In this book, we start by exploring the basics of blockchain technology and its potential applications in various industries. We then dive into the concept of glocal policy and how it can be applied to blockchain ecosystems. We also examine the challenges and opportunities of building sustainable blockchain ecosystems, including issues of governance, regulation, and social impact. Throughout the book, we feature case studies and examples of successful blockchain initiatives from around the world. By the end of the book, readers will have a comprehensive understanding of how to build sustainable and inclusive blockchain ecosystems that benefit both local and global communities.

We believe that blockchain technology has the potential to create a more equitable and sustainable world, but it will take all of us working together to make this a reality. We invite readers to join the conversation and contribute to the development of sustainable blockchain ecosystems that benefit both local and global communities. Together, we can build a better future with blockchain.

Gülsün Kurubacak
Anadolu Üniversitesi Açikögretim Fakültesi, Turkey

Ramesh Chander Sharma
Dr. B.R. Ambedkar University, Delhi, India

Hakan Yıldırım
Eskişehir Osmangazi University, Turkey

REFERENCES

Ahiabenu, K. (2023). Unpacking Central Bank Digital Currency (CBDC) Implementation Challenges and Risks. In S. Ahmed, J. Addae, & K. Ofori (Eds.), *Exploring the Dark Side of FinTech and Implications of Monetary Policy* (pp. 207–231). IGI Global. doi:10.4018/978-1-6684-6381-9.ch010

Al-khateeb, B. A. (2021). Regulatory Standards and Measures: Panacea for Blockchain Technology Acceptability. In A. Sghari & K. Mezghani (Eds.), *Influence of FinTech on Management Transformation* (pp. 54–72). IGI Global. doi:10.4018/978-1-7998-7110-1.ch003

Alisawi, M., Al-Dawoodi, A., Mohammed Wahab, Y., Hammood, L., Yaseen Nawaf, A., & Ghazi, A. (2022). Developing the Real Estate Rental Sector in Third World Countries Using Blockchain Technology: Iraq as Case Study. In P. Swarnalatha & S. Prabu (Eds.), *Blockchain Technologies for Sustainable Development in Smart Cities* (pp. 87–109). IGI Global. doi:10.4018/978-1-7998-9274-8.ch006

Alketbi, A., Nasir, Q., & Talib, M. A. (2018). Blockchain for government services-Use cases, security benefits and challenges. *2018 15th Learning and Technology Conference, L and T 2018*, 112–119. 10.1109/LT.2018.8368494

Asharaf, S., & Adarsh, S. (2017). *Decentralized Computing Using Blockchain Technologies and Smart Contracts: Emerging Research and Opportunities*. IGI Global. doi:10.4018/978-1-5225-2193-8

Bansal, R., Malyadri, P., Singh, A., & Pervez, A. (Eds.). (2021). *Blockchain Technology and Applications for Digital Marketing*. IGI Global. doi:10.4018/978-1-7998-8081-3

Bhattacharjee, P., Saha, V., & Chatterjee, P. (2022). Cryptocurrency: A Detailed Study. In N. Dewani, Z. Khan, A. Agarwal, M. Sharma, & S. Khan (Eds.), *Handbook of Research on Cyber Law, Data Protection, and Privacy* (pp. 161–183). IGI Global. doi:10.4018/978-1-7998-8641-9.ch010

Boily, É. (2022). Smart Contract and Collaborative Platforms: Exploring the Impacts of the Computerized Transaction Protocol on the Collaborative Economy. In M. Ertz (Ed.), *Handbook of Research on the Platform Economy and the Evolution of E-Commerce* (pp. 104–127). IGI Global. doi:10.4018/978-1-7998-7545-1.ch005

Casino, F., Dasaklis, T. K., & Patsakis, C. (2019). A systematic literature review of blockchain-based applications: Current status, classification and open issues. *Telematics and Informatics, 36*, 55–81. doi:10.1016/j.tele.2018.11.006

Grech, A., & Camilleri, A. F. (2017). *Blockchain in Education*. doi:10.2760/60649

Joshi, S., & Choudhury, A. (2022). Tokenization of Real Estate Assets Using Blockchain. *International Journal of Intelligent Information Technologies, 18*(3), 1–12. doi:10.4018/IJIIT.309588

Madakam, S., & Harshita. (2021). Blockchain Technology: Concepts, Components, and Cases. In Z. Mahmood (Ed.), *Industry Use Cases on Blockchain Technology Applications in IoT and the Financial Sector* (pp. 215-247). IGI Global. doi:10.4018/978-1-7998-6650-3.ch010

Mahmood, Z. (Ed.). (2021). *Industry Use Cases on Blockchain Technology Applications in IoT and the Financial Sector*. IGI Global. doi:10.4018/978-1-7998-6650-3

Manzoor, A. (2023). Privacy and Security Under Blockchain Technology: Challenges and Solutions. In A. Naim, P. Malik, & F. Zaidi (Eds.), *Fraud Prevention, Confidentiality, and Data Security for Modern Businesses* (pp. 176–195). IGI Global. doi:10.4018/978-1-6684-6581-3.ch008

Mason, N., Halgamuge, M. N., & Aiyar, K. (2021). Blockchain and Cryptocurrencies: Legal and Ethical Considerations. In Z. Mahmood (Ed.), *Industry Use Cases on Blockchain Technology Applications in IoT and the Financial Sector* (pp. 132–159). IGI Global. doi:10.4018/978-1-7998-6650-3.ch007

Nakamoto, S. (2008). *Bitcoin: A Peer-to-Peer Electronic Cash System*. https://bitcoin.org/bitcoin.pdf

Ølnes, S., Ubacht, J., & Janssen, M. (2017). Blockchain in government: Benefits and implications of distributed ledger technology for information sharing. *Government Information Quarterly*, *34*(3), 355–364. doi:10.1016/j.giq.2017.09.007

Önkan, Ö., & Arikan, Z. (2022). The Impact of Blockchain Technology on Tax and Accounting Practices. In P. Nijalingappa & M. Ghonge (Eds.), *Blockchain Technologies and Applications for Digital Governance* (pp. 1–36). IGI Global. doi:10.4018/978-1-7998-8493-4.ch001

Patel, H., & Thakur, G. S. (Eds.). (2021). *Blockchain Applications in IoT Security*. IGI Global. doi:10.4018/978-1-7998-2414-5

Pina, P. (2022). Blockchain and Copyright: Challenges and Opportunities. In S. Khan, M. Syed, R. Hammad, & A. Bushager (Eds.), *Blockchain Technology and Computational Excellence for Society 5.0* (pp. 125–145). IGI Global. doi:10.4018/978-1-7998-8382-1.ch007

Polas, M. R., Saha, R. K., & Ahamed, B. (2022). Leveraging Green IoT and Blockchain Technology in the Era of Transformative Digitalization: A Green Energy Usage Perspective. In P. Lai (Ed.), *Handbook of Research on Social Impacts of E-Payment and Blockchain Technology* (pp. 115–135). IGI Global. doi:10.4018/978-1-7998-9035-5.ch008

Popescu, A. D. (2022). Understanding FinTech and Decentralized Finance (DeFi) for Financial Inclusion. In M. Anshari, M. Almunawar, & M. Masri (Eds.), *FinTech Development for Financial Inclusiveness* (pp. 1–13). IGI Global. doi:10.4018/978-1-7998-8447-7.ch001

Sahani, A., Priyadarshini, S. B., & Chinara, S. (2021). The Role of Blockchain Technology and Its Usage in Various Sectors in the Modern Age: Various Roles of Blockchain and Use Cases in Different Sectors. In S. Pani, S. Lau, & X. Liu (Eds.), *Blockchain and AI Technology in the Industrial Internet of Things* (pp. 221–245). IGI Global. doi:10.4018/978-1-7998-6694-7.ch014

Sharma, R. C., Yildirim, H., & Kurubacak, G. (Eds.). (2020). *Blockchain Technology Applications in Education*. IGI Global. doi:10.4018/978-1-5225-9478-9

Skiba, D. J. (2017). The Potential of Blockchain in Education and Health Care. *Nursing Education Perspectives, 38*(4), 220–221. doi:10.1097/01.NEP.0000000000000190 PMID:28622267

Teles, D., & Azevedo, I. (2019). Blockchain: Past, Present, and Future. In H. Lum (Ed.), *Critical Issues Impacting Science, Technology, Society (STS), and Our Future* (pp. 21–52). IGI Global. doi:10.4018/978-1-5225-7949-6.ch002

Thomason, J. (2022). Fintech and Blockchain: Maximizing Benefit and Minimizing Harm. In I. Vasiliu-Feltes & J. Thomason (Eds.), *Applied Ethics in a Digital World* (pp. 185–198). IGI Global. doi:10.4018/978-1-7998-8467-5.ch013

Thomason, J., Bernhardt, S., Kansara, T., & Cooper, N. (2019). *Blockchain Technology for Global Social Change*. IGI Global. doi:10.4018/978-1-5225-9578-6

Üçoğlu, D. (2022). Blockchain Technology and Future Banking: Opportunities and Challenges. In S. Gupta, P. Kansra, & G. Kukreja (Eds.), *Applications, Challenges, and Opportunities of Blockchain Technology in Banking and Insurance* (pp. 43–68). IGI Global. doi:10.4018/978-1-6684-4133-6.ch003

Varadharajan, V., Bansal, D. J., Nair, S., & Leevinson, J. R. (2021). Blockchain Reinventing the Healthcare Industry: Use Cases and Applications. In Z. Mahmood (Ed.), *Industry Use Cases on Blockchain Technology Applications in IoT and the Financial Sector* (pp. 309–328). IGI Global. doi:10.4018/978-1-7998-6650-3.ch013

Wang, Y., Singgih, M., Wang, J., & Rit, M. (2019). Making sense of blockchain technology: How will it transform supply chains? *International Journal of Production Economics, 211*, 221–236. doi:10.1016/j.ijpe.2019.02.002

Williams, I. (2020). Cross-Chain Blockchain Networks, Compatibility Standards, and Interoperability Standards: The Case of European Blockchain Services Infrastructure. In I. Williams (Ed.), *Cross-Industry Use of Blockchain Technology and Opportunities for the Future* (pp. 150–165). IGI Global. doi:10.4018/978-1-7998-3632-2.ch010

Zakaria, R. B., Zainuddin, M. N., & Mohamad, A. H. (2022). Distilling Blockchain: Complexity, Barriers, and Opportunities. In P. Lai (Ed.), *Handbook of Research on Social Impacts of E-Payment and Blockchain Technology* (pp. 89–114). IGI Global. doi:10.4018/978-1-7998-9035-5.ch007

Acknowledgment

We want to express our sincere gratitude and appreciation to everyone who has supported and contributed to our book 'Glocal Policy and Strategies for Blockchain: Building Ecosystems and Sustainability.'

Gulsun Kuruback thanks her parents from the core of her heart, Öznur, and Cemalettin Kurubacak. She expresses her profound thanks to her spouse Prof. Dr. Serhat Çakir.

Ramesh thanks his parents, Sh Kidar Singh Sharma and Mrs. MohanVati Devi, for their unwavering love, support, and guidance. Their encouragement and motivation have been instrumental in shaping him into the person he is today. He is profoundly grateful to his wife Madhu Sharma, daughter Aakanksha Sharma, and son Apoorv Sharma for their love, patience, and understanding during the book-writing process. Their unwavering support and encouragement have been his source of inspiration. Special thanks go to his friends, Mr. Veer Batra, Dr. Aras Bozkurt, Dr. Vandna Guliya, Dr S K Pulist, Dr Ashish Awadhiya, Dr Deepika Kohli, and Dr Siran Mukerji, for their valuable feedback and insights, which helped shape the book's content. Their suggestions and comments have been invaluable in the book-writing process.

Hakan expresses his thanks to his mom Naile Yildirim, dad Muzaffer Yildirim, brother Kadir Yildirim and wife, Meltem Eryilmaz Yildirim. Also, he would like to thank TÜBİTAK (Scientific and Technological Research Council of Turkiye) for granting a scholarship during his Ph.D. studies, and his doctoral thesis advisor at Anadolu University, Prof. Dr. Gülsün Kurubacak, and his host advisor at Purdue University Assoc. Prof. Dr. Muhsin Menekşe, and his team.

We want to express our sincere gratitude to Professor Junhong Xiao for writing the foreword for this book which has added immense value. His expertise and knowledge in the field of Open and Distance Education are unparalleled, and his thoughtful comments and insights in the foreword have helped to contextualize the book's content and its relevance to the field.

The insightful and thought-provoking chapters presented in this book would not have been possible without the authors' hard work, dedication, and expertise. We extend our heartfelt gratitude and appreciation to each author for their contributions, which have added immense value to the book.

We also thank the chapters' reviewers for their excellent feedback and valuable suggestions. Their meticulous review process has helped us ensure the book's quality and that it is a valuable resource for the blockchain community. We are grateful for the authors' and reviewers' commitment and contribution to making this book a reality. Working with such a talented group of individuals whose insights and expertise have enriched the book and made it a valuable resource for researchers, practitioners, and policymakers is truly an honor.

We express our sincere gratitude and appreciation to the IGI Global staff who published our book meticulously. First and foremost, we extend our heartfelt thanks to Jan Travers, Director of Intellectual Property & Contracts, and Melissa Wagner, Vice President of Editorial, for their invaluable contribution to the publication process. Your expertise, professionalism, and attention to detail have ensured the smooth publication of the book.

We would also like to thank Elizabeth Barrantes and Gianna Walker, Assistant Development Editors, who have been with us from the beginning of the book-writing stage. Your invaluable support, guidance, and feedback have been critical in shaping the book's content and ensuring its quality.

We are grateful for the support and encouragement provided by IGI Global, which has made this book possible. We are honored to have been associated with a publishing house that upholds the highest standards of excellence in academic publishing.

Section 1
Policy Implications

Chapter 1
Blockchain:
A Legal Perspective

Kalpana Muthireddi
Independent Researcher, India

ABSTRACT

This chapter looks at blockchain technology from a legal perspective and discusses the opportunities, risks, and challenges that surround the use of this technology. It adopts a three-pronged approach to discussing blockchain technology, how it can be leveraged, and the challenges that must be addressed before it can be adopted. The three areas of discussion are 1) legal and regulatory compliance issues; 2) the various business applications, possible risks, and how they are addressed, and 3) the possibilities for harnessing this technology for well-being and welfare of the general public.

INTRODUCTION

The advantage with Blockchain, a type of DLT (Distributed Ledger Technology), is its decentralized, immutable, tamper-proof feature. As we already know, it is a decentralized database of transaction data that is shared, replicated, and synchronized by multiple member participants across multiple nodes of a network (Oh and Shong,2017). And every single transaction is recorded with an immutable cryptographic signature called a hash, containing a timestamp, a link to the previous block, and transaction data (Mahankali & Chaudhary, 2020).

It provides verifiable integrity for transactions, as well as transactional security, all of which is conducted within a resilient system that is believed to be impervious

DOI: 10.4018/978-1-6684-4153-4.ch001

Copyright © 2023, IGI Global. Copying or distributing in print or electronic forms without written permission of IGI Global is prohibited.

to hacking, thereby preventing false and counterfeit transactions. Given that it is unalterable, keeping permanent records of transactions is possible (Koldemir, 2020).

It can be private/permissioned network with a closed ecosystem requiring permission to join and run on proprietary software; or could be a public (permissionless) network running on opens source software such as Hyperledger Fabric and Ethereum; or a hybrid platform. All these have varying levels of participation and transaction access. They also have their limitations of scalability, privacy, access and security, and centralization.

Different types of data viz., medical records, insurance records, real estate records, Intellectual Property documentation, supply chain details and any kind of data that requires a time stamp system and transaction validation can be stored on the network (Oh et al., 2017). Every entity on the Blockchain plays a part in maintaining the network and the data it hosts (Kommadi, 2022).

In this chapter, we look at the opportunities and challenges that Blockchain and other distributed ledger technologies have introduced, the regulatory challenges they present and how we can harness the potential of this technology while managing possible risks.

Blockchain technology interests lawyers and policy makers for its ability to create authentic records of transactions with little or no risk of vulnerability to exploitation. However, there are concerns and complexities involving practical applications of and the risks inherent in this technology (Iansiti & Lakhani, 2017). This chapter looks at Blockchain technology from a legal perspective and discusses the opportunities, risks and challenges that surround the use of this technology.

This chapter adopts a three pronged approach to discussing Blockchain technology; how it can be leveraged and the challenges that must be addressed before it can be adopted. The three areas of discussion are:

1. Legal and regulatory compliance issues;
2. The various business applications; possible risks and how they are addressed; and
3. The possibilities for harnessing this technology for well-being and welfare of the general public.

LEGAL AND REGULATORY COMPLIANCE

With emergence of Bitcoin on the scene, the underlying technology on which it runs, viz., Blockchain has generated tremendous interest in the business world in general and the financial sector in particular. Cryptoassets, Blockchain and smart contracts are set to change how we do business and provide services, across various sectors including the legal industry (European Economic and Social Committee, 2019).

The initial euphoria of being able to conduct transactions without need for trusted intermediaries has given way to a more cautious approach, recognizing the need for regulated markets in which Blockchain and other DLT can function effectively.

Blockchain and DLT, is disruptive technology at its best, offering exponential possibilities for growth, but given that it is still at an incipient stage, there is a call for regulation and building effective governance models (Tiwari and Pal, 2022). Only then can we harness its immense potential effectively, without adversely affecting the interests of stakeholders and the general public.

But new technologies introduce new legal issues. Here are a few of them:

Jurisdictional Issues

Since the decentralized set up of the DLT spans different countries and jurisdictions worldwide, it is difficult to identify when, where and how a dispute might arise on the system and what jurisdiction and laws will be apply.

Given the presence of multiple participants across multiple jurisdictions worldwide, it is critical for the parties to decide before hand, the applicable law, venue and forum should a dispute arise. If that is not done beforehand, and a dispute arises, the courts of the jurisdiction in which the dispute has arisen and to which it is subject, will determine the matter resulting in unexpected and sometimes unwanted outcomes.

Due care must be taken to decide whether you want arbitration, litigation or other alternate dispute resolution systems; or if you want to arbitrate or other pursue other means of dispute resolution before proceeding to litigation. Other points to consider are: the applicable law, the venue of the litigation/arbitration, and if you choose arbitration, would it be ad hoc or under institutional rules; where would be the seat of the arbitration, and the composition of the tribunal.

Care must be taken to avoid multiple and conflicting decisions on a single matter so that enforcement does not become a problem. Parties must be cautious about choosing a single option and venue and ensuring that it applies to all participants on the DLT network.

Questions still remain about the acceptance, enforceability and the binding nature of these arbitral awards or decisions of a court in other jurisdictions. Since an award applies to all the participants, they must ensure beforehand that any awards and decisions are binding on all the participants outside the jurisdiction of the courts/ arbitration tribunal which has granted the respective decision or award.

Data Protection, Privacy, and Security

Different types of data are stored and processed on the block chain including sensitive and personal information such as medical and financial records, which are subject to different data privacy and protection laws such as the GDPR (Europe), CAA (USA) the Data Protection Rules, 2011 (India) etc. depending on the geography to which they are related.

By putting such data on the DLT which by its nature is decentralized and accessible from anywhere in the world, the transactions violate these data protection and privacy laws.

We will look at it taking into account, some of the GDPR regulations relating to Data Privacy, Data Subject Rights, International Data Transfer and Rule of Minimization.

Concerns relating to data privacy are heightened when data is put on a shared ledger, where the data is immutable, accessible across borders, and cannot be erased.

By putting personal data or personal identifying information relating to individuals on the DLT, it gets challenging to comply with the GDPR regulation relating to data privacy as it provides access to personal data of individuals without their consent and by using it for purposes beyond the scope for which it was collected. Further, this data will remain on the ledger in perpetuity. This violates the key principle of Minimization, set out in the GDPR.

Similarly, this also conflicts with various data subject rights such as right to erasure, because the data is immutable and perennial. It cannot be amended or erased unless every single node on the network erases the data, which means even more computations across every system on the network if you wish to scrub the data of even one individual.

Each block on the chain includes a hash containing the underlying information, and a link to the previous block of data, and these blocks could be in different countries or regions outside the European Economic Region. When data is shared on a public chain, which may have multiple nodes criss-crossing countries, the data has, by default crossed the home jurisdiction where it was generated, thereby violating the regulation of the GDPR relating to international data transfer. When every participant is called upon to validate a transaction containing a certain piece of data, who would be considered a Data Controller on the chain, for the purposes of ascertaining obligation under the GDPR. These and several other questions arise when processing data on the Blockchain.

This calls for a deeper understanding of the technology in conjunction with the applicable law and clarity of the activities that are being conducted on the chain.

This does not mean that data must not be shared on block chains or that DLTs are in violation of data privacy or protection laws, but that there is a need to reconcile the use of the DLT with data protection and privacy regulations.

One way to achieve this is by using a closed chain or off-chain, permissible and private DLTs, where the data is restricted to a few participants and there is a greater measure of control over how the data is handled. If you do intend to use block chains, make sure that you do not put personal data or any personal identifying information on the Blockchain.

Data Security Concerns

There is a common perception that data stored on the Blockchain is more secure and there is almost nil likelihood of data being compromised. It is true that there are various checks and measures built into the system by the architecture and the decentralized way the DLT is structured. Compared to traditional centralized systems, a decentralized set up provides more robust data security and poses less risk, due to the network architecture, high level access controls, and protocols surrounding transaction authenticity.

But all things considered, DLT merely comprises of servers, systems, databases and platforms all of which are vulnerable to system failures and hacking just like any other computer. A few incidents of hacking attest to this.

In 2018, a 15 year old hacked the Ledger Hardware Wallet through the back door by developing a proof of concept code that allowed him entry. Later, Ledger admitted to a complex vulnerability that was deep in the architecture of their system.

Before that, in 2016, DAO a Decentralized Autonomous Organization, launched on the Ethereum blockchain was hacked due to vulnerabilities in its code base. A hacker found a loophole in the coding that allowed him to drain funds from The DAO. It had raised $150 million USD worth of ether and the hacker, moved approximately $50 million in investor funds to a sub-contract that the hacker controlled.

More recently, the Poly Network hacking incident in 2021 has turned out to be the largest crypto currency heists so far, with around $610 million stolen in August of 2021, where hackers exploited vulnerability in its system.

These incidents demonstrate that weaknesses in the technology, be it architecture or data validation or any other can seriously jeopardize the integrity of the Blockchain. This calls for well thought out and thorough validation of outside technologies, architectures, and the data before integration into the DLT. Security of the Blockchain and safety of the data housed in it is critical to the success of a Distributed Ledger Transaction before we can safely introduce it to the public for large scale application/ consumption.

Money Laundering and Related Concerns

Money laundering activities are a major concern on digital platforms where transactions are made and recorded on distributed ledgers. The confidential nature of the transactions, the lack of an overseeing authority and the difficulty of regulating these technologies makes it easy for proliferation of illegal transactions.

One of the earliest and most famous cases in point was the Silk Road. The Silk Road was a digital black market platform operating between 2011 and 2013 that was popular for hosting money laundering activities and illegal drug transactions using Bitcoin. In 2015 the founder Ross Ulbricht, was successfully prosecuted for distributing narcotics and money laundering; and 69,000 Bitcoins worth over one billion US Dollars (1bn USD) in today's value was seized.

In addition to the above, money laundering activities, arms deals, narcotic deals, terrorist funding and other illegal transactions can be conducted easily and much more efficiently on blockchain and other DLTs away from the scrutiny of legal and regulatory authorities. This calls for regulatory frameworks that can effectively monitor such platforms; and transactions on Blockchain (and other distributed ledgers) that endanger national safety and security (Karisma, 2022).

Intellectual Property Rights

Block chain and other DLTs provide immense possibilities for recording, storing, protecting and managing intellectual property rights. It could make it easier for calculating and claiming royalties, creating new IP deals and asserting one's IP rights. Digital content can be managed more efficiently using DLT.

We require further guidance on how the precedent on the topic of "communication to the public will be applied, and how the rights and obligations of all parties will be considered. This also ties in closely with the copyright infringement issues.

The technology of the DLT such as decentralized applications, peer to peer network and pseudonymous nature facilitates easier management, sharing and transfer of digital files. But the same technology could also encourage copyright infringement. To avoid this, one may resort to storing full data in an off-chain database while using a hash on the chain with a hyperlink to the stored data.

Immutability, a key feature of DLTs like Block chain, conflicts with the requirement of taking down infringing material upon a claim of copyright infringement. How does one address these requirements without violating the law?

How about the *right of erasure* under the GDPR? How can a DLT platform comply with that request of a data subject when the data on the chain is supposed to be immutable? How can a platform operator claim exemption and to what extent is that exemption allowed.

There is no clarity yet on database rights, and the rights of developers with regard to protection for their creations on the DLT. These and several other questions and concerns surrounding IP rights still remain, such as:

"Who owns the IP of data that is hosted on a block chain?

"Will the platform operator/owner be responsible for any breach or violation of IP that is hosted on the block chain?

"Has the platform permitted, encouraged or facilitated illegal sharing of protected content?

"Will every participant be considered liable in the event of a copyright or trademark infringement?"

These questions must be addressed accurately if we must be assured of safe block chain and distributed ledger transactions. There are no simple answers to these scenarios and it becomes important to understand the business models and the actions pursued by such platforms, and contributing factors that must be considered if one were to determine questions of IP infringement, liability, and the extent of responsibility, and whether there was a breach of rights and obligations; use or misuse of the systems in place.

There is no need for special laws to address these, but it calls for a deeper understanding of the technology, the infringing activity, and the actors (the operators, the core developers, the miners, the participants etc.) responsible for such infringement and the extent of their liability.

Regulation of Crypto Assets

The driving factors behind regulatory frameworks for crypto assets and the block chain transactions in general are twofold –Firstly, protecting the interests of the stakeholders viz., the retail, enterprise and professional users of the various platforms; and secondly, concern over the use of crypto currencies for illicit purposes, as was the case with the Silk Road platform we discussed a little while ago.

In layman terms, Crypto assets are tokenised digital assets created by using cryptographic techniques and are digital representations of assets of value and connected rights (Spafford, Stanaway and Chung, 2019). They can be stored, transferred or traded on distributed ledger technology (DLT).

There are different types of crypto assets ranging from Crypto currencies (Bitcoin, Ethereum, Binance coin, Cardano etc.), Central bank digital currencies (CBDC), utility coins, e-money tokens, security tokens, platform tokens, to stable coins and NFTs, among others.

One of the major challenges with regulating crypto assets and crypto currencies is their classification. How does one classify them? Are they property, money, securities, commodities or other kinds of financial instruments?

At the moment, the authorities are trying to regulate them using the existing regulatory framework based on the different categories into which the crypto assets fall into depending on their characteristics.

In the US, the sentiment seems to be that the IRS considers them as "Property", the Treasury Department considers them as "Money", and the CFTC considers them as "Commodity" (it is no longer considered as "Securities") and is subject to the jurisdiction of all these authorities.

The Financial Crime Enforcement Network (FinCEN), the enforcement arm of the Department of Treasury (DoT), USA, was one of the first regulators to claim jurisdictions over cryptocurrecy transactions and issued guidelines in 2013 and has since brought cases and issued administrative rulings in this regard.

The key legal issues focus mainly around anti-money laundering and sanctions, securities, commodities and derivative regulations, advertising and consumer protection, investor protection laws, state money transmitter regulations and other related issues.

If you are using a token, you have to consider all related issues such as, for instance, Money Transmitter Licenses (MTL). Obtaining such licenses is not a one-and-done thing. There are compliance requirements related to MTL maintenance which are on-going obligations. That is not an easy thing especially with lack of regulatory clarity and may also not be cost effective, if you are not sure what you are doing, because as is the case with all legal compliances, you must be aware of and comply with the obligations across regimes and industries relating to the crypto currencies you are dealing with.

In the UK, crypto assets are regulated based on the substantive characteristics of the crypto asset and then subject it to the relevant regulatory framework.

For instance, depending on whether the crypto asset is characterised as a security, asset, deposit, e-money or any other type of regulated financial instrument, the regulations of that asset classification will apply to the crypto asset in question.

The regulatory authorities depend on a case-by-case analysis to determine the nature of the crypto asset and the applicable regulation. The UK government has also introduced new regulations for monitoring the Virtual Asset Service Providers (VASP) providers and other developments in the DeFi space.

It introduced new registration requirements in January 2020, for "crypto asset exchange providers" and "custodian wallet providers" as set out in Regulation 14A of the Money Laundering, Terrorist Financing and Transfer of Funds (Information on the Payer) Regulations 2017 (MLRs) and FCA rules. The term "crypto asset" has been defined broadly to cover both regulated as well as unregulated types of crypto assets.

While the government of India has not banned crypto currencies, (in March 2020, the Supreme Court reversed the order of the RBI banning cryptos), it is likely to regulate them strongly. The proposed crypto currency bill, the "Crypto currency and Regulation of Official Digital Currency Bill, 2021' has been deferred in the winter session of the Parliament, and is likely to feature in the February 2022 budgetary session.

Crypto ICOs

Organizations must be cognizant of the laws when launching crypto in the market. Initial Coin Offerings of Crypto Assets (ICOs) are regulated under relevant applicable laws. When an entity is offering an ICO of cryptos, it must be clear whether it is offering it as an investment or as a service. It must clarify to the purchasers whether it is raising investment or if it is selling a service. If it is raising investment, it must comply with the applicable investor protection laws in every jurisdiction it is offering the ICO. If it is offering as a service, it must comply with all applicable, laws relating to selling services.

Also when offering it across jurisdictions, it must take care to see what regulatory regimes it is entering, and ensure that sale of tokens is limited to those jurisdictions where cryptos are allowed. It may run the risk of violating the law if it extends the offering to jurisdictions where cryptos are heavily regulated or completely banned.

More challenges will emerge, as this technology gets widespread, but with a controlled and regulated marketplace, we can leverage the benefits of this technology to its fullest.

Providing legitimacy to blockchain based operations including crypto currencies, while safeguarding the interests of the various stakeholders seems to be the focus of the regulatory regimes.

Other legal and Regulatory Concerns

There are several other regulatory concerns that will need to be addressed as the technology gains traction in general and the business world in particular. A couple of them are discussed below:

Competition and anti-trust laws

Would competitors be considered to be in violation of anti-trust and competition laws simply by being on the same block chain together?

Is there a possibility of them working in collusion on a DLT without the knowledge of the regulatory authorities; without invoking anti-monopoly laws? Can they get away with potentially monopolistic actions and without the regulatory authorities monitoring their actions?

While the current competition laws in India and elsewhere are very comprehensive and cover all sectors, how the presence of competitors on Single blockchain or DLT will play out remains to be seen. In all likelihood, it may be taken on a case-by-case basis to prevent any miscarriage of justice.

Liability

The question of apportionment of liability assumes importance in a DLT scenario where every single node on the network is responsible for the maintenance, if not introduction of the data that is stored on the chain. The question of joint and several liability assumes importance when deciding the liability in a transaction on the DLT.

In the event of a breach, how far down the chain will the liability extend? Is there a way to clearly demarcate the liability of each of the participants and their contribution if there is a breach or a breakdown of the transaction? There are no easy answers to this, as several factors have to be taken into consideration and matters have to be decided on a case-by-case basis.

Taxation

When it comes to tax matters there are two main categories of questions that arise: one is taxation of the block chain and crypto assets, and two - where in the chain of the transactions does the Tax location situate and which tax law has jurisdiction over the transactions? How can we address the problems of double taxation and tax avoidance?

These lead to a third question about how to harness the block chain technology and the ability of DLTs to store data and validate transactions to reduce false and fraudulent transactions and use it to revamp the taxation system. This calls for a tax system that understands this technology, and utilizes it to bring in a better taxation regime.

These are some of the current potential challenges and benefits if we can address the challenges and leverage the potential. Providing legitimacy to Blockchain based operations including crypto currencies, while safeguarding the interests of the various stakeholders seems to be the focus of many regulatory regimes. There is significant impact of blockchain on accounting and taxation, including the accounting entry process, tax practices, accounting professionals, and taxpayers. By applying blockchain to accounting and tax auditing, it is believed that errors and fraud will be prevented, resulting in more effective and concurrent taxation (Önkan & Arikan, 2022).

THE VARIOUS BUSINESS APPLICATIONS OF BLOCKCHAIN; POSSIBLE RISKS, AND HOW THEY ARE ADDRESSED

Blockchain (and DLT) is one of the few technologies that have presented game changing potential across business, social, and cultural domains. While the potential is immense, the risks are yet to be fully understood and managed before it can be fully utilized. Here we will discuss three common topics that are discussed and show immense promise for business applications across industries. - NFTs, DeFi and Smart contracts.

NFTs (Non-Fungible Tokens)

A non-fungible token (NFT,) is a non-divisible digital token that is unique, and linked to an object such as music, art or other asset and uses the blockchain technology to record and certify its authenticity and ownership.

The Merriam Webster dictionary defines a non-fungible token (NFT) as - : "a unique digital identifier that cannot be copied, substituted, or subdivided, that is recorded in a blockchain, and that is used to certify authenticity and ownership (as of a specific digital asset and specific rights relating to it)."

To understand the concept of non-fungible tokens, (NFTs), we must first understand what fungibility means. Fungiblity is the characteristic of being interchangeable with something of similar value. Fungible items have value but are not unique. For example: Currency.

NFTs are those items that have value and are also unique, in addition to being indivisible and amenable to digital recording on a blockchain, For example – A piece of art, music or movie. Sotheby's successfully conducted its curated online NFT art sale in 2021. A Picasso painting for instance, has value and is also unique because it is the only one of its kind in the world and is not transferable in value. In other words it is not fungible.

NFTs are digital assets created on DLT and are freely tradeable. They use smart contracts to enable the transactions relating to their creation, transfer and all other transactions (Pranata & Tehrani, 2022). NFTs have of late, become popular and are used largely in art, interactive entertainment, and other digital collectables. The NFT metadata contains information about the original object, the smart contract address which manages the ownership, and the transferability of the NFT (Saeidnia and Lund, 2023).

Without going too much into detail as to all the aspects of NFTs, we will examine here, how NFTs are gaining popularity in the world of business, and the opportunities and risks they present.

We are seeing a spurt in the popularity of NFTs, and everyone wants to benefit from either the creation, or investments relating to NFTs. NFTs are not regulated at the moment, but current laws relating to securities, commercial trading exchanges and assets will be used to handle the issues on a case by case basis. It is clearly an evolving area and there will be developments in the legal sphere to address the issues surrounding NFTs.

In this scenario, we see a few issues occurring commonly, that one must be aware of when handling NFTs.

Copyright and Other IP Issues

Most people purchasing an NFT may not be aware that NFTs do not grant a copyright in the underlying asset. When you buy a book, piece of music, or art, you are only granted a right of use and enjoyment in the object, not a right to make multiple copies or sell it to third parties. That right is reserved with the creator or author. Same is the case with NFTs most of the time.

You have to do a full and thorough due diligence to see whether you are buying a right to use the asset or a right to full ownership in the asset including copyright. Ascertain the rights of the seller; verify their authenticity and whether any third party IP rights are present in the NFT, hindering your rights as the owner of the asset.

Authenticity of the NFT

Since an NFT is hosted on a distributed ledger, it is connected to its underlying digital asset via a link. There is a possibility of the link between the asset and its NFT being broken or otherwise compromised due to technical or other more nefarious reasons. It is crucial that a purchaser must do their due diligence to check and verify the authenticity of the NFT link and the server hosting the asset to which the NFT is connected.

Due to any of the above reasons, the link between the asset and its corresponding NFT may break. Therefore, it is advisable for purchasers of NFT to check the reliability of the digital asset as well as the security and authenticity of the link. A reliable NFT seller will greatly minimise the related risks and you will be assured of getting a verified and authentic token with full understanding of the rights you are purchasing (be they use or ownership rights) in the underlying asset.

Royalty Pay-Outs and Other Transactions

Since NFTs are as asset whose creation, trade and exchange happen on the block chain; their transfers are governed by smart contracts. Most transactions including royalty pay outs are made using smart contracts. These are contracts written in code and programmed to execute pre-determined outcomes, once a set of conditions are fulfilled. The code is embedded into the token on the block chain and is immutable and cannot be replaced, amended or deleted.

With smart contracts enabled, certain automatic actions are executed without need for human intervention and facilitate automatic calculations such as royalties to the creator of the NFT. This is convenient but since these depend on by pre-determined terms absent human intelligence, any unexpected changes in the marketplace cannot be foreseen and may lead to unwanted results.

DeFi (Decentralized Finance)

Decentralized Finance, "DeFi" for short, is a comprehensive term used for financial services that are provided on public blockchains, such as Ethereum. DeFi facilitates all transactions that a bank or financial institution delivers, and does it faster, without need for paperwork or a third party intermediary. All activities such as saving, lending, trading, insurance are supported with DeFi. Like other digital ledger transactions, it is peer-to-peer (or a peer-to-platform), decentralized, global, pseudonymous and immutable (Thomason, 2022).

Similar to all transactions on the DLT, DeFi exhibits all features and functionalities of decentralised, and immutable transparency. The challenge here is in understanding the technology, the risk involved in using decentralised exchanges, identifying protocols to monitor the application and manage, avoid or mitigate risks. A substantial amount of the value of cryptocurrency was received from DeFI protocols via retail transactions and visits to DeFi platforms worldwide. One of the fundamental questions concerning DeFi is whether it is a 'software service' or a financial service, a "virtual asset based service" coming within the purview of financial regulations and regulations monitoring activities of virtual asset based service providers (VASP) .

The strongest argument in favour of DeFi is its financial inclusion, that is, making available financial services to anyone with a computer and access to the platform, without need for any authentication by an intermediary such as a bank. The biggest concern is its potential for illegal transactions, mainly money laundering.

This has prompted equal amounts of interest as well as caution from governments worldwide. Some have outright banned cryptomarkets, while China, US, UK and the EU have seen a humungous amount of business transacted in the past couple of years. China initially accounted for a majority of these transactions between 2020-2021, but it has steadily been moving towards its own solutions. It started testing its own central bank digital currency (CBDC), the digital yuan, which has been used for $8.3 bn worth of transactions as of the second half of 2021.

The biggest challenge for regulators when it comes to DeFi is the possibility of money laundering using peer to peer platforms. Regardless the growth of DeFi transactions has been phenomenal with China leading the list in the crypto markets, closely followed by North America, and UK.

In UK, as of the moment, there are no "tests" for decentralization as a legal concept, but the Financial Action Task Force (on Money Laundering), (FATF), has noted that peer to peer trading platforms may be categorised as an exchange when it conducts the activity as a business or on behalf of another person. But there are several other factors that must be considered before arriving at a conclusion of a these providers as virtual asset based service providers (VASP).

Under the UK MLR (Money Laundering Regulations), any firm or sole sole practitioner who by way of business provides services relating to exchanging or arranging or making arrangements with a view to the exchange of one crypto asset for another, crypto assets for money or money for crypto assets, or operates a machine which utilises automated processes to exchange crypto assets for money or money for crypto assets, shall be considered a "crypto asset exchange provider". Software developers and providers who do not earn any income or benefit from consequent transactions fall outside the purview of the definition.

This seems to be the approach in other regions as well. The EU also seems to focus on the "facilitating" aspect of the transactions. The same sentiment seems reflected in Thailand's approach to regulating DeFi. The Indian government has not given a go ahead yet, and there are indications that DeFi will be heavily regulated similar to cryptocurrencies.

This indicates that the challenge for the regulatory authorities seems to be one of preventing fraudulent activities, and protect the interest of the investors rather than rather than blocking genuine financial transactions on the DeFi platforms. The concerns are not without cause. The DeFi industry is still nascent and but has already seen a few crashes such as the so called "Black Thursday" incident at *MakerDAO,* when a sudden price crash triggered an automatic liquidation of portfolios." This was followed by China's DeFi protocol *dForce* when cyber criminals first stole and then stealthily reinstated US$25 million worth of assets into the exchange.

But the overall approach seems to be to avoid any set of test factors and to go with the individual elements of the transactions and trigger factors that could indicate a "trade", "facilitation" or "exchange" of "digital assets". How these regulations and protocols will translate into law remains to be seen given the many challenges, including and a "lack of expertise and understanding" regarding virtual assets and VASPs, as well as ground level constraints (Gaur, 2020).

For all the potential the DeFi promises, it is also fraught with technical and operational risks. Besides, potential for misuse of the technology to defraud investors, there is a major risk of use for illegal purposes such as money laundering. Given the high stake risks, governments are understandably taking a cautious approach towards it.

SMART CONTRACTS

Another area that has exploded on the blockchain and distributed ledger scene, is smart contracts. From a legal perspective, few technological innovations have had the kind of impact the blockchain revolution has caused, because no innovation has impacted the legal industry as directly as blockchain and smart contracts have.

While the concept of smart legal contracts using smart contract code is still at a nascent stage, there is a lot of excitement surrounding smart contracts aka Smart Legal Contracts (SLC) and their potential to do away with the trusted intermediary, in this case, the lawyer. But how far are smart contracts capable of creating truly effective contracting.

By now, we have all heard the common refrain, that smart contracts are neither contracts, nor are they necessarily smart. Why has this phrase become so synonymous with smart contracts and what does it actually mean? If a smart contract is not a contract, and if it is not even a smart thing, why is everyone looking at it as next revolution in the legal industry?

Here, we will try to understand what the term "smart contract" actually means; the true potential of this new development, and if it can really deliver on the hype that has been generated around it.

At its simplest, a smart contract is a code or program that executes certain actions based on pre-determined inputs, commands and upon occurrence of certain fixed trigger factors. As such, it does not need an intermediary.

Going by the above understanding, smart contracts have been around for longer than we assume. The earliest examples of such pre-programmed transactions could be vending machines where you can put in money and get a bottled beverage or snacks of your choice; and railway weighing machines, where you could get your weight and fortune told for a few paise.

If the amount paid into the vending machine is more than the value of the item you choose, the excess change returns to you along with the item you ordered, and if you paid less than the value, the transaction does not occur and you get your money back.

Similar to this, were the card-ticket weighing machines. A common sight at railway stations in India until a few years ago, these machines told your weight and fortune when you put in a coin of the exact denomination. Put it a coin of higher value, you would get your card and the excess change back, put in a coin of lower value, the transaction would fail and your coin would be returned.

These were smart contracts at its simplest and they have been around since at least the 1950s.

Smart contracts of today are a far cry from the above and consist of highly sophisticated transaction codes that are being used for increasingly complex transactions, but their potential is only just being addressed.

Before we get too enamoured by the idea of having contracts drawn by predetermined codes, we must understand how smart contracts work, and how realistic it is to have our legal obligations drawn out by a prearranged set of codes and fixed set of outcomes. What we refer to smart contracts is code in contract form, but depending on how it is structured and its functions, a smart contract may not even be a true contract in the real sense of the term, meeting all requirements of a valid and legally binding agreement.

A smart contract is an agreement between two parties and is in the form of code. They run on blockchain, are stored on a shared database and cannot be erased. The transaction does not need a third party intermediary and it works with a set of pre-determined commands, trigger factors and outcomes. They are useful to automate processes that are standard and do not need customization. For example, contracts that calculate and pay royalties on NFTs. This does not change with every contract and is therefore amenable to being coded as a smart contract.

But what about more complex agreements which have may variables? The biggest advantage of a regular contract with natural language is the ability to accommodate unforeseen events by making relevant amendments to the contract. In that sense, can a smart contract, anticipate everything and be completely autonomous? Or does it require human intervention to be more effective?

Are We Smart Contract Ready?

One question that arises is whether we are smart contract ready at the moment. There are arguments in favour of embracing smart contracting in the world of business and that technology will make contracting more efficient and faster without human intervention.

The real question is, are smart contracts truly capable of replacing human contracting ability in their current stage of complexity?

Let us examine some of the pros and cons of using smart contracts and how effective technology can be in creating a valid, legally binding agreement without negative repercussions. Some of the arguments in favour of smart contracts are:

1. As long as the parties are willing, they do not need an intermediary such as lawyers to execute the document, thereby saving time.
2. They are efficient for automated processes which are high on volume and where the transactions are all of a similar nature.
3. They lead to accurate and transparent terms since they are pre-coded, are visible to all concerned parties, and keep confidential information out hidden from the other participants.
4. Since the terms are pre-coded, there is less scope for multiple interpretation of the same terms and saves time that might otherwise be spent in negotiations.
5. Have more evidentiary value since it is an automatically generated document with little to no human involvement. For the same reason, it also facilitates, faster and more effective dispute resolution.

On the other end of the spectrum, we still have certain practical issues to address before we can trust computer code to decide our rights and obligations. While it is tempting to do away with an intermediary, in their current avatar, smart contracts present limited benefit to a segment of contacts.

A lot of questions and technical issues must be addressed before we can see widespread and effective utilisation of this technology. Some of these are:

Complete Automation. The main feature of smart contracts is complete automation. This means low to nil flexibility, which is not a good thing. One of the main reasons, a traditional contract with natural language is successful in effective execution, is because it makes room for amendments to accommodate unexpected and unforeseen changes. As long as the parties agree to incorporate the changes, any kind of change or amendment to the contract is possible.

Full Automation Is Not Always Possible. A contract is a subjective document requiring human intelligence to exploit its full value; to understand situations and respond accordingly. Current levels of automaticity and code do not have the capability to fully understand and bring to fruition the desire of the parties as they truly intend it to because there are subtle nuances to human interaction and transactions that code cannot capture.

Limited Reach of Smart Contracts. Smart contract code is still in its nascent stage and it is limited to minor, simple and routine transactions that can be automated because there are no changes in the execution. A transaction with some level of complexity, variety and subjective or external requirements and external dependencies will not do well with pre-determined code and a smart contract will not work for these types of transactions.

Inflexibility. The immutable nature of the smart contract may be a disadvantage where changes or amendments to the contract are required due to many reasons ranging from force majeure events, (for instance, the current pandemic threw entire supply chain systems into loop), change in laws to change of market circumstances, all of which require amendments, to change, reverse or sometimes even prevent the execution of the contract.

An inability to change the contract to reflect the changing circumstances affecting the parties could mean that the rights of the parties may be adversely affected.

Oracle Problems. This brings us to the next problem which is called the "Oracles Problem", ie the problem one encounters when external data is required to execute certain actions but that external data cannot be validated as it comes from an "untrusted source". This means that the information must come from a third party source whose reliability is undisputed for all nodes on the chain. This is where Oracles step in. Oracles are systems that act as a bridge that can provide blockchain with information coming from the real world in a way that the blockchain can understand.

For example, in a smart contract of royalty payments for a book, the amount of annual royalty is calculated based on the sales data of the book. The sales data is external data based on which the amount of royalty is calculated. Unless that information is entered into the smart contract, the exact royalty pay-outs to the author cannot be calculated. An "Oracle" provides this information. Sometimes these Oracles are not available and that can lead to issues in the contracts.

Contract Terms and Obligations. In addition to what we discussed above, we also encounter certain practical issues, such as how confidentiality terms in the contract are addressed and affected when a contract is put on the Blockchain. Some of the standard exceptions to a confidentiality obligation are when that confidential information comes into public domain, or when it is shared with a third party. By putting this data or information on the blockchain, we have triggered these exceptions to the obligation and have made it available free of obligations of confidentiality, especially if it is put up platform that uses open source platforms.

Another issue is how do you handle the scope of different terms such as indemnities, or what would be considered a material breach of the contract. Most often, human understanding and subjectivity is required to address issues which could have lasting impact on the rights and obligations of the parties, in a way pre-determined code cannot.

Terms such as confidentiality, IP and other such obligations will continue as long as the agreement survives. A smart contract on the chain will survive in perpetuity, making these obligations perpetual. This is not preferable since they come with a lot of risk attached to them.

As we can see, smart contracts are at a nascent stage and these concerns must be answered before they become commonplace. For now, we may see simple transactions in certain segments of business activity being more amenable to smart contracting. More complex agreements still require human intervention at this stage of the technological innovation.

Rather than take extreme stands on the point, we can embrace the possibility of change in the way contracts are drafted but it must be tempered with a practical outlook on current realities and challenges (Tang, Xiong, Becerril-Arreola and Iyer, 2020).

We should not hesitate to choose the various options open to us in order to leverage this technology and gain as much traction as we can with technical advances. It could be using smart contracts for simple agreements, or choosing the right platforms viz., open source or proprietary, and private (permissioned) or public (permissionless) to provide a combination of security as well technical excellence.

As we can see from the above, this is a technology that is ripe with immense possibilities and the only drawback seems to be that it is in its nascent stage and we need to understand how we can leverage this potential while addressing the legal and other technical issues that arise along the way.

THE POSSIBILITIES FOR HARNESSING THIS TECHNOLOGY FOR WELL-BEING AND WELFARE OF THE GENERAL PUBLIC

Regardless of technical innovation and proficiency, perhaps the biggest test of any technology seems to be its impact from environmental, social, and governance (ESG) perspective (Wharton School of the University of Pennsylvania, 2018).

Blockchain (and other DLT) is at its core, secure and verifiable public register with multiple access points and an almost perpetual existence. These features can be transferred to many social, cultural, and general welfare aspects in society. There are several areas of general social living that can benefit from distributed ledger platforms (Kuhail, Mathew, Hammad, & Bahja, 2022). We will examine here, how this technology can be harnessed in areas of general and public interest that go beyond mere economic advancement. If the above two topics discussed the hard-core technical and business aspects of the blockchain technology, this part of the chapter is more a wishful thinking about leveraging block chain for the benefit of society.

Land Records

Real estate transactions depend heavily on registration of documents and authentication of land and ownership rights. Fraudulent transactions and multiple documentation are common because of the difficulty in obtaining land records from the local land registrar. Most often, lack of transparency leads to corruption and fraudulent deals where the genuine buyer has to undertake a lot of effort to check the authenticity and antecedents of a land transaction. By putting land records on the block chain, the government can effectively reduce corruption because the middleman is removed; it can also eliminate fraud and multiple transactions relating to the same piece of land (G Hiremath, 2022). Fee based access will bring additional revenue to government and add more transparency and a stamp of government approval to the entire operation.

Education

This is another area that will greatly benefit from using the block chain. Starting from maintaining student records, granting certificates, ensuring safety of the certificates in digital format, to using block chain to store all the student records and course related information on the chain, Blockchain can be successfully harnessed in the schools and universities.

Courses can be delivered via digital platforms, tests can be conducted and students can be evaluated online. The pandemic has brought into sharp focus the need for online education to go mainstream. Blockchain can effectively make that happen.

This also makes background verification of students easier and faster. It also helps to check the educational background of potential employees faster and efficiently (Bhaskar, Tiwari, and Joshi, 2021, Dash, Panda, Kumar and Luthra, 2022, Chen, Bohloul, Ma, and Li, 2022).

Medical Records

Hospitals have been using technological means to maintain medical records and histories of their patients for quite some time now. By leveraging DLT like Blockchain, old records of patients can be maintained and transferred between hospitals and doctors so that the entire history of a patient is available in one shot without loss of time or misplacing patient data (Parlakkılıç, 2022). Since medical institutions are working with privacy regulations like data privacy, HIPAA and GDPR, they can easily and seamlessly integrate those compliance requirements into any regulations that monitor block chain transactions (Dhagarra, Goswami, Sarma, and Choudhury, 2019, Hasan, Deng, Sultana, and Hossain, 2021).

Background Verification

Background verification has become de rigueur across all kinds of engagements. Employers in organizations have traditional background verification services that they can avail of, but non-traditional employers such as people employing nannies for their children or care givers for their elderly, have no such facility. People who need healthcare and social assistance services, where certain temperament is called for along with a specific subject matter expertise, can greatly benefit from a data base on block chain which provides verifiable information about a potential hire (Sharma & Gupta, 2021, Zaidi, 2022).

Intellectual Property

In the field of Intellectual property; the security, transparency and immutable nature of Blockchain serves an important purpose. Besides facilitating prior art searches, it helps with identifying proof-of ownership with accuracy. It gives the owner of copyrightable content, a measure of control over their creation as all downloads and transactions relating to their Intellectual property are transparent and smart contracts will help with timely royalty payments Pina, 2022).

Owners of IP can license their IP to potential licensees and market their IP, all in a culture of transparent and secure IP transactions. The current trend seems to encourage an IP marketplace where Intellectual property can be traded. Creation of smart IP registries, enabling digital transfer of IP and IP rights and effective payment mode, will encourage R&D and innovation. This will also bring the world closer to a more unified patent system worldwide, resulting in more effective IP rights management.

Tracking Donations and Other Funds Granted to Non-Profits and Charities

With the lack of accountability, inefficient utilization, and corruption are common with many non-profits and charitable organizations. Donations can be tokenized and tracked in their Blockchain-based platform. This will help with increasing transparency and tracing its movement across the chain starting from receipt of donations, to the ultimate beneficiary. Donations of all manner, digital, tangible and intangible can be tracked across the chain.

Voting Records

This is another area that can benefit from the immutable and verifiable records. Complaints of missing and fraudulent voter information can be effectively addressed by recording and storing voter information on the Blockchain. Once data is on the Blockchain, details belonging to a voter can be kept secure and changes can be made only as and when they occur. This is helpful to both the voters as well as the authorities as it will curtail misuse of voter data during the voting process (Verma & Sheel, 2022), Lin, Wang, Li, Lu, Qi, & Xie, 2022).

Other Uses

- Any document that can benefit from transparency and security with relevant time stamps will find block chain useful. One such area is wills, where there is need for a security and time stamping to ensure that the draft of the will is indeed the original and the latest one drafted by the testator. By harnessing the power of the block chain to the storing, verification and accessing of wills, it will prevent a lot of problems regarding the authenticity and the proof of time of the drafting of the will.
- Co-operatives in close knit societies or in fields like agriculture, horticulture, and crafts and arts can ensure that the goods and products sold by their co-operatives are identifiable and traceable on a Blockchain-based platform after the sale has been made. This will prevent fraudulent and counterfeit products that could impact small groups which may not have the resources to take preventive or protective action against fraudulent transactions.

CONCLUSION

As we can see, there are a lot of questions, concerns surrounding the use of Blockchain and relatively fewer instances of large scale adoption of the technology both in the spheres of business as well as social and welfare measures.

It is not to think that this technology does not have social advantages, but that there are a few threshold issues that must be addressed both at the corporate as well as societal levels before this technology can be safely embraced across all spheres (Martínez-Ríos, Marmolejo-Saucedo, and Abascal-Olascoaga, 2020).

Providing legitimacy to Blockchain based operations including cryptocurrencies, while safeguarding the interests of the various stakeholders seems to be the focus of many regulatory regimes. These provide challenges esp. since Blockchain is still in a nascent stage and it is difficult to predict how it will develop. It is also the right time to introduce technological solutions and frameworks. These inbuilt solutions will then grow and morph with changing technology.

REFERENCES

Bhaskar, P., Tiwari, C. K., & Joshi, A. (2021). Blockchain in education management: Present and future applications. *Interactive Technology and Smart Education, 18*(1), 1–17. doi:10.1108/ITSE-07-2020-0102

Chen, W., Bohloul, S. M., Ma, Y., & Li, L. (2022). A blockchain-based information management system for academic institutions: A case study of international students' workflow. *Information Discovery and Delivery, 50*(4), 343–352. doi:10.1108/IDD-01-2021-0010

Dash, M. K., Panda, G., Kumar, A., & Luthra, S. (2022). Applications of blockchain in government education sector: A comprehensive review and future research potentials. *Journal of Global Operations and Strategic Sourcing, 15*(3), 449–472. doi:10.1108/JGOSS-09-2021-0076

Dhagarra, D., Goswami, M., Sarma, P. R. S., & Choudhury, A. (2019). Big Data and blockchain supported conceptual model for enhanced healthcare coverage: The Indian context. *Business Process Management Journal, 25*(7), 1612–1632. doi:10.1108/BPMJ-06-2018-0164

European Economic and Social Committee. (2019, July 31). *Blockchain can have multiple applications in social economy but must not create new digital economy elite, says EESC*. EESC. https://www.eesc.europa.eu/en/news-media/news/blockchain-can-have-multiple-applications-social-economy-must-not-create-new-digital-economy-elite-says-eesc

Gaur, N. (2020). Blockchain challenges in adoption. *Managerial Finance, 46*(6), 849–858. doi:10.1108/MF-07-2019-0328

Hasan, M. R., Deng, S., Sultana, N., & Hossain, M. Z. (2021). The applicability of blockchain technology in healthcare contexts to contain COVID-19 challenges. *Library Hi Tech, 39*(3), 814–833. doi:10.1108/LHT-02-2021-0071

Hiremath, G. A. K., G. M., R., & K. R., N. K. (2022). A Secure Blockchain-Powered Land Record Modernization System for Documenting Land User's Rights: Land Transparency Initiative. In P. Nijalingappa & M. Ghonge (Eds.), Blockchain Technologies and Applications for Digital Governance (pp. 37-59). IGI Global. doi:10.4018/978-1-7998-8493-4.ch002

Iansiti, M., & Lakhani, K. R. (2017). The Truth About Blockchain. *Harvard Business Review*. https://hbr.org/2017/01/the-truth-about-blockchain

Karisma, K. (2022). A Pragmatic Regulatory Framework for Artificial Intelligence. In P. Tehrani (Ed.), *Regulatory Aspects of Artificial Intelligence on Blockchain* (pp. 21–39). IGI Global. doi:10.4018/978-1-7998-7927-5.ch002

Koldemir, B. (2020). A Digital Transformation in International Transport and Logistics: Blockchain. In G. Ceyhun (Ed.), *Handbook of Research on the Applications of International Transportation and Logistics for World Trade* (pp. 425–453). IGI Global. doi:10.4018/978-1-7998-1397-2.ch023

Kommadi, B. (2022). Blockchain Open Issues, Research Gaps, and Road Map of Future Blockchain for 2030: Charting the Digital Future. In P. Nijalingappa & M. Ghonge (Eds.), *Blockchain Technologies and Applications for Digital Governance* (pp. 200–229). IGI Global. doi:10.4018/978-1-7998-8493-4.ch009

Kuhail, M. A., Mathew, S. S., Hammad, R., & Bahja, M. (2022). Blockchain Primer: Introduction to Blockchain Foundations and Implementation. In S. Khan, M. Syed, R. Hammad, & A. Bushager (Eds.), *Blockchain Technology and Computational Excellence for Society 5.0* (pp. 28–47). IGI Global. doi:10.4018/978-1-7998-8382-1.ch003

Lin, R., Wang, L., Li, B., Lu, Y., Qi, Z., & Xie, L. (2022). Organizational governance in the smart era: the implications of blockchain. Nankai Business Review International. doi:10.1108/NBRI-02-2021-0014

Mahankali, S., & Chaudhary, S. (2020). Blockchain in Education: A Comprehensive Approach – Utility, Use Cases, and Implementation in a University. In R. Sharma, H. Yildirim, & G. Kurubacak (Eds.), *Blockchain Technology Applications in Education* (pp. 267–293). IGI Global. doi:10.4018/978-1-5225-9478-9.ch014

Martínez-Ríos, F. O., Marmolejo-Saucedo, J. A., & Abascal-Olascoaga, G. (2020). A New Protocol Based on Blockchain Technology for Transparent Operation of Corporate Social Responsibility. In Strategy, Power and CSR: Practices and Challenges in Organizational Management. Emerald Publishing Limited. doi:10.1108/978-1-83867-973-620201012

Oh, J., & Shong, I. (2017). A case study on business model innovations using Blockchain: Focusing on financial institutions. *Asia Pacific Journal of Innovation and Entrepreneurship*, *11*(3), 335–344. doi:10.1108/APJIE-12-2017-038

Oh, S.-C., Kim, M.-S., Park, Y., Roh, G.-T., & Lee, C.-W. (2017). Implementation of blockchain-based energy trading system. *Asia Pacific Journal of Innovation and Entrepreneurship*, *11*(3), 322–334. doi:10.1108/APJIE-12-2017-037

Önkan, Ö., & Arikan, Z. (2022). The Impact of Blockchain Technology on Tax and Accounting Practices. In P. Nijalingappa & M. Ghonge (Eds.), *Blockchain Technologies and Applications for Digital Governance* (pp. 1–36). IGI Global. doi:10.4018/978-1-7998-8493-4.ch001

Parlakkılıç, A. (2022). Blockchain Use Cases in Healthcare. In P. Nijalingappa & M. Ghonge (Eds.), *Blockchain Technologies and Applications for Digital Governance* (pp. 85–104). IGI Global. doi:10.4018/978-1-7998-8493-4.ch004

Pina, P. (2022). Blockchain and Copyright: Challenges and Opportunities. In S. Khan, M. Syed, R. Hammad, & A. Bushager (Eds.), *Blockchain Technology and Computational Excellence for Society 5.0* (pp. 125–145). IGI Global. doi:10.4018/978-1-7998-8382-1.ch007

Pranata, A. R., & Tehrani, P. M. (2022). The Legality of Smart Contracts in a Decentralized Autonomous Organization (DAO). In P. Tehrani (Ed.), *Regulatory Aspects of Artificial Intelligence on Blockchain* (pp. 112–131). IGI Global. doi:10.4018/978-1-7998-7927-5.ch006

Saeidnia, H. R., & Lund, B. D. (2023). Non-fungible tokens (NFT): A safe and effective way to prevent plagiarism in scientific publishing. *Library Hi Tech News*, *40*(2), 18–19. doi:10.1108/LHTN-12-2022-0134

Sharma, A., & Gupta, P. (2021). Blockchain Revolution: Adaptability in Business World and Challenges in Implementation. In S. Singh & A. Jurcut (Eds.), *Revolutionary Applications of Blockchain-Enabled Privacy and Access Control* (pp. 189–218). IGI Global. doi:10.4018/978-1-7998-7589-5.ch009

Spafford, M. L., Stanaway, D. F., & Chung, S. (2019). Blockchain and cryptocurrencies: A cross-border conundrum. *Journal of Investment Compliance*, *20*(3), 10–19. doi:10.1108/JOIC-05-2019-0027

Tang, Y., Xiong, J., Becerril-Arreola, R., & Iyer, L. (2020). Ethics of blockchain: A framework of technology, applications, impacts, and research directions. *Information Technology & People*, *33*(2), 602–632. doi:10.1108/ITP-10-2018-0491

Thomason, J. (2022). Fintech and Blockchain: Maximizing Benefit and Minimizing Harm. In I. Vasiliu-Feltes & J. Thomason (Eds.), *Applied Ethics in a Digital World* (pp. 185–198). IGI Global. doi:10.4018/978-1-7998-8467-5.ch013

Tiwari, C.K., & Pal, A. (2022). Using blockchain for global governance: past, present and future. *South Asian Journal of Business Studies*. doi:10.1108/SAJBS-07-2022-0252

Verma, S., & Sheel, A. (2022). Blockchain for government organizations: Past, present and future. *Journal of Global Operations and Strategic Sourcing*, *15*(3), 406–430. doi:10.1108/JGOSS-08-2021-0063

Wharton School of the University of Pennsylvania. (2018, November 28). *How the Blockchain Brings Social Benefits to Emerging Economies*. Wharton School of the University of Pennsylvania. https://knowledge.wharton.upenn.edu/article/blockchain-brings-social-benefits-emerging-economies/

Zaidi, T. (2022). Blockchain: Emerging Trends, Applications, and Challenges. In S. Khan, M. Syed, R. Hammad, & A. Bushager (Eds.), *Blockchain Technology and Computational Excellence for Society 5.0* (pp. 189–203). IGI Global. doi:10.4018/978-1-7998-8382-1.ch009

Chapter 2
A Blockchain-Based Tourism Industry:
How Promising Can It Be?

Stelios A. K. Ioannidis
University of the Aegean, Greece

ABSTRACT

Blockchain technology is highly regarded for its potential to revolutionize various economic sectors and meet consumer needs. In the tourism industry, blockchain has attracted investor attention due to its characteristics, such as decentralisation, disintermediation, security, transparency, and immutability. Blockchain applications like smart contracts, tokens, cryptocurrencies, decentralised applications (Dapps), and decentralised autonomous organisations (DAOs) have the potential to transform the relationship between tourism providers and tourists by improving trust, automating processes, and reducing costs. However, widespread adoption of blockchain in tourism is limited due to several challenges, including a lack of awareness and expertise, high energy consumption, and a lack of regulatory frameworks for data management and income taxation. Despite these challenges, blockchain technology's potential benefits make it an essential area to watch in the future.

INTRODUCTION

The advent of the Internet has accelerated the pace of change in tourism industry, as it has empowered customers to browse and reserve their travel products through online channels (Önder & Treiblmaier, 2018). Tourism and travel sectors are witnessing radical changes nowadays and are set to undergo an even more substantial

DOI: 10.4018/978-1-6684-4153-4.ch002

Copyright © 2023, IGI Global. Copying or distributing in print or electronic forms without written permission of IGI Global is prohibited.

transformation in the coming years, driven by the preferences and requirements of tourism consumers, with digitalization playing a pivotal role (Puri et al., 2023; Erceg et al., 2020; Ozdemir et al., 2020). Traditionally, technology has been at the core of business operations, with automating processes to reduce costs (Menon & Mady, 2021). From the demand side, new trends have been established, reshaping the entire tourism industry (Calvaresi et al., 2019), with modern tourists being actively engaged in crafting their travel experiences, assuming the role of brand ambassadors as they share their first-hand encounters and consequently enhance the appeal of a destination. In other words, "they become protagonists of a process of co-creation of value that can affect both the destination in general and individual companies operating in the sector" (Caddeo & Pinna, 2021, p. 1). It is exactly for that reason, according to Susanto et al. (2020), that tourism service providers actively develop smart tourism: the enhancement of the tourist experience.

As Büttgen et al. (2021) and Cobanoglu et al. (2021) explain, both academics and practitioners are deeply interested in novel technologies and the impacts of innovations on industry players and consumers, with blockchain technology being one of the most discussed disruptive and revolutionary innovations. Since 2016 and 2018 the World Economic Forum and Gartner have included blockchain in top 10 emerging technologies (Calvaresi et al., 2019). The potential of this technology to disrupt economic sectors and lead digital transformation with the ultimate goal of satisfying consumers' needs, along with other disruptive technologies such as artificial intelligence, 5G and IoT, is widely recognized by numerous researchers (Fragnière et al., 2022; Hashimy et al., 2022; Filimonau & Naumova, 2020; Ertemel, 2018). Industries such as tourism, banking, marketing, and advertising are expected to gain significant benefits through the implementation of blockchain technology (Baudier et al., 2022; Yan et al., 2021; Kwok & Koh, 2019; Harvey et al., 2018). Blockchain technologies demonstrate a disruptive impact on value chain and value co-creation (Novianti et al., 2022; Cobanoglu et al., 2021), a fact that have already attracted the interest of investors, as from a marketing perspective, investments on such disruptive technologies "are dependent on the added value derived from them" (Menon & Mady, 2021, p. 2). For the tourism industry in particular, innovations stemmed from the usage of blockchain technology for accommodation facilities or communication with customers can grant significant advantages over competitors (Demirel et al., 2021). In overall, blockchain is considered a disruptive technology that offers novel and unique value propositions to the market. Irrelevantly to its current adoption extent or whether it appears crucial to customers currently, there exists a correlation between technical advancement and shifts in customer demands (Aghaei et al. 2021). Disruptive technologies, though seemingly inferior to existing products, introduce fresh values for customers, such as affordability, convenience, size variations, and more.

In the current desk research, a comprehensive articles review has been conducted with the aim of shedding light on the potential changes that the usage of blockchain technology is expected to bring to the tourism industry. Considering the extremely low adoption rate of the particular technology, as it is still in its infancy (Puri et al., 2023; Raluca-Florentina, 2022; Arbatskaya & Khoreva, 2021; Alazab et al., 2021; Ozdemir et al., 2020; Dogru et al., 2018) and the acknowledgment of its great potential for tourism industry (Muharam et al., 2023; Baudier et al., 2022; Antoniadis et al., 2020; Melkić & Čavlek, 2020), the current study aspires to contribute to identifying the beneficial role of blockchain technology in tourism. To this end, a detailed description of its core characteristics along with its mode of operation is provided. Benefits and drawbacks for the tourism sector derived from the nature of blockchain technology are described, constituting the current study a significant source of guidelines for academics and practitioners in the field of tourism. In addition, this study responds to calls for further research on blockchain and tourism (Chang et al., 2022; Aghaei et al., 2021; Ahmad & Shah, 2021; Arbatskaya & Khoreva, 2021; Caddeo & Pinna, 2021; Treiblmaier, 2020; Benduch, 2019) customer experience (Fragnière et al., 2022), service management (Büttgen et al., 2021) and product design and distribution (Filimonau & Naumova, 2020). As stated by Treiblmaier (2021) about first publications on blockchain potential, they "triggered a hype that, more often than not, praised blockchain technology as a silver bullet for numerous business problems" (p. 1). Characteristically, Waldo (2019) confidently declares that it will change the way we work and live in general. Hence, clarification of real potential of blockchain technology for the tourism industry becomes of great significance before broad adoption takes place.

DEFINING BLOCKCHAIN

Blockchain is a portmanteau of the words "block" and "chain", indicating by default its operational process (Chang et al., 2022; Vacchio & Bifulco, 2022). "As the name suggests, a blockchain can be considered as a database where digital information (ie, "Blocks") is stored in a distributed network as a chain of blocks" (Hassija et al., 2021, p. 1). Consequently, a blockchain is a series of blocks that are chained with one another (Das et al., 2022). It is a digital (Abooleet & Kinnett, 2023), decentralized (Demirel et al., 2021), distributed (Ahmad & Shah, 2021) database with no central authority for the verification of the transactions (Das et al., 2022; Cobanoglu et al., 2021). Instead, all the executed transactions are shared and visible from all participants (Demirel et al., 2021). It is also described as a ledger (AlShamsi et al., 2022; Balasubramanian et al., 2022; Antoniadis et al., 2020; Ertemel, 2018), with the capability of recording all transactions (entries) among the members of the network. It has specific rules for building the chains of transactions and obtaining information (Arbatskaya & Khoreva, 2021). "We can imagine the blockchain network as the accounting books kept by a central bank to record all transactions, with the difference that the blockchain does not belong to any central authority but it is public and accessible to everyone. Everyone can have a "copy of this book" at any time while the user's personal information is not obvious but encrypted" (Varelas et al., 2019, p. 886). Ahmad and Shah (2021) explain that encryption offers anonymity to all participants involved.

According to the Cambridge Dictionary webpage (2023), blockchain is defined as a system "to record all the occasions a cryptocurrency is bought or sold and that is constantly growing as more blocks are added". As it concerns the blocks, each of them contains registries (transactions) into a Merkle Tree, with the hash of the previous block and the hash of the genesis block (Abooleet & Kinnett, 2023). Genesis block is called the initial block created in every blockchain (Menon & Mady, 2021). The corresponding technology refers to an add only repository of transactions, in the form of a list of linked blocks (Gökal et al., 2023) where every user participating can make an entry validated by other users. This is the major characteristic of a group of technologies named Distributed Ledger Technologies (DLTs) (Daskalakis & Georgitseas, 2020). DLTs are included in disruptive technologies and are characterized by decentralization, in other words, by the complete absence of a central authority for handling transactions (Antal et al., 2021). Blockchain is the most well-known DLT nowadays, although the two terms are often used interchangeably (Li & Kassem, 2021). In a blockchain context, the term "transaction" is attributed to any kind of record inside of a block and not necessarily a financial purchase. In fact, blockchain's conceptual roots can be traced back to the public ledgers of ancient or medieval societies, which documented significant events such as inheritance ownerships, births, deaths, judicial rulings and election results (Menon & Mady, 2021). However, unlike the old town ledger maintained by a human overseer, blockchain achieves a secure, transparent, and immutable repository by leveraging cryptography and distributed networks. From this angle, Tapscott & Tapscott (2018) define blockchain as a digital ledger, programmable to record everything of value in addition to financial transactions. The sum of transactions (a certain number of them) is what is referred as a "block" (Hewa et al., 2021). Accordingly, a blockchain is a decentralized ledger of all transactions (and all blocks) across a peer to peer (P2P) network (Himeur et al., 2022; Melko & Uvarova, 2021; Khan et al., 2021).

Based on Ozdemir et al., (2020), there is a specific pseudonym attached to every transaction (or a user code), implying that what blockchains (or DLTs) in essence offer to their users is pseudonymity and not anonymity. Pseudonymity is achieved through cryptographic techniques and guarantees data security, whilst consensus mechanisms in a P2P network play the role of the central authority or the commonly known trusted third party (Chen, 2018). Specifically, it is achieved with the use of public keys (Hassija et al., 2021). A public key is the unique address of a user within the network, is visible and known to everyone and is used by the user to receive funds (or digital assets), resembling the traditional bank accounts numbers (Daskalakis & Georgitseas, 2020). Public key encryption act as a shield to various cyber threats as ransomwares, trojans, botnets and worms, whilst ensures data authenticity and users privacy, according to Alazab et al. (2021) and Calvaresi et al. (2019). In this context, several researchers claim that blockchain should be rather considered as a combination of underlying technologies, such as cryptography, hashing, P2P network, distribution ledger and consensus protocols (Cobanoglu et al., 2021; Ertemel, 2018). P2P network means that participants directly interact with each other, and they all have the same rights and obligations, while distributed ledger and consensus algorithms ensure an immutable history log of all entries (Antal et al. 2021; Varelas et al., 2019).

S. Haber and W. S. Storrentta initially introduced the concept of blockchain in 1991, but it wasn't until Satoshi Nakamoto's publication of Bitcoin's whitepaper in 2008 and Bitcoin's genesis block creation in 2009 that the technology was associated with Bitcoin and gained in popularity (Baudier et al., 2022; Vacchio & Bifulco, 2022; Aghaei et al., 2021; Li and Kassem, 2021; Antoniadis et al., 2020; Önder & Treiblmaier, 2018). V. Buterin expanded Bitcoin's concept in the Ethereum project in 2014, allowing the development of any decentralized application with the deployment of smarts contracts apart from a payments system (Muharam et al., 2023; Buraga, 2021; Varelas et al., 2019), with a great number of cryptocurrencies and tokens been launched ever since (Puri et al., 2023). Although cryptocurrencies have been the initial application of blockchain technology (Kapil & Kapil, 2022), the common notion that blockchain is merely the underlying technology for them is an oversimplification. A thorough analysis of the key concepts that facilitated Bitcoin's success and led to the development of more cryptocurrencies and further applications highlights the great potential of blockchain components, such as linked timestamping, consensus protocols, public keys as IDs and smart contracts (Treiblmaier, 2020).

THE ROLE OF SMART CONTRACTS

The term "smart contract" was initially presented in 1994 and updated later in 1997 by the cryptographer Nick Szabo (Ante, 2021; Antoniadis et al., 2020; Benduch, 2019) with the aim of boosting trust and dedication to the terms among all parties involved in a contract (Dogru et al., 2018). "Many kinds of contractual clauses (such as collateral, bonding, delineation of property rights, etc.) can be embedded in the hardware and software we deal with, in such a way as to make breach of contract expensive (if desired, sometimes prohibitively so) for the breacher" (Szabo, 1997). Simply put, smart contracts are pieces of code that run on a blockchain network and impose terms of an agreement, without any human intervention, but with full traceability, limited transaction costs and limited execution time (Fragnière et al., 2022; Hewa et al., 2021; Khan et al., 2021). "Actually, they are software scripts, just like scripts that run on non-blockchain applications" (Ante, 2021, p. 4). They contain a set of rules that the parties agree to interact with each other, facilitating, verifying, and enforcing the negotiation or execution of an agreement or transaction (Ozdemir et al., 2020). Once activated by a blockchain transaction and confirmed by the network, predetermined actions are carried out (Ante, 2021).

Smart contracts are self-enforcing and self-executing pieces of code (Hewa et al., 2021; Tapscott & Tapscott, 2018). Their fundamental programming logic follows the structure of "if...then...else...", in which "if" denotes the satisfaction of preconditions, and "then" or "else" represents the corresponding execution section. "Then" is applied when the conditions are met, whereas "else" pertains to the opposite scenario. In the case of financial transactions and payments between two individuals, smart contracts can adequately serve in the simple form described above (Buterin, 2022). However, in cases of more complex agreements, where the judgment of a third party is needed to determine whether the terms of the agreement have been fulfilled or not, the above logic changes as follows: "if" preconditions are met is defined by the judge, which triggers the "then" or "else" commands. In real life, the role of the judge is played by any individual or authority or, in general, an entity trusted by all participants. In smart contracts, that person can be replaced by applications called Oracles, as "People are susceptible to collusion, bribery, error, and fraud" (Albizri & Appelbaum, 2021, p. 1). An oracle is a trusted data source that feeds the smart contract with information from the real world. The term derives from the ancient Greek mythology, where "certain individuals, usually women, were given that title. They were believed to have the power to channel messages from the gods to mortals) Now, oracles channel messages from data sources to smart contracts" (Hedera, 2023). In brief, smart contracts offer advantages across different sectors by offering disintermediation, ensuring transaction transparency, providing resistance to tampering, and enabling automated and accurate execution (Buterin, 2022; Caddeo & Pinna, 2021; Hewa et al., 2021; Daskalakis & Georgitseas, 2020; Tapscott & Tapscott, 2018).

BLOCKCHAIN CHARACTERISTICS

Several characteristics have been attributed to blockchain technology, as a result of the way it operates and the nature of its underlying technologies. Antoniadis et al., (2020) identifies six features, namely decentralization, data transparency, open source access, autonomy, immutability and anonymity. The above features are discussed by several researchers in blockchain literature. Decentralization is the main feature to support disintermediation (Marikyan et al., 2022; Hassija et al., 2021; Menon & Mady, 2021; Treiblmaier, 2020). Data transparency is achieved by the way blockchain operates, since once a piece of information is entered into the system, is recorded, stored and becomes visible to every user. Nobody can change or delete this piece of information or modify its timestamp (Kapil & Kapil, 2022; Arbatskaya & Khoreva, 2021). Data transparency leads to trust and auditability (Aghaei et al., 2021; Büttgen et al., 2021; Dogru et al., 2018) or traceability (Sharma et al., 2021; Ozdemir et al., 2020;), while for some researchers it's disintermediation that creates trust (Balasubramanian et al., 2022). Vacchio et Bifulco (2022) highlight decentralization, immutability, anonymity and transparency through traceability as the main 4 features of blockchain, in line previous researchers. For Hewa et al. (2021) cryptographic link is one more essential feature of blockchain, apart from decentralization and immutability. In an attempt to clarify the notion of cryptographic link, the authors explain that it refers to the link between each block in the chain which is cryptographic, meaning that each block contains a hash of the previous block in the chain. This creates an unbroken chain of records that cannot be altered or deleted without detection. The digital signature verifies the integrity of each record using hashing techniques and asymmetric key cryptography. When a record or transaction is added to a block, it is hashed, which means that a unique digital fingerprint is created for that record. Any attempt to alter a block or transaction record will violate the integrity of the blockchain.

Dogru et al. (2018) describe four features of blockchain in a more technical-oriented approach, namely shared ledger, security, efficiency and smart contracts. Shared ledger refers to the concurrent availability of every record to all members of the network, providing a "single source of truth" (p. 3). Security refers to consensus mechanisms, which act as a shield against data manipulation, efficiency is achieved by disintermediation and smart contracts enforce governing rules whilst "eliminate costs and delays associate with ordinary contracts" (p. 4). All the above features are acknowledged by Aghei et al. (2021), in addition to three more: peer-to-peer exchange, transparency with anonymity and immutability of data. Transparency and immutability are adequately explained by Treiblmaier (2020): "The immutability of the blockchain results from the fact that new data can only be appended, such that existing data remains unaltered in the chain. Blockchain can help to create transparency in systems since all participants can access the same data" (p. 5). In this way, all records (or transactions are visible by all members, while personal details of users are covered by the public key, which offers anonymity (or pseudonymity, as explained earlier). Immutability is also discussed by Irannezhad and Mahadevan (2021), who use the term 'irreversibility" of the original content, a feature that protects data records from fraudulent alterations.

BLOCKCHAIN IN HOSPITALITY AND TOURISM

The tourism and travel industry, among many other industries, may soon be significantly impacted by blockchain, which represents the latest in a series of technological advancements (Treiblmaier, 2020; Treiblmaier & Onder, 2019). The tourism industry is constantly seeking for novel approaches to enhance competitiveness, improve performance and increase customer satisfaction. All viewpoints about the numerous ways blockchain can benefit the tourism industry, sourced in tourism articles review of the current desk research, are illustrated in Table.1

Table. 1 Contribution of blockchain in travel and tourism industry

Authors	Feature/Contribution	Benefits in Tourism Industry
Aghei et al. (2021)	DLT P2P Encryption Smart contracts Transparency with anonymity Consensus mechanisms Decentralisation	Trust. Sense of security. Nice emotional experience for tourists. Enhance tourists' loyalty. Reduce costs. Disintermediation. Secure transfer of tourist data.
Ahmad & Shah (2021)	Online customer reviews Disintermediation Disintermediation Decentralized architecture Smart contracts	Trust in reviews. Reduce costs. Reliance on OTAs. Immutability against hacks. Avoidance of overbookings, baggage tracking, travelers' ID verification, coordination among hotels and transport system.
Antoniadis et al. (2020)	Online reviews Cryptocurrencies adoption Smart contracts Smart contracts ICOs ICOs	Trust in reviews. Comparative advantage. Speed of transactions. Security of transactions. Cash flow, funding. New investments, new experiences offered.
Ozdemir et al. (2020)	Online customer reviews Digitalisation of tourism Smart contracts Cryptocurrencies, Tokens Creation of DAOs	Trust in reviews Enhance tourist experience, Meet modern tourists' demands. Disintermediation, Rewards, Loyalty programs, currency exchange.
Balasubramanian et al. (2022)	Smart Contracts – Pre trip Smart Contracts – During trip Cryptocurrencies – During trip Smart Contracts – Post trip	Planning, Booking. ID verifications, E-gates, baggage tracking, insurance claims, payments, refunds. Transparent Reviews, Personalized offers for next trip, planning next trip.
Benduch (2019)	Smart Contracts Smart Contracts Decentralized Traceability Disintermediation	Transaction transparency, user friendly loyalty programs, real time luggage tracking, traveler's ID verification, contracts simplification, cooperation among all parties. Data verification and security. Trusted, secure, up to date data for the firms. Reduce costs.
Dogru et al. (2018)	Tracking guests Tracking food Loyalty programs Digital ID Smart contracts	Reduce wait time, check-in process. Quality control. Hotels, Airlines. Airports, Hotels, Restaurants. Cooperation hotels-travel agents, travel insurance, check-in process, inventory management.
Bulut (2021)	ICOs	Crowdfunding.

Continued on following page

Table 1. Continued

Authors	Feature/Contribution	Benefits in Tourism Industry
Caddeo & Pinna (2021)	Smart contracts & Dapps	Inventory management. Collaboration hotels-travel agencies. Traceability of food products. Baggage tracking. Reservations & ticketing. Travelers' loyalty, Privacy. Eliminate paperwork, store biometrics .
Calvaresi et al. (2019)	Loyalty programs ID management Tracking guests, food All transactions Rewards program	
Çapar (2020)	Attract tech savvy tourists	Destination choice criterion for medical tourists.
Chen & Tham (2023)	Cryptocurrencies	Feel of security, destination loyalty.
Cobanoglu et al. (2021)	Disintermediation Online reviews Gamification Baggage tracking Digital Identity Cryptocurrencies Loyalty programs	Reduce costs, own customers' data. Eliminate fake ratings. Manage tourism flows, enhance experience. Eliminate travelers' annoyance. Guests' data security. Reduce costs and time of payments. Rewards for guests.
Demirel et al. (2021)	Smart contracts Disintermediation Cryptocurrencies	Reduce costs about sustainability, access to rooms without physical contact, transparency in the terms of the contract. Democratization of the industry. Speed, efficiency, interoperability.
Daskalakis & Georgitseas (2020)	Smart contracts	Hotel and tickets reservation without human intervention.
Dogru et al. (2018)	Shared Ledger Security Efficiency Smart Contracts	Automation, Trust, Reduce costs, Disintermediation.
Erceg et al. (2020)	Blockchain use	Secure and transparent transactions. Ease and speed of retrieving past transactions. Promote a level of trust and confidence. Improve guest satisfaction. Promote a higher level of sophistication.
Ertemel (2018)	Smart contracts DAOs	Decentralization. Privacy. Reliability. Trust enabling, transparency. Personal Identity Ownership. Disintermediation. Value co-creation. Cost reduction. Democratization of supply.

Continued on following page

Table 1. Continued

Authors	Feature/Contribution	Benefits in Tourism Industry
Filimonau & Naumova (2020)	Smart contracts DLT Cryptocurrencies	Safe and secure payments. Rapid feedback. Better demand forecasting. Accurate billing and data. Knowledge sharing. Business cooperation. Carbon audit. Loyalty programs. Personalized marketing. Unbiased reviews.
Fragnière et al. (2022)	Smart contracts Cryptocurrencies	Reservations and ticketing. ID management. Loyalty programs. Digital payments. Credential management. Inventory management. Collaboration among all stakeholders.
Hewa et al. (2021)	Decentralization Immutability Cryptographic link	Disintermediation. Integrity of transactions. Transparency. Automation. Accuracy.
Melko & Uvarova (2021)	Smart contracts Cryptocurrencies	Automate transactions. Reduce time and preparations for agreements. Reduce paperwork. Reduce costs. Automate ID verifications. Loyalty programs.
Ilyas & Sukhrob (2022)	Tokens Cryptocurrencies	Tokenization of tourism services. Loyalty programs. Currency exchange. Reduce costs. Eliminates monopolies.
Irannezhad & Mahadevan (2021)		Reduce costs. Speed efficiency. Transparency. Tracking and tracing. Interoperability. Trust and authenticity. Enabling sharing economy. Risk reduction. Prevent human errors. Disintermediation.
Kapil & Kapil (2022)	Rating Systems Loyalty programs Transactions	Decentralized. Immutable – Resilient against cyber-attacks. Fault tolerant. Enhance guest experience.

Continued on following page

Table 1. Continued

Authors	Feature/Contribution	Benefits in Tourism Industry
Melkić & Čavlek (2020)	Smart contracts	Real time interactions. Greater personalization. Automation. Disintermediation.
Muharam et al. (2023)	Smart contracts	Enables P2P accommodation. Disintermediation. Enhances privacy. Data protection.
Öndern & Treiblmaier (2018).	Online Review systems Cryptocurrencies Disintermediation	Trustworthiness. New C2C markets. Increase providers' market power.
Ozdemir et al. (2020)	Consensus mechanisms Cryptocurrencies Governance model Smart contracts Tokens Blockchain platforms	Criteria for evaluating Dapps in tourism.
Puri et al. (2023)	Decentralization Efficiency of tourism system	Trustworthy. Transparent. Loyalty programs. Reduce costs.
Tham & Sigala (2020)	Smart contracts Cryptocurrencies Disintermediation ICOs	Manage corporate procurement. Manage B2B relations. Security protocols. Accessibility of SMEs to tourism markets. Fundraising. Sustainable development goals. Transparency and evaluation of tourism providers. Co-creating sustainable value.
Vacchio & Bifulco (2022)	Decentralization Immutability Anonymity Traceability and Transparency	Guarantee authenticity of work arts of cultural heritage, Tokenization of assets, rights management, and protection of intellectual property.
Varelas et al. (2019)	Smart contracts	Re-booking feature when rates are lowered. Real time booking – Overbooking prevention. Resell tickets. Best offers through integration of all travel mediums

Source: The Authors

According to Aghaei et al (2021), blockchain integration offers travel agencies exclusive opportunities to efficiently cater to customer preferences and meet their requirements, creating more personalized peer-to-peer interactions, and elevate value by strengthening customer loyalty. Peer-to-peer engagements foster trust and a feeling of security, which results in a positive emotional experience for tourists. This emotional experience, in turn, has a favorable impact on tourists' loyalty. Based on Valeri & Baggio, (2020) and Erceg et al. (2020)'s approach, blockchain technology is poised to transform the tourism industry by improving cost efficiency, speed, transparency of transactions, trust among all parties and promotion of innovation and sophistication in the sector. Hotels and accommodation facilities can now safely execute digital deals without intermediaries, democratizing participation between consumers and providers and creating a more comprehensive future for tourism (Demirel et al., 2021). Such a future includes features such as booking suggestions for hotel rooms when rates are lowered and eliminating potential double booking with smart contracts, real-time booking confirmations, peer-to-peer ticket reselling, and the integration of all major travel mediums into a single platform to offer the cheapest package deals (Varelas et al., 2019). In essence, smart contracts will be the core element of a variety of decentralized tourism applications (Dapps), which will eventually enhance smart tourism experience (Antoniadis et al., 2020), customer experience (Ozdemir et al., 2020), provide solutions to the tourism logistics industry (Arbatskaya & Khoreva, 2021) and, finally, make the entire industry more transparent, eliminating the current state of oligopolization or monopolization (Benduch, 2019).

Table 1 makes it evident that most authors emphasize the significance of disintermediation to reduce costs by eliminating commissions charged by third parties. Additionally, disintermediation grants tourism providers the advantage of obtaining and managing tourist data, which was previously controlled by intermediaries such as OTAs. The role of smart contracts in automating procedures, creating trust and reducing costs and time for both providers and tourists is also highlighted in blockchain literature. Usage of cryptocurrencies is also proposed as a solution to tackle foreign exchange barriers and interoperability in transactions (Ilyas & Sukhrob, 2022; Antoniadis et al., 2020; Ozdemir et al., 2020). Further, use of tokens is esteemed essential for optimization of loyalty programs and rewards distribution (Kapil & Kapil, 2022; Filimonau & Naumova, 2020). Tham and Sigala (2020) analyze the potential of blockchain and cryptocurrencies in advancing sustainable development goals. They highlight that this technology empowers small and medium-sized tourism businesses, which constitute the majority of the industry, as well as tourists. By removing intermediaries and enabling evaluation of tourism providers, blockchain creates a conducive environment for a smarter and more sustainable tourism ecosystem. The automation of processes for large corporations using blockchain technology can also contribute to a more sustainable environment. By enabling secure and transparent tracking of supply chain activities, blockchain can help to identify inefficiencies, reduce waste, and minimize the environmental impact of production and distribution processes. This can lead to more efficient resource management, lower carbon emissions, and greater sustainability in the long term. Additionally, the use of blockchain can provide greater accountability and transparency in corporate practices, promoting responsible business conduct and increasing consumer trust in the sustainability of the products and services being offered.

ADOPTION BARRIERS IN TOURISM INDUSTRY

Despite the potential benefits of blockchain technology, there are various obstacles to the adoption of blockchain technology that need to be addressed. One of these is the lack of awareness and expertise in blockchain (Marikyan et al., 2022; Aghaei et al., 2021; Ozdemir et al., 2020; Calvaresi et al., 2019; Harvey et al., 2018), which leads to various misunderstandings (Chang et al., 2022). This could be addressed by organizations such as chambers of commerce, tourism education institutions, universities, and academies of business and tourism by providing specialized training in blockchain for the tourism industry. It is essential to note that while blockchain-based technologies may have significant practical implications, the field is still relatively new and requires exploration of the technical and cryptographic aspects. This presents a particular obstacle for those who are interested in the technology (Benduch, 2019). Apart from this, a major finding of Büttgen et al. (2021)'s study is that "financial service organizations' managers do not prioritize blockchain implementation because they cannot see a clear path to value creation" (p. 74). Thus, from a managerial perspective, blockchain integration may seem a waste of sources under the current circumstances. After all, not a few times the hospitality industry, at the decision-making level, tends to be risk-averse, conservative and resistant to the embracement of disruptive innovations (Filimonau & Naumova, 2020). Another dimension is also added by Treiblmaier & Špan (2022) in the context of poor knowledge and misunderstanding. The authors argue that those two factors can lead to higher expectations or exaggerations about blockchain, making the current state of the technology seem insufficient. Accordingly, proponents of blockchain technology may seem as heralds of a fictional and unrealistic concept, and, as a result, not been taken seriously. In addition to that, the fact that blockchain technology does not constitute a technology per se, but the combination of a variety of underlying technologies, with most people having very limited knowledge on them, makes the entire blockchain solution as a blur and inconsistent concept.

The second obstacle is the absence of an appropriate regulatory framework regarding blockchain use by individuals and organizations (Kwok & Koh, 2019). Governments should collaboratively enact laws that facilitate the implementation of blockchain in travel and hospitality businesses and clarify grey areas (Aghaei et al., 2021). Specifically, as it concerns the crypto economy, Çapar (2020) points out the lack of a regulatory framework to ensure stability in the cryptocurrency market and prevent money laundering. The reluctance of tourists to utilize cryptocurrencies (or other forms of crypto assets) has an impact on the supply side, as well, because "until cryptocurrencies are accepted as an official financial instrument, it is unlikely that tourists will use them, and if they do not use them, then they cannot be used by tourism intermediaries either" (Melkić & Čavlek, 2020, p. 135).

The consensus approach of blockchain makes it challenging to scale (AlShamsi et al., 2022), as it constitutes a relevantly time-consuming mechanism, which can lead to significant delays when there are many users on the network. Some applications, also, require high energy consumption and integration of blockchains with multiple other systems (i.e., oracles). The technology gap has proved to be critical in such cases so far. This is the main reason, according to Arbatskaya & Khoreva (2021), that only 1% of organizations are using the blockchain technology in their supply chain operations by 2021 and only 35% are currently exploring the possibility of using blockchain technology. According to Ozdemir et al. (2020), the Amadeus platform attributed the limited adoption of blockchain technology in the tourism industry in 2017 to technological gaps, and expressed optimism that decentralized applications (Dapps) will gain broader usage in the sector once these gaps are filled by advancements in technology.

Additionally, one of the significant constraints on popularizing smart contracts in the tourism industry is that other technological solutions, particularly cloud solutions, can provide a similar level of security at lower organization costs. Furthermore, using a public network of blockchain such as Ethereum requires significant computing power and often results in unsatisfactory running speed. Another major risk associated with smart contracts is the possibility of errors, which can occur when they are programmed by humans. According to Szczebrowski (2019), as cited in Benduch (2019), if a mistake is found in a smart contract that has been embedded in blockchain, there are significant technical limitations that allow for it to be fixed, unlike traditional programs. However, there have been instances of malpractice in economics where errors in the code of smart contracts were exploited, emphasizing the need for thorough testing and consideration of potential flaws and risks associated with software code and hacker attacks when using smart contracts for contracting purposes. This viewpoint comes in line with Calvaresi et al. (2019)'s argument that "blockchain technology itself does not guarantee security and privacy, therefore, more research / building new mechanisms on top has to be studied and applied to real-word use-cases" (p. 11).

Hassija et al. (2021) added pseudo-anonymity as a factor to the list of drawbacks for adopting blockchain. According to the authors, while blockchain users can benefit from anonymity by using public keys, the decentralized nature of the blockchain makes every transaction visible and accessible to everyone. As a result, the public key associated with each transaction is also visible, which reduces anonymity. This is particularly true in the scenario that a user knows the public key of another user. The first can easily track and access the entire transaction log of the latter, since the public key serves as their identity. Efforts have already been made to address the issue of pseudo-anonymity, with the usage of new cryptographic techniques.

CONCLUSION AND DISCUSSION

The goal of the present articles review study is to elucidate the unique characteristics of blockchain technology and conceptualize how it can be implemented in the travel and tourism industry. In this regard, a comprehensive literature review on blockchain in tourism hospitality has been conducted, shedding light on the available blockchain applications today and their distinct characteristics. The use of smart contracts is generally acknowledged to have the potential to revolutionize the tourism industry, by showcasing significant benefits for tourism enterprises, such as disintermediation, reduce costs, automation in processes, real time transactions, data management, more efficient customers targeting, integrity and trust in collaborations among all stakeholders, and a novel channel of income. Additionally, smart contracts can benefit tourists by reducing costs, reducing waiting time, ensuring anonymity, personal data security, enhancing tourist experience and trust in every provider-guest interaction. This conclusion is consistent with Aghaei et al.'s (2021) findings which suggest that implementation of blockchain may built new business models and new structures in tourism industry. According to the authors, disintermediation and peer-to-peer interactions seem to better meet stakeholders and tourists' needs compared to the current state of the tourism system.

The decentralized nature of blockchain with consensus protocols being in charge of authorizing every transaction, in addition to data traceability, seem to have a great impact on users' attitude towards blockchain, according to Muharam et al. (2023) findings. The authors add that "In this context, self-sovereignty is a form of user empowerment in which users have more control over how their data is used in the system" (p. 8). From a broader standpoint, blockchain as a new technology can affect tourists' perception of a destination. According to Novianti et al. (2022) smart tourism applications have a significant impact on the attitudes, perceived behavioral control, beliefs, and subjective norms of tourists, ultimately influencing their intention to select and visit various tourist attractions. The use of cryptocurrencies can also contribute to reduced costs for overseas travelers, as they minimize commissions and foreign currency exchange rates. Apart from using cryptocurrencies for financial transactions, tourism firms can also utilize tokens as rewards for loyalty programs or for governance purposes. Additionally, by tokenizing tourism assets or services and integrating them with a peer-to-peer (P2P) system, a new market model is created where every user can sell, resell or trade tickets, accommodation, or activities. In essence, blockchain technology lays the foundation for every user to function as a travel agent and generate extra income by selling goods or services in a P2P network. This feature can prove useful in cases of cancellations, where a tourism package or activity can be resold from one user to another, even at a higher rate with a profit margin, instead of adhering to the provider's cancellation policy and potentially losing a portion of the initial payment.

The present study examined adoption barriers, and it was found that the tourism industry's reluctance to adopt blockchain technology on a large scale is primarily due to inadequate knowledge about the technology. A lack of knowledge about the technology could potentially hinder user experience, as highlighted by Kamble et al. (2019) research, which found that perceived ease of use plays a critical role in shaping perceived usefulness of blockchain. This result is in line with Büttgen et al. (2021) assertion that blockchain has yet to find its place in terms of value creation at the managerial level. Furthermore, the technology's lagging development is a significant factor contributing to its slow performance at times and challenging the claims of proponents regarding its security. The lack of regulations has also been recognized as a significant factor that raises concerns about blockchain technology being used as a facilitator for money laundering and tax evasion. The current study fully agrees with Treiblmaier (2020)'s statement that "blockchain is not a silver bullet for all organizational problems and several shortcomings exist" (p. 7), the present study acknowledges the importance of showcasing the current state of this technology, including its advantages and disadvantages, for the benefit of both academic research and practical implementation.

REFERENCES

Abooleet, S., & Kinnett, S. (2023). A Systematic Review of Blockchain-based Loyalty Programs. *Proceedings of the 56th Hawaii International Conference on System Sciences*, 5323-5332. https://hdl.handle.net/10125/103284

Aghaei, H., Naderibeni, N., & Karimi, A. (2021). Designing a tourism business model on block chain platform. *Tourism Management Perspectives*, *39*, 100845. doi:10.1016/j.tmp.2021.100845

Ahmad, M. S., & Shah, S. M. (2021). Moving Beyond the Crypto-Currency Success of Blockchain: A Systematic Survey. *Scalable Computing: Practice and Experience*, *22*(3), 321–346. doi:10.12694cpe.v22i3.1853

Alazab, M., Alhyari, S., Awajan, A., & Abdallah, A. B. (2021). Blockchain technology in supply chain management: An empirical study of the factors affecting user adoption/ acceptance. *Cluster Computing*, *24*(1), 83–101. doi:10.100710586-020-03200-4

Albizri, A., & Appelbaum, D. (2021). Trust but Verify: The Oracle Paradox of Blockchain Smart Contracts. *Journal of Information Systems*, *35*(2), 1–16. doi:10.2308/ISYS-19-024

AlShamsi, M., Al-Emran, M., & Shaalan, K. (2022). A Systematic Review on Blockchain Adoption. *Applied Sciences (Basel, Switzerland)*, *12*(9), 4245. doi:10.3390/app12094245

Antal, C., Cioara, T., Anghel, I., Antal, M., & Salomie, I. (2021). Distributed Ledger Technology Review and Decentralized Applications Development Guidelines. *Future Internet*, *13*(3), 62. doi:10.3390/fi13030062

Ante, L. (2021). Smart contracts on the blockchain – A bibliometric analysis and review. *Telematics and Informatics*, *57*, 101519. doi:10.1016/j.tele.2020.101519

Antoniadis, I., Spinthiropoulos, K., & Kontsas, S. (2020). Blockchain applications in tourism and tourism marketing: A short review. Strategic Innovative Marketing and Tourism. In A. Kavoura, E. Kefallonitis, & P. Theodoridis (Eds.), *Strategic Innovative Marketing and Tourism. Springer Proceedings in Business and Economics* (pp. 375–384). Springer. doi:10.1007/978-3-030-36126-6_41

Arbatskaya, E., & Khoreva, L. (2021). Blockchain as an innovative digitalization achievement in logistics of eco-tourism. *E3S Web of Conferences, 258*. https://doi. org/ doi:10.1051/e3sconf/202125802016

Balasubramanian, S., Sethi, J. S., Ajayan, S., & Paris, C. M. (2022). An enabling Framework for Blockchain in Tourism. *Information Technology & Tourism*, *24*(2), 165–179. doi:10.100740558-022-00229-6

Baudier, P., Chang, V., & Arami, M. (2022). The Impacts of Blockchain on Innovation Management: Sectoral Experiments: *Journal of Innovation Economics & Management, 37*(1), 1–8. doi:10.3917/jie.037.0001

Benduch, D. (2019). *Risks And Opportunities for Tourism Using Smart Contracts. 6th Geographic Information Systems Conference and Exhibition "GIS ODYSSEY 2019"*. Retrieved from https://www.researchgate.net/publication/344340600

Bouraga, S. (2021). A taxonomy of blockchain consensus protocols: A survey and classification framework. *Expert Systems with Applications*, *168*, 114384. doi:10.1016/j.eswa.2020.114384

Buhalis, D., & Law, R. (2008). Progress in information technology and tourism management: 20 years on and 10 years after the Internet—The state of eTourism research. *Tourism Management, 29*(4), 609–623. doi:10.1016/j.tourman.2008.01.005

Bulut, E. (2021). Blockchain-based entrepreneurial finance: Success determinants of tourism initial coin offerings. *Current Issues in Tourism*, 1–15. doi:10.1080/13683500.2021.1980505

Buterin, V. (2022). *Proof of Stake*. Seven Stories Press.

Büttgen, M., Dicenta, J., Spohrer, K., Venkatesh, V., Raman, R., Hoehle, H., De Keyser, A., Verbeeck, C., Zwienenberg, T. J., Jorgensen, K. P., ... Schar, F. (2021). Blockchain in Service Management and Service Research—Developing a Research Agenda and Managerial Implications. *Journal of Service Management Research, 5*(2), 71–102. doi:10.15358/2511-8676-2021-2-71

Caddeo, F., & Pinna, A. (2021). *Opportunities and challenges of Blockchain-Oriented systems in the tourism industry*. doi:10.15358/2511-8676-2021-2-71

Calvaresi, D., Leis, M., Dubovitskaya, A., Schegg, R., & Schumacher, M. (2019). Trust in Tourism via Blockchain Technology: Results from a Systematic Review. In J. Pesonen & J. Neidhardt (Eds.), *Information and Communication Technologies in Tourism 2019* (pp. 304–317). Springer International Publishing. doi:10.1007/978-3-030-05940-8_24

Cambridge Dictionary. (2023). *Blockchain*. Retrieved from https://dictionary.cambridge.org/dictionary/english/blockchain

Çapar, H. (2020). Using cryptocurrencies and transactions in medical tourism. *Journal of Economic and Administrative Sciences., 37*(4), 677–693. doi:10.1108/JEAS-07-2019-0080

Chang, M., Arachchilage, C. S. M. W., Kim, M., & Lim, H. (2022). Acceptance of Tourism Blockchain Based on Utaut and Connectivism Theory. SSRN *Electronic Journal*. doi:10.2139/ssrn.4017406

Chen, S.-H., & Tham, A. (2023). A crypto-tourism case study of agnes water/seventeen seventy, Australia. *Tourism and Hospitality Research, 23*(1), 108–112. doi:10.1177/14673584221085472

Chen, Y. (2018). Blockchain tokens and the potential democratization of entrepreneurship and innovation. *Business Horizons, 61*(4), 567–575. doi:10.1016/j.bushor.2018.03.006

Cobanoglu, C., Dogan, S., Berezina, K., & Collins, G. (2021). *Hospitality and Tourism Information Technology*. USF M3 Publishing, LLC. https://doi.org/doi:10.5038/9781732127593

DasS.RoutJ.PriyadarshiniR.MishraM. (2022). A Comparative Analysis of the Consensus Algorithms in Blockchain Technology. SSRN. Retrieved from doi:10.2139/ssrn.4091413

Daskalakis, N., & Georgitseas, P. (2020). *An Introduction to Cryptocurrencies*. Routledge. doi:10.4324/9780429352584

Demirel, E., Karagöz Zeren, S., & Hakan, K. (2021). Smart contracts in tourism industry: A model with blockchain integration for post pandemic economy. *Current Issues in Tourism*, 1–15. doi:10.1080/13683500.2021.1960280

Dogru, T., Mody, M., & Leonardi, C. (2018). Blockchain Technology & its Implications for the Hospitality Industry. *Boston Hospitality Review*, 1-12. Retrieved from https://www.bu.edu/bhr/files/2018/02/Blockchain-Technology-and-its-Implications-for-the-Hospitality-Industry.pdf

Erceg, A., Damoska Sekuloska, J., & Kelić, I. (2020). Blockchain in the Tourism Industry—A Review of the Situation in Croatia and Macedonia. *Informatics (MDPI)*, *7*(1), 5. doi:10.3390/informatics7010005

Ertemel, A. V. (2018). Implications of Blockchain Technology on Marketing. *Journal of International Trade. Logistics and Law*, *4*(2), 35–44.

Filimonau, V., & Naumova, E. (2020). The blockchain technology and the scope of its application in hospitality operations. *International Journal of Hospitality Management*, *87*, 102383. doi:10.1016/j.ijhm.2019.102383

Fragnière, E., Sahut, J.-M., Hikkerova, L., Schegg, R., Schumacher, M., Grèzes, S., & Ramseyer, R. (2022). Blockchain technology in the tourism industry: New perspectives in Switzerland. *Journal of Innovation Economics Management*, *37*(1), 65–90. doi:10.3917/jie.pr1.0111

Gökal, H., Buzun, M., & Adalier, A. (2023). Blockchain: A Content Analysis of Graduate Theses in Turkey. *The Turkish Online Journal of Educational Technology*, *22*(1). http://www.tojet.net/articles/v22i1/2215.pdf

Harvey, C. R., Moorman, C., & Castillo Toledo, M. (2018). How Blockchain Will Change Marketing As We Know It. SSRN *Electronic Journal*. doi:10.2139/ssrn.3257511

Hashimy, L., Jain, G., & Grifell-Tatjé, E. (2022). Determinants of blockchain adoption as decentralized business model by Spanish firms – an innovation theory perspective. *Industrial Management & Data Systems*. Advance online publication. doi:10.1108/IMDS-01-2022-0030

Hassija, V., Zeadally, S., Jain, I., Tahiliani, A., Chamola, V., & Gupta, S. (2021). Framework for determining the suitability of blockchain: Criteria and issues to consider. *Transactions on Emerging Telecommunications Technologies*, *32*(10). Advance online publication. doi:10.1002/ett.4334

Hedera. (2023). Blockchain Oracles: How Smart Contracts Talk to the World. *Hedera website*. Retrieved from https://hedera.com/learning/decentralized-finance/blockchain-oracle

Hewa, T., Ylianttila, M., & Liyanage, M. (2021). Survey on blockchain based smart contracts: Applications, opportunities and challenges. *Journal of Network and Computer Applications*, *177*, 102857. doi:10.1016/j.jnca.2020.102857

Himeur, Y., Sayed, A., Alsalemi, A., Bensaali, F., Amira, A., Varlamis, I., Eirinaki, M., Sardianos, C., & Dimitrakopoulos, G. (2022). Blockchain-based recommender systems: Applications, challenges and future opportunities. *Computer Science Review*, *43*, 100439. doi:10.1016/j.cosrev.2021.100439

Ilyas, A., & Sukhrob, B. (2022). Advantages of Using and Developing Blockchain Technology in Tourism. *Journal of Advanced Research and Stability*, *2*(1), 525–526.

Irannezhad, E., & Mahadevan, R. (2021). Is blockchain tourism's new hope? *Journal of Hospitality and Tourism Technology*, *12*(1), 85–96. doi:10.1108/JHTT-02-2019-0039

Kamble, S., Gunasekaran, A., & Arha, H. (2019). Understanding the Blockchain technology adoption in supply chains-Indian context. *International Journal of Production Research*, *57*(7), 2009–2033. doi:10.1080/00207543.2018.1518610

Kapil, S., & Kapil, K. N. (2022). Blockchain in Hospitality and Tourism Industry Way Forward. *International Journal of Business and Economics*, *6*(2), 289–298. doi:10.5281/zenodo.5839071

Khan, S. N., Loukil, F., Ghedira-Guegan, C., Benkhelifa, E., & Bani-Hani, A. (2021). Blockchain smart contracts: Applications, challenges, and future trends. *Peer-to-Peer Networking and Applications*, *14*(5), 2901–2925. doi:10.100712083-021-01127-0 PMID:33897937

Kwok, A. O. J., & Koh, S. G. M. (2019). Is blockchain technology a watershed for tourism development? *Current Issues in Tourism*, *22*(20), 2447–2452. doi:10.1080/13683500.2018.1513460

Li, J., & Kassem, M. (2021). Applications of distributed ledger technology (DLT) and Blockchain-enabled smart contracts in construction. *Automation in Construction*, *132*, 103955. doi:10.1016/j.autcon.2021.103955

Marikyan, D., Papagiannidis, S., Rana, O. F., & Ranjan, R. (2022). Blockchain adoption: A study of cognitive factors underpinning decision making. *Computers in Human Behavior*, *131*, 107207. doi:10.1016/j.chb.2022.107207

Melkić, S., & Čavlek, N. (2020). The impact of blockchain technology on tourism intermediation. *Tourism (Zagreb)*, *68*(2), 130–143. doi:10.37741/t.68.2.2

Melko, M., & Uvarova, G. (2021). Blockchain Technology in the Tourist Sphere: Realities of Today. *International Conference on Economics, Accounting and Finance*. Retrieved from https://library.krok.edu.ua/media/library/category/materiali-konferentsij/melko_0034.pdf

Menon, M., & Mady, A. (2021). Blockchain: A Non-Technical Primer For Marketing Academics. *Marketing Education Review*, *32*(3), 1–19. doi:10.1080/10528008.2021.1943445

Muharam, I. N., Tussyadiah, I. P., & Kimbu, A. N. (2023). A theoretical model of user acceptance of blockchain-based peer-to-peer accommodation. *Current Issues in Tourism*. doi:10.1080/13683500.2022.2164485

Novianti, S., Susanto, E., & Rafdinal, W. (2022). Predicting Tourists' Behaviour Towards Smart Tourism: The Case in Emerging Smart Destinations. *Journal of Tourism Sustainability*, *2*(1), 19–30. doi:10.35313/jtospolban.v2i1.30

Önder, I., & Treiblmaier, H. (2018). Blockchain and tourism: Three research propositions. *Annals of Tourism Research*, *72*(C), 180–182. doi:10.1016/j.annals.2018.03.005

Ozdemir, A. I., Ar, I. M., & Erol, I. (2020). Assessment of blockchain applications in travel and tourism industry. *Quality & Quantity*, *54*(5), 1549–1563. doi:10.100711135-019-00901-w

Puri, V., Mondal, S., Das, S., & Vrana, V. G. (2023). Blockchain Propels Tourism Industry—An Attempt to Explore Topics and Information in Smart Tourism Management through Text Mining and Machine Learning. *Informatics (MDPI)*, *10*(1), 9. doi:10.3390/informatics10010009

Raluca-Florentina, T. (2022). The Utility of Blockchain Technology in the Electronic Commerce of Tourism Services: An Exploratory Study on Romanian Consumers. *Sustainability (Basel), 14*(2), 943. doi:10.3390u14020943

Sharma, M., Sehrawat, R., Daim, T., & Shaygan, A. (2021). Technology assessment: Enabling Blockchain in hospitality and tourism sectors. *Technological Forecasting and Social Change, 169*, 120810. doi:10.1016/j.techfore.2021.120810

Susanto, E., Novianti, S., Rafdinal, W., Prawira, M. F. A., & Septyandi, C. B. (2020). Visiting Tourism Destination: Is It Influenced by Smart Tourism Technology? *Journal of Indonesian Tourism and Development Studies, 8*(3), 145–155. doi:10.21776/ub.jitode.2020.008.03.04

Szabo, N. (1997). *The Idea of Smart Contracts.* Retrieved from https://nakamotoinstitute.org/literature/the-idea-of-smart-c
ontracts

Tapscott, D., & Tapscott, A. (2018). *Blockchain Revolution.* Portfolio Penguin. doi:10.1108/JHTT-05-2019-0069

Tham, A., & Sigala, M. (2020). Road block (chain): Bit (coin) s for tourism sustainable development goals? *Journal of Hospitality and Tourism Technology, 11*(2), 203–222. doi:10.1108/JHTT-05-2019-0069

Treiblmaier, H. (2020). Blockchain and Tourism. In Z. Xiang, M. Fuchs, U. Gretzel, & W. Höpken (Eds.), Handbook of e-Tourism (pp. 1–21). Springer International Publishing. doi:10.1007/978-3-030-05324-6_28-1

Treiblmaier, H. (2021). *Need some inspiration for blockchain research? Try tokenization!* Academia Letters. doi:10.20935/AL2937

Treiblmaier, H., & Onder, I. (2019). The impact of blockchain on the tourism industry: A theory-based research framework. In H. Treiblmaier & R. Beck (Eds.), *Business Transformation through blockchain* (pp. 3–21). Palgrave Macmillan. doi:10.1007/978-3-319-99058-3_1

TreiblmaierH.ŠpanŽ. (2022). Will Blockchain Really Impact Your Business Model? Empirical Evidence from Slovenian SMEs. *Economic and Business Review, 24*(2), 132–140. doi:10.15458/2335-4216.1302

Vacchio, E. D., & Bifulco, F. (2022). Blockchain in Cultural Heritage: Insights from Literature Review. *Sustainability (Basel), 14*(4), 2324. doi:10.3390u14042324

Valeri, M., & Baggio, R. (2021). A critical reflection on the adoption of blockchain in tourism. *Information Technology & Tourism*, *23*(2), 121–132. doi:10.100740558-020-00183-1

Varelas, S., Georgitseas, P., Nechita, F., & Sahinidis, A. (2019). Strategic Innovations in Tourism Enterprises Through Blockchain Technology. In A. Kavoura, E. Kefallonitis, & A. Giovanis (Eds.), *Strategic Innovative Marketing and Tourism* (pp. 885–891). Springer International Publishing. doi:10.1007/978-3-030-12453-3_102

Waldo, J. (2019). A Hitchhiker's Guide to the Blockchain Universe? Blockchain remains a mystery, despite its growing acceptance. *ACM*, *16*(6), 38-42. Retrieved from https://queue.acm.org/detail.cfm?id=3305265

Yan, C., Zhu, J., Ouyang, Y., & Zeng, X. (2021). Marketing Method and System Optimization Based on the Financial Blockchain of the Internet of Things. *Wireless Communications and Mobile Computing*, *2021*, 1–11. doi:10.1155/2021/9354569

Chapter 3
Strategic Planning and Policy Framework for Implementation of Blockchain Technology in Education in India

Anjana
Indira Gandhi National Open University, India

Siran Mukerji
https://orcid.org/0000-0001-9861-3013
Indira Gandhi National Open University, India

ABSTRACT

Blockchain technology is considered to be one of the most significant innovations in information technology of the 21st century. There is a wide spectrum of application of this technology in sectors including education. In view of the increasing popularity, greater acceptance, and merits of this technology, the Government in India has initiated steps towards adopting this technology. Blockchain technology as an important research area has the potential for its application in various domains (i.e., governance, banking and finance, education, cyber security, etc.). The objectives of this chapter are to showcase the strategic planning and policy framework for implementation of this technology in India and provide an insight into the international scenario and current status of implementation of blockchain technology across sectors in India. It deliberates upon the multidimensional use of blockchain and analyses the challenges in its implementation and the way ahead for this technology in education in India.

DOI: 10.4018/978-1-6684-4153-4.ch003

Copyright © 2023, IGI Global. Copying or distributing in print or electronic forms without written permission of IGI Global is prohibited.

Blockchain technology, synonymous with distributed ledger technology, which is considered to be one of the most significant innovations in information technology of the 21st Century, has been widely recognized across the globe as a network of safe data storage and transactions. Given the design and development of this network, it enables data sharing amongst multiple entities, without compromising on the data security with a layer of trust without having to validate the transactions by a third party. With the passage of time, there has been a wide spectrum of application of this distributed ledger based technology in varied sectors including education.

In view of the increasing popularity, greater acceptance worldwide and relative merits of using this technology in the organizations, the Central as well as the state Governments in India have also initiated a number of steps towards adopting this blockchain technology. In this context, according to the Ministry of Electronics and Information Technology (Meity), Government of India, Blockchain Technology as an important research area, has the potential for its application in various domains such as governance, banking & finance, education, cyber security amongst many other sectors.

One of the objectives of this paper is an attempt to showcase the strategic planning and policy framework for implementation of blockchain technology in India. While providing an insight into the international scenario, it delves in detail on the National Strategy of Blockchain by Government of India and also the Current Status of Implementation of Blockchain Technology across different sectors in India. The paper deliberates at length the multidimensional use of blockchain presently in India, beginning with certification management by CBSE, logistics supply chain management of drugs (Aushada), certification chain management, supply chain tracking of commodities and database management under the PDS in the country, blood bank management system, and land records management. Thereafter, in the subsequent section, an analysis of the challenges that exist in the implementation of blockchain technology in the education sector in India has been done and the way ahead for Blockchain Technology in the Education Sector in India has been discussed.

BLOCKCHAIN TECHNOLOGY: A BRIEF INTRODUCTION

As is widely known, Blockchain which is an individually connected array of blocks with each block comprising several transactions yielding a distributed, incontrovertible data store that can be used in a broad range of applications (Fanning and Centers, 2016) is one of the most significant innovations of the 21st century in the IT industry. In technical parlance, it is distributed ledger technology, introduced by Satoshi Nakamoto in 2009 for the design and development of cryptocurrency "Bitcoin". In business terms, it allows for sharing of a ledger among a number of business entities for conducting a transaction, thus, enabling a layer of trust and removing the necessity for validating the transaction by a third party. Therefore the data and transactions are stored as blocks in the Blockchain and cannot be tampered with, thus adding the much needed security to the transaction. The inherent features namely, tamper-evident, consensus based validation of transactions, irreversibility, openness and accountability for all transactions carried out in a blockchain network and secure data storage has made this technological innovation popular for its adoption worldwide.

In fact due to these vital inherent features, the Blockchain is being widely used in e-governance also and it is expected to usher in major transformations in the services that e-governance can provide. With the application of this technology, the governance data remains secure and adds transparency, vigilance, trust and accountability. Thus, in other words, it promotes more trust within the various stakeholders in the form of a collaborative and consortium environment, suiting it for implementation across multiple departments and agencies where they are mandated to work on a particular assignment, making the system more efficient and trustworthy. The blockchain technology enables the creation of decentralized applications programmed to run on a network and records sets of data in encrypted form that can be shared securely without third-party mediation.

IMPORTANCE OF BLOCKCHAIN TECHNOLOGY IN EDUCATION SECTOR

In the field of education, the University of Nicosia in Cyprus has been reported to be the first University in the world that issued academic certificates with authenticity verification through the blockchain technology (Sharples and Domingue, 2016). In addition to issuing certificates, it has been suggested that this technology can be applied in educational institutions for various purposes such as administering academic activities, administering MOOCs all the way through actual degrees, recording learners' acquired competencies, learner evaluation and assessment (Clark, 2016), effective and efficient recognition and management of certification and accreditation as well as monitoring the mobility of students, educators and professionals (Gatteschi et al., 2020). Apart from these areas, three potential domains of learning and instruction for future use of blockchain technology include Formative assessment, learning activities design/implementation, and records of learning processes (Chen et al., 2018). Also, blockchain technology can execute other tasks in educational administration and management through its shared ledgers such as staff hiring via smart contracts, continuous professional development (Clark, 2016), performance assessment and payments right from the first day of work to retirement pensions.

This technology due to its capacity in allowing stakeholders to validate learning records and identity management makes it suitable for applications in higher education where a huge amount of data is exchanged among the institutions in terms of certificates. In addition, its distributed ledger without the need for a trusted third party can improve smart contract based protocols that automatically enforce a contract in students throughout multiple levels of administration, which is a major advantage of blockchain technology for higher education (Arndt, 2018).

It is expected that blockchain technology will accelerate the automation of the award, recognition and transfer of credits, thereby ending the paper-based system for certificates and storage and verification of complete record of formal and non-formal achievements of students, while the qualifications and records of achievement be permanently and reliably secured. It will also enable automatic verification of the validity of certificates, eliminating the need to contact the issuing organization for validating credentials.

In the near future, certificates of accreditation being issued to institutions by quality assurance bodies, or licenses to teach being issued to educators may be publicly available and verifiable by any user against a blockchain. It also has the applicability in intellectual property management for tracking of publication and citations, without the need for management of databases by a central authority enabling the possibility of automatic tracking of open educational resources use and re-use. With the implementation of blockchain technologies for creating data management structures with users having increased ownership and control over their own data, data management costs of educational organizations' as well as liability resulting from data management issues can be significantly reduced (Grech and Camilleri, 2017).

As is commonly known, Blockchain technology is now being increasingly applied in many sectors due to inherent characteristics of maintaining the data security and adding transparency, vigilance, trust and accountability to the inter-departmental, inter-organizational and multi-locational transactions and data processing. In the following section, an attempt is made to discuss and delve into the aspects of its application in the global context as well as in the Indian context.

APPLICATIONS OF BLOCKCHAIN TECHNOLOGY

As has been discussed earlier, the application of blockchain technology is widespread across nations, sectors and functions. While being considered for use in cryptocurrency only earlier, its acceptability and usage is gaining ground with the passage of time. The application of this distributed ledger technology is being discussed here under two heads, one, the international scenario and second, the national perspective.

Application of Blockchain Technology: The International Scenario

In principle, blockchain technology and cryptocurrency have almost become synonymous for general parlance, understanding and practice. However, the usage of this technology in wider areas globally is gradually gaining ground and the relative advantages of its utilization is being recognized by everyone. Both public and private organizations are devising function-specific platforms by using blockchain technology including providing it as a Service to the customers. A brief account of the major developments taking place worldwide in this sector is being highlighted in the ensuing paragraphs.

1. **European Blockchain Services Infrastructure (EBSI):** The European countries through European Blockchain Partnership (EBP) aims to develop an effective, secure and resilient EBSI with the use of blockchain and distributed ledger technology. The aim of this infrastructure is to meet the highest standards of IT related functions in terms of interoperability, privacy, cyber security and regulation in applying the policies. Further towns such as Zug in Switzerland are using this technology in the form of Bitcoin payments for tax. Infact, it has also successfully developed Blockchain based voting systems, apart from creating Blockchain TaskForce, for providing guidance to the companies towards implementing Blockchain framework. Likewise, the Estonian government is using blockchain based Keyless Signature Infrastructure (KSI) for proving the authenticity of electronic records. They have also implemented X-Road, an open source data exchange layer that enables secure exchange of information within different organizations.
2. **Use of Ethereum Platform in South America:** Countries such as Brazil and Chile have been using the Ethereum platform for various purposes. Brazil is using Ethereum platform for developing various applications including for voting purposes, student certificates and tracking student performance, apart from focusing on public bidding of contracts with the government's on-line bid solution for ascertaining secure and transparent deals for agriculture applications. Besides Brazil, Chile is also using the Ethereum platform in the energy grid to enable transparency by tracking data and finances related to it.
3. **Use of Blockchain in Food Standards:** Both USA and UK are using blockchain technology for ensuring food standards. In the UK, this is being put to use to keep a track on the distribution of meat to enhance food traceability. Further, in the US, blockchain technology is being applied for ensuring transparency and security in health data processing during food and drug inspection. Similarly,

Canada is using the Ethereum platform to bring transparency in the spending of government grants.

4. **Blockchain Technology in the Middle East Region:** United Arab Emirates has developed its policy related to Smart Dubai initiative with the aim of becoming the first city fully powered by Blockchain and enhancing services related to healthcare, education, traffic management and environmental sustainability.
5. **Blockchain-Based Service Network (BSN) of China:** China has developed a BSN initiative with the objective of helping companies and individuals deploy faster and economical Blockchain applications. This network has a number of tools for developers that help them in developing applications that mainly concentrate on smart cities and the digital economy. Besides this initiative, China has also planned a blockchain based notary system and an identification system for smart city infrastructure which will assign a unique, global digital ID to Chinese smart cities, aimed at improving the connectivity and data sharing between these cities.
6. **Use of Blockchain-Based Technology in Singapore:** The Monetary Authority of Singapore (MAS) is aiming to use Blockchain technology in the financial sector for enabling cross-border payments and also for adopting smart contracts by Insurance companies. Besides, work is in progress in the healthcare sector also for designing a solution for securing the healthcare data.
7. **Initiatives in the Private Sector:** Companies such as Samsung, LG, Amazon, Microsoft, Oracle etc. are applying this distributed ledger based technology services for various functions such as Amazon Managed Blockchain, Microsoft Azure Workbench, IBM Blockchain, Oracle Blockchain and BlockappsStrato. LG has developed Blockchain platform for applications related to digital authentication, community token and supply chain management. Further, Samsung Blockchain Wallet powered by COSMOCHAIN Blockchain has developed CosmeeDApp for purchase of contents using cryptocurrency.

BLOCKCHAIN IN INDIA: NATIONAL STRATEGY OF BLOCKCHAIN BY GOVERNMENT OF INDIA

Globally, the Blockchain technology has become widely accepted for carrying out distributed transactions, prompting the launch of numerous digital transaction platforms and services by using this technology for its use in different sectors. In the Indian context, a number of initiatives have been taken at the Central as well as at the state Government level. In this context, according to the Ministry of Electronics and Information Technology (Meity), Government of India, Blockchain Technology as an important research area, has the potential for its application in various domains such as governance, banking & finance, education, cyber security amongst many other sectors. The following section highlights the strategic framework and efforts initiated by various departments of the Government of India:

1. **MeitY:** The Ministry in consultation with a number of other agencies has launched a multi-institutional project namely "Distributed Center of Excellence in Blockchain Technology". The agencies involved are C-DAC, Institute for Development & Research in Banking Technology (IDRBT), Hyderabad and Veermata Jijabai Technological Institute (VJTI), Mumbai. These agencies have jointly developed Proof-of-Concept solutions which are being applied in various sectors such as for property registration being led by Shamshabad District in the state of Telangana. Additionally, Proof-of-Concept solutions are also in place for Cloud Security Assurance, Central Know Your Customers (CKYC) and trade finance. Generic Blockchain based Proof-of-Existence (POE) Framework have been developed and are being used for digital artifacts, for checking the integrity of academic certificates, MOUs, etc.
2. **NIC:** A Center of Excellence (CoE) in Blockchain technology has been established by NIC in collaboration with NICSI, with multiple objectives such as to speed up adoption and deployment of Blockchain technology in Government, ensure execution of different projects, pilot deployment, provide consultancy services and capacity building. All these projects are expected to be through inter-organizational collaborations within government and public and private sectors.
3. **NeGD:** National e-Governance Division has been established as an independent division by the MeitY, for supporting the ministry in the programme management and implementation of e-governance projects of the central and state governments. This Division has been working on developing a national strategy for Blockchain related to Blockchain application domains, challenges in adoption of the Blockchain technology, its societal impact, and role of the government and principles of the government that guide the national strategy on Blockchain.
4. **C-DAC:** MeitY has yet another important organization namely, Centre for Development of Advanced Computing (C-DAC) which is responsible for undertaking R&D in IT, Electronics and other related areas. "C-DAC's mission on Blockchain technology includes: (a) National Blockchain Framework for cross domain applications addressing the performance, scalability, interoperability, security and privacy challenges; (b) eSign integration with Blockchain based proof-of-existence (PoE); and (c) development and large-scale deployment of applications."

5. **Education and Academia:** A number of educational institutions are conducting research on Blockchain technology that would contribute to e-governance systems and platforms as for instance IIT-Kanpur is working on project that would be accomplished in three stages, first, the developing a feasibility report on the use of this technology, second development of protocols and finally conducting research on validating the property of tamper resistance. Ripple, the leading Blockchain company, unique for its collaborative vision and powerful vision, has launched University Blockchain Research Initiative (UBRI), has forged partnerships with IIIT-Hyderabad for promoting Blockchain research in India and for this purpose, it has set up a Center of Excellence in Blockchain in IIIT-H wherein research is being conducted use of blockchain technology on security, use of game-theoretic techniques and evolution of next generation Blockchain.
6. **NITI Aayog:** NITI Aayog is playing an equally important role in conducting research and development on the use of blockchain, such as its collaboration with Gujarat Narmada Valley Fertilizers & Chemicals Limited for fertilizer subsidy. It has also released a detailed document titled "Blockchain: The India Strategy (Towards Enabling Ease of Business, Ease of Living and Ease of Governance)" in January 2020 wherein it firmly believes that although this technology very fast evolving but it is very important for the policy makers, regulators, industry and citizens to have an in-depth understanding of the Blockchain technology and its uses in various sectors, along with a detailed study on the regulatory framework for implementation in the country.
7. **Banking and Financial Services:** The Government is also mulling on the use of this technology in the banking and financial sector. Reserve Bank of India is looking into the prospects of utilizing the Blockchain technology for banking purposes. Various banks both public as well as private sector banks such as SBI, Yes Bank, Axis Bank, and ICICI Bank are adopting Blockchain for their business.
8. **IDRBT:** The Institute for Development and Research in Banking Technology (IDRBT) is also involved in conducting research and development on the use of this technology in banking and financial sector and it envisages this in three essential steps: first, Intrabank: a private Blockchain for the banks internal purpose; second, Interbank: Implementation and testing of POCs; and third, Central Bank: for interventions related to digitizing the Indian Rupee through Blockchain technology. IDRBT has also been awarded a major research project entitled "Blockchain and Machine Learning Powered Unified Video KYC Framework" (to be carried out in collaboration with IIIT Hyderabad, and IIT Bhilai), sponsored by the Ministry of Electronics and Information Technology (MeitY), Government of India, IIT Bhilai Innovation

and Technology Foundation (IBITF) and RBI Innovation Hub on 14th February 2022. The overall coordination and execution of the project will be done by IDRBT. The objective of the project is to develop a Blockchain and Machine Learning powered unified video KYC framework for the financial sector.

9. **NIELIT:** National Institute of Electronics & Information Technology (NIELIT), formerly known as DOEACC Society, an Autonomous Scientific Society under the administrative control of MeitY, was set up to carry out Human Resource Development and related activities in the area of Information, Electronics & Communications Technology (IECT). NIELIT due to its vast experience in the domain of providing both Formal & Non-Formal Education in the area of IECT besides development of industry oriented quality education and training programmes, in collaboration with C-DAC, has been entrusted with the planning and implementation of Future Skills Prime Program for upskilling and reskilling in the domain of Blockchain so that a pool of skilled manpower in this specialized area can be developed in the country.
10. **Industry:** Development of Blockchain technology stacks by collaborative efforts of Government, industry, startups and research institutes, for encouragement of entrepreneurial development. It is planned to support the industry, academia and startups for working on application oriented research and technological challenges in the area of Blockchain technology through public private partnership, with the objective of facilitating and evolving techniques for innovative platforms that would contribute towards NBF. Encourage the industry to use the existing government schemes for the promotion of Blockchain technology.
11. **Legal and Regulatory Services:** Use of blockchain technology in the domain of legal and regulatory services is yet another important area that needs to be planned and a comprehensive framework needs to be developed. To begin with, primary focus should be on using the technology for enhancing transparency or openness in these services. Adequate consultations with the stakeholders need to be initiated so as to address the challenges that could be faced while admitting smart contracts in the courts. As the national blockchain framework evolves, updation of the prevailing policies and regulations need to be taken up. To begin with, initial use cases can be taken so that regulatory obligations and legal aspects are not required to be altered.
12. **State Governments:** The central government acting as a catalyst, has the important function of involving different stakeholders including industry, education, banking and finance, research institutions and state governments in developing a national blockchain strategic framework. It should act as a facilitator in encouraging the state governments for adopting the blockchain technology in its regular use and also creating a synergy amongst various inter-

governmental departments and institutions. For its successful development and implementation, the government needs to form committees, councils and working groups that could ensure expeditious strategy formulation and execution of the national blockchain framework. Specific technical parameters should be clearly defined before involving these stakeholders who need to undergo comprehensive security audit and assessment by appropriate agencies before coming onboard the national government blockchain infrastructure.

13. **Security and Privacy:** Initiating efforts through research for evolving guidelines pertaining to security audit and assessment processes. Hosting of Blockchain infrastructure, data and smart contracts in India for data localisation. Use of infrastructure of various organizations and state data centers in the country for execution of Blockchain technology framework. Enforcing identity, authentication and role based access control for in permissioned Blockchain framework.

From the aforementioned discussion, it is evident that several steps have been initiated by the Government for the design and development of a national strategic framework for the implementation of blockchain technology in various sectors in India. The subsequent section discusses in detail the status of nation-wide implementation of blockchain technology in different areas and sectors in the country.

CURRENT STATUS OF IMPLEMENTATION OF BLOCKCHAIN TECHNOLOGY ACROSS DIFFERENT SECTORS IN INDIA

National Informatics Centre (NIC) under the Ministry of Electronics and Information Technology (MeitY) is the technology partner of the Government of India, established in the year 1976 with the objective to provide technology-driven solutions to Central and State Governments has set up a Centre of Excellence (CoE) in Blockchain Technology in order to facilitate and promote adoption of this technology. This CoE is responsible for enabling an ecosystem of Blockchain Technology in the country and facilitates the building of required capacity for adoption of Blockchain alongwith testing viability of the use cases through development of Proof of Concepts and provide Blockchain as a Service (BaaS). Baas is an offering that allows Government Departments to leverage Blockchain services to build, host and use their own Blockchain applications, smart contracts and functions of the blockchain while the CoE manages all the necessary tasks and activities to keep the infrastructure agile and operational.

The inherent features of Blockchain technology makes it usable for almost all fields, however there are certain specific areas that involve identity management, workflow management, supply chain management, verification, etc. that may derive maximum benefit from this technology. Some of these areas where it is being implemented by government authorities in India include Certificate Chain, CBSE Certificate Chain, Drugs Logistics Supply Chain, Public Distribution System, Goa GST, Goa Excise, Blood Bank, Digidhan and State Excise Supply Chain System (eAbgari). In this section, let us have a look at some of the areas where blockchain technology is being implemented in India.

1. **CBSE Certificate Chain:** Blockchain technology has been adopted by the Central Board of Secondary Examination (CBSE), an autonomous organization under the Ministry of Education, Government of India that advocates continuous and comprehensive evaluation with an emphasis on holistic development of learners. It prescribes conditions for examinations, conducts public examinations at the end of Class 10 and 12 and grants qualifying certificates to successful candidates of the affiliated schools.

Once the result is declared, challenges are faced by the board in terms of making digital academic documents instantly available and verification of these documents for higher education as well as selection of the students for the jobs. In order to handle the challenge of verification of the correctness and genuineness of the documents, CBSE in collaboration with NIC has come up with a solution using Blockchain Technology, named as "Academic {BlockChain} Documents (ABCD)". It ensures the storage of academic documents in secure, tamper-proof and immutable manner and allows online access of these documents in a trustable manner by the educational institutes for admission in higher education programmes, job providers for offering employment, financial institutions for sanctioning educational and other loans, etc. making the verification process transparent, tamper proof and paperless. Initially, the academic documents of the years 2019 to 2021 for Class 10 and 12 have been provided, however it intends to append that of previous years as well.

2. **Drug Logistics Supply Chain:** The Government of the State of Karnataka, India has an Online Drug Supply Management System (Aushada) through which it procures and supplies free drugs for the patients across the state. Under this scheme, there are around 2911 hospitals and medicines worth more than 300 crores are procured and supplied every year. The system involves key challenges that include poor visibility i.e. the genuineness of the drug, information about its manufacturer, quality and expiry cannot be ascertained by the consumers that leads to possibility of counterfeit drugs in the supply chain, drug shortages, etc.. Another challenge is in terms of difficulty in tracking and tracing of the drugs as the stakeholders of the supply chain maintain their own records that

are inaccessible to others and collection of data to trace the history of drugs leading back to the manufacturer is a cumbersome process.

In order to handle these challenges, Drug Logistics chain using Blockchain Technology is integrated with the Aushada for recording the transactions in the block chain . This Drug Logistics Chain records all the transactions that take place in the supply chain of the procurement and distribution of the drugs in a shareable manner and can be accessed by all the stakeholders. Blockchain ensures complete surveillance of the process and makes the process transparent, trackable, traceable, tamper proof with no dependency on third parties for quality check. It also helps in avoiding delays and eliminates fraudulence, thereby enhancing the consumer/supplier trust.

3. **Certificate Chain:** Throughout the life of an individual, a number of certificates are required at various stages for the purpose of admission, scholarship, employment, availing concessions and government schemes, etc. and these include birth certificate, educational certificates, transfer certificate, income certificate, caste certificate and so on. There are few certificates which have limited validity and need to be issued again and again to continue getting the benefits which is a tedious process not just for an individual but also the issuing authority.

Integration of blockchain technology in the system eliminates the issue pertaining to fake certificates by building a trusted environment enabling the issuing authorities to store certificate details in a tamper proof manner and the validating authorities may login the Blockchain ledger for verification as and when required.

The stakeholders that may be a part of this Blockchain network include municipalities or revenue department that issues birth, caste, income and rural area certificates; department of education that issues academic, school completion, transfer certificates, etc.; welfare departments that offer scholarships and fee concessions; employers for verification of educational certificates as well as the ones required for implementation of reservation policy of Government of India. Therefore, this system brings a change to the certificates verification process by various stakeholders for the purpose of admission, recruitment, scholarships, social security pensions and various other government initiatives.

Recently, during the 35th Convocation ceremony of Indira Gandhi National Open University (IGNOU), New Delhi, India on 26th April 2022, more than 60 thousand students were awarded blockchain-based digital degrees with the help of Indian Institute of Technology.

4. **Public Distribution System:** Under the Public Distribution System (PDS) in India, food grains are distributed to the ration card holders, i.e., population below poverty line (almost two-third of total population) at a subsidized price. This process involves a number of entities including food grain producing farmers; registered millers that collect commodities from the farmers, process and transport them to the State godowns; transporters; fair prices shop owners, schools, colleges and charitable institutions that distribute the commodities to the beneficiaries; and central and state government agencies.

Due to the complex nature of PDS and involvement of many entities, there are a number of challenges involved in the process, i.e., lack of accountability down the supply chain leading to leakages, unpredictable supply, wastage due to logistic issues and administrative obstacles.

The objective of Blockchain technology in PDS is to track the commodities and record all the key data throughout the supply chain, i.e., price of the commodity, minimum support price at which it is procured from the farmers, quantity delivered at each stage, etc., thereby effectively managing the entire supply chain through distributed ledger technology, enabling fool proof end to end public distribution system. Since the movement of stocks through the supply chain will be tracked and traced, it will ensure identification of inefficiencies and inaccuracies and also ensure payment to the farmers and millers in a timely manner.

The proof of concept application of PDS has three stakeholders i.e. millers, farmers and admin and it covers the procurement process that involves registration of farmers with the miller and that of millers with the government.

5. **Blood Bank Management System:** This system is developed for facilitation of workflow management at the blood banks that carry multiple critical tasks that include donor record management, tests result management, component management, crossmatching, billing and inventory management, etc. to which blood bank management system is the solution. There are a number of challenges involved in this process mainly transfusion of unsafe blood due to human error, ineffective supply and demand management, non-availability of the medical history of the donor, lack of transparency and traceability, excessive human intervention leading to high operational costs, etc. These challenges may be adequately addressed while maintaining a tamper-proof repository of blood using blockchain technology that facilitates integration with the blood bank application for recording transactions in the chain and connecting all the key stakeholders involved in blood supply management i.e. donor, test centre, blood bank and hospital. It is envisaged that application of blockchain technology will lead to supply of quality blood to the patients, visibility of availability of

blood to all the stakeholders, trackability and verification of donor history, traceability of blood from donor to recipient and elimination of wastage of blood due to mismanagement. It is also expected to have a significant impact on reducing the HIV rates in developing countries due to transfusion of infected blood, thereby protecting the future of people.

6. **Land Records Management:** Land is considered to be one of the most valuable assets in the world and thus maintenance of land records requires high security and accuracy with immutable storage of data, however the ownership of a property in India is proven through presumptive land titling Record of Rights, i.e., a chain of documents providing history of transaction property from one person to another. One of the major challenge in the system is that the intermediate transactions are liable to be challenged due to undertaking of deed registration by the office of sub-registrar under Central Registration Act, 1908, however the ownership of the land is not verified as the custodian of the land records is Revenue Department/Revenue and Panchayat Raj that maintain land records and all other transactions such as change of ownership through sale, loan, mortgage, etc. and gets the Record of Rights document updated. Since the Registration departments in the country use software independent of the land record system, it leads to a number of challenges, i.e., increase in fraudulent transactions and land disputes, lack of a single source of truth for land ownership, time consuming and cumbersome process, high turn-around time for service delivery, possibility of data tampering, etc.

The benefits expected from the implementation of blockchain technology in land record management include availability of data at central location with accessibility to all departments making it immutable by spurious persons, easy and quick processing of loan applications, ensuring availability of benefits offered by the government to eligible persons only, verifiability of the integrity of documents by the registration department, elimination of fake documents, reduction in the pendency of property dispute cases in the courts and elimination of dependence of land purchaser on non-reliable personnel/agency for verification of authenticity of the documents provided by seller.

As is clear from the deliberations here, there has been a multidimensional use of blockchain presently in India, beginning with certification management by CBSE, logistics supply chain management of drugs (Aushada), certification chain management, supply chain tracking of commodities and database management under the PDS in the country, blood bank management system, and land records management. What follows here in the subsequent section is an analysis of the challenges that exist in the implementation of blockchain technology in the education sector in India.

CHALLENGES IN THE IMPLEMENTATION OF BLOCKCHAIN TECHNOLOGY IN EDUCATION SECTOR

Although blockchain technology is being widely adopted in different sectors due to its relative advantages including in the education sector, yet there are numerous challenges that the educational institutions may experience in the adoption of this technology. An attempt has been made here to review these challenges in the education sector:

1. **Technical Constraints and Scalability in Implementation:** The first and foremost challenge faced is of blockchain scalability. Technically, blockchain is yet to be adopted by the mainstream industries due to its scalability issues and still limited technical familiarity and skills (Gatteschi et al., 2020). It has been seen that the increase in the number of transactions in the blockchain network leads to an increase in block sizes. This will eventually increase the transaction latency i.e it causes lag or delay in communication over a network.
2. **Security and Privacy Concerns:** Use of blockchain entails managing large online networks. This raises serious security concerns that include threats such as malicious attacks and data leakage.
3. **Cost of Adopting the Technology:** One of the major challenges is pertaining to the cost incurred to educational institutions for adopting blockchain technology in various functional areas. These costs include the cost incurred in transformation of the presently used infrastructure, calculating the power cost, cost of time, cost due to slow transactions, and cost of large database management.
4. **Skepticism Amongst the Stakeholders:** Another important factor affecting the use of this technology is trust. Educational institutions are yet to instill the much needed trust on the implementation of blockchain, significant skepticism and reluctance exists about sharing the data on this platform.
5. **Selection of Data and Its Management:** Decisions related to the choice or selection of data is yet another important dimension of concern for educational institutions. In this process of transformation, the data rests with the stakeholders which might raise the issue of proper management of the database.
6. **Immutability:** Although immutability is one of the inherent characteristics of blockchain, but it is seen that this particular feature causes hindrance in data processing as it does not allow correction of inaccurate data and makes it difficult to apply new information storage laws.
7. **Nascent Stage of Technology Adoption in Education:** The implementation of blockchain technology in the education sector is in the nascent stage wherein this technology is being used only for issue of digital certificates to the learners

by institutions. It is yet to be implemented for managing other areas in the education sector.

CONCLUSION

As is seen from the discussion here in this paper, blockchain is now being implemented across the sectors in various functional areas which is also the case for educational institutions. As such, blockchain technology in the education sector in India has a long way to go, it is now being adopted at a larger scale for managing various activities in educational institutions as for instance, CBSE is using Blockchain Technology, named as "Academic {BlockChain} Documents (ABCD)" wherein it ensures the storage of academic documents in secure, tamper-proof and immutable manner and allows online access of these documents in a trustable manner by the educational institutes for admission in higher education programmes, job providers for offering employment, financial institutions for sanctioning educational and other loans, etc. making the verification process transparent, tamper proof and paperless. Similarly yet another public higher education institution has recently awarded digital certificates by utilizing this technology.

To conclude, blockchain in the domain of education, is presently at its developmental phase in India, has immense potential and promising future for ushering in mega transformation that is expected to encompass various levels of education, be it K-12, undergraduate or higher education, research and development. It is anticipated to be a leading force in online education database management, educational administration and governance.

REFERENCES

Alammary, A., Alhazmi, S., Almasri, M., & Gillani, S. (2019). Blockchain-Based Applications in Education: A Systematic Review. *Applied Sciences (Basel, Switzerland)*, *9*(12), 2400. doi:10.3390/app9122400

Arndt, T. (2018). Empowering university students with blockchain-based transcripts. *Proceedings of the 15th International Conference on Cognition and Exploratory Learning in the Digital Age*, 399–400.

Chen, G., Xu, B., Lu, M., & Chen, N.-S. (2018). Exploring Blockchain technology and its potential applications for education. *Smart Learning Environments*, *5*(1), 1–10. doi:10.118640561-017-0050-x

Clark, D. (2016). *10 ways Blockchain could be used in education*. Retrieved from https://oeb.global/oeb-insights/10-ways-blockchain-could-be-used-in-education/.

Fanning, K., & Centers, D. P. (2016). Blockchain and its coming impact on financial services. *Journal of Corporate Accounting & Finance*, *27*(5), 53–57. doi:10.1002/jcaf.22179

Gatteschi, V., Lamberti, F., & Demartini, C. (2020). Blockchain technology use cases. In S. Kim & G. C. Deka (Eds.), *Advanced applications of Blockchain technology* (pp. 91–114). Springer. doi:10.1007/978-981-13-8775-3_4

Grech, A., & Camilleri, A. F. (2017). Blockchain in Education. EUR 28778 EN. doi:10.2760/60649

Sharples, M., & Domingue, J. (2016). The Blockchain and Kudos: A distributed system for educational record, reputation and reward. In K. Verbert, M. Sharples, & T. Klobučar (Eds.), *Adaptive and adaptable learning* (Vol. 9891, pp. 490–496). Springer. doi:10.1007/978-3-319-45153-4_48

ADDITIONAL READING

Blockchain for Government. (2020). National Informatics Centre. Retrieved from https://blockchain.gov.in/assets/docs/Whitepaper_30jan.pdf

Central Board of Secondary Education (CBSE). (2021). Department of School Education & Literacy. Retrieved from https://dsel.education.gov.in/cbse

Chaturvedi, A. (2021, December 28). PM Modi launches blockchain-based digital degrees at IIT Kanpur. All you need to know about the technology. Hindustan Times. Retrieved from https://www.hindustantimes.com/india-news/blockchainbased-digital-degrees-all-you-need-to-know-about-the-technology-101640677737420.html

Daley, S. (n.d.). Blockchain in Education: 8 Examples to Know. BuiltIn. Retrieved from https://builtin.com/blockchain/blockchain-education

How blockchain could impact the education sector in the next five years. (n.d.). India Today. Retrieved from https://www.indiatoday.in/education-today/featurephilia/story/how-blockchain-could-impact-the-education-sector-in-the-next-five-years-1874422-2021-11-08

How Blockchain is Used in Education. (n.d.). Maryville University. Retrieved from https://online.maryville.edu/blog/blockchain-in-education/

Park, J. (2021). Promises and challenges of Blockchain in education. *Smart Learning Environments*, *8*(1), 1–13. doi:10.118640561-021-00179-2

Steiu, M. F. (n.d.). Blockchain in education: Opportunities, applications, and challenges. First Monday. Retrieved from https://firstmonday.org/ojs/index.php/fm/article/download/10654/9726/71482

Section 2
Industry Implications

Chapter 4
Investigation of Purchasing Applications After Digital Change in Industrial Markets

Ayla Avcı
https://orcid.org/0000-0002-1871-4090
Antalya Belek University, Turkey

ABSTRACT

In the constantly developing and changing global markets, the demands and demands of consumers also change. There have been changes in the industrial markets where too many products and services are bought and sold. World countries have become a common market thanks to the internet usage, which is widely used all over the world. The needs and wishes of consumers have also differentiated with cultural interaction. With the rapid increase in the online shopping method, internet shopping has started to rise with the fast and timely delivery options of the products. Demand for fast purchasing and product diversity have also increased in complex industrial markets where many people make purchases. Recently companies have entered the period of using online web pages in supplier selection. In this research, purchasing processes in industrial markets and new area digital transformation processes are examined.

DOI: 10.4018/978-1-6684-4153-4.ch004

Copyright © 2023, IGI Global. Copying or distributing in print or electronic forms without written permission of IGI Global is prohibited.

INTRODUCTION

When the markets are evaluated according to their activities those who buy products and services for their own needs form the seller's markets those who buy goods and services for production purposes form the product market and those who buy and sell the products in order to resell them form the seller's market. This markets can be added to regional, national, international and global markets (İslamoğlu, 2009).

People who are called sellers, consumers who buy raw materials, semifinished tools and equipment from other institutions in order to produce raw materials, semifinished products, goods and services for organizations and generally form and organizational structure are called industrial consumers. Industrial consumers earn a certain profit as a result of reselling the products, goods and services they have purchased or they create a new product they buy by evaluating them in the production processes. Industrial purchasing rates of small, medium and large enterprises and public institutions and organizations that are geographically distributed in various regions vary. Therefore, doing business with one or more of the industrial consumer is like selling the whole product. Industrial consumers to provide product benefits. They have expert knowledge about the technical features of the products. Since the relations in products purchases are based on trust hard bargains are made in purchasing transactions in these markets (Özdemir, 2006). Since continuous shopping gains more importance than a one-time purchase for businesses their interest in the consumer continues after the sale (Tengilimoğlu and Öztürk, 2011).

Organizations have people in charge of purchasing materials such as raw materials and semi-products (Tekin and Şenol, 2011). They are in the purchasing department of organizastions that buy a product or service. The number of employees in this role varies between 1 and 35. Increasing numbers of executives involved in larger and more important acquisitions (Bonoma, 2008). Purchasing unit employees are the center that coordinates with other departments. Members working in purchasing are required the obtain full approval from all units by cooperating with all departments in the organization from the R&D to logistics, in case a new products is developed or a new supplier joins. On the contrary, developing a product only in terms of cost or changing suppliers creates a disadvantageous situation contrary to what is expected. That's why the concept of "looking at the total picture" which is frequently mentioned in business life, is important. Especially in the purchase of food based products this situation causes various problems. Effective units in the development of a product or supplier; marketing, purchasing, R&D, production, logistics and finance units. This situation takes place as purchasing other departments depending on product and supplier change (Aksoy, 2013). Therefore industrial buyers are in the purchasing center position (Can, 2016).

Purchasing is one of the primary responsibilities of companies. Firms rely on suppliers to produce products, good and services that meet the needs of their customers. Purchasing accounts for a significant percentage of firms financial expenditures. For example materials constitute 58% of industrial enterprises. Therefore that see the purchasing transactions have some important responsibilities. These (Demir and Gümüşoğlu, 2009):

- Determining and evaluating the goods and service resources needed by the company.
- Ensuring business relations with these resources on issues such as quality, delivery, payment, and product replacement.
- Investigation and evaluation of potential resources such as new material product.
- Purchasing the materials required fort eh company at the best price according to the quality requirements.
- Conduct cost reduction programs value analysis buy or manufacture applications and market research.
- Maintaining effective communication with interdepartmental and external suppliers and potential suppliers in the company.
- Inform senior management about the cost of acquisitions in the firm, changes in the market that may affect the firm's profit and market share.

Determination of needs takes in cooperation with the parties in the organization, such as management, R&D, IT and production departments. Needs are determined by evaluating the information coming from inside and outside the organization (Karabıyıkoğlu, 2018). In the researches, it is seen that the purchasing unit is effective in searching, evaluating and choosing suppliers. Mogee and Been (1978), stated that employees in this unit are very effective in determining suppliers in criteria such as delivery times, cost and relations with costumers. Technical personnel on the other and play an important role in determining the technical specifications of the suppliers. In the uncertainties (suppliers and products) experienced in alternative purchases, purchasing units are effective in external information gathering processes (Upah, 1980).

Information Resources for Industrial Buyers

Some researchers have proposed some conflicting theories about how industrial buyers seek information. For example, Webster (1965) stated that the search process begins with the evaluation of goals and there is no need to seek information when the achievement of an existing goal is satisfactory. If there is a new target the search process will be activates. However information search efforts are costly and time constrained. webster, recommends stopping the search when the value is less than the information collection costs. Ammer (1968), stated that buyers are limited in the information search process but they are in an effort to develop new supplier in order to reduce price, insure supply and quality continuity and establish positive relations with suppliers. Bauer (1960), on the other hand, emphasized that unlike webster and Ammer, people seek information that comfirms the wisdom of decisions going to ways to reduce risk after reaching a decision (Copley and Callom, 1971).

Industrial markets are divided into two parts: personal and non-personal. Personal sources of business information: Salespeople, trade shows, interactive website, email, personal experiences, senior management, users, external consultants, colleges, purchasing unit, chat rooms (Özçelik and Akyol, 2017). Non-personal sources of business information are a wide variety of sources, such advertising, news information from trade associations and direct mail brochures (Dawn et al., 2002). Industrial ads can lead to technology transitions. For example, thanks to a publication showing the developments in the chemical industry, a Teflon section can be used in jet engine or another innovation can be seen in a digital wristwatch. Industrial manufacturers have a target audience range from manufacturers whose annual sales averages do not reach 100 thousand dollars to 35 million dollars. For this reason it is difficult to access measurements expressed in millions as in consumer advertisements. Reaching buyers through means such as sectoral magazine publications and direct mailing is an effective and economical method (Peltekoğlu, 2019).

Table 1. Industrial procurement information resources

	Personal	Impersonal
Commercial	Sales People Trade Shows Interactive Websites E-Mail	Advertisements in commercial publications Sales literature Direct mail brochures Information Websites Mass e-mails (spam)
Non-Commercial	Personal Experiences Senior Management Users Non business consultants Colleagues Purchasing Unit Email Chat Rooms	Newspaper publications Trade Associations Rating Services Internet newspapers Address lists News boards

Source: Yılmaz and Erçiş (2012)

PURCHASING APPROACHES AFTER DIGITAL CHANGE

Units where institutions buy and sell raw materials/products are called suppliers. Collaborative and positive behavior towards suppliers makes a significant contribution to the timely delivery of the desired raw materials and the timely completion of the works (Okay, 2013). Starting from the 1990s, it gained a new direction with the "Supply Chain" approach, which provides integrating supply sources and buyers. With this approach, suppliers, manufacturers, wholesalers, retailers and costumers in the chain have formed a single and integrated structure (Öz, 2011). In the research, it has been revealed that companies trust their suppliers in the buyer-buyer-seller relationship and they maintain their relationship and they maintain their relationship in the long term (Kılvan & Barutçu, 2018).

Today's digital transformations have enabled industrial consumers to turn to modern communication tools. In B2B purchasing processes, industrial buyers were communicating with sales representatives to purchase or research products in 2017. As of 2019, buyers preferred to search for product from companies websites through online searches. The data in Figure 1 shows the proportions of the methods buyers used to search for suppliers in 2017 and 2019 (Saha, 2019).

Figure 1. Methods used to search for suppliers in 2017-2019
Source: Saha (2019). UPS 2019 Industrial Buying Dynamics Study.

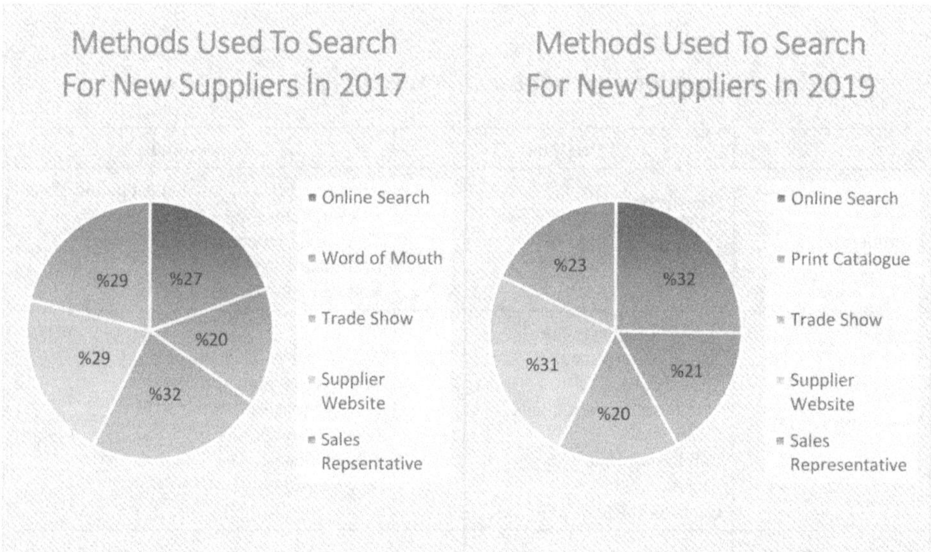

The benefits for digital content for the user are (Penpece, 2013).

- **Sincerity:** Since consumers have easy access to information about companies, it is a necessity for their digital content to be sincere.
- **Intelligibility:** Clear understanding of the message about the brand and product given to the consumer
- **Topicality:** Providing accurate and up-to-date information about the product to the consumer is effective in creating a perception of trust.
- **Consistency:** Provides brand trust wheen the content of the products offered to the consumer is consistent with each other
- **Mobilize the Target Audience, Attract Attention:** Content prepared in accordance with the target consumer attracts attention. Ekstras like discount coupons activate.

The most important agenda of companies recently is the digital transformation in purchasing and supply chain. According to the results of the PwC EMEA Digital Procurement Research conducted with more than 400 participants in 30 countries in the EMEA Region in 2020, it shows that digitalization and the rapid rise in purchasing will intensify in the coming periods (Karakış, 2020). B2B companies aim to increase their investments in marketing and IT technologies, which make up %12 of their budgets to %21 in the next two years in order to integrate new technologies into their companies (Mourtzis et al, 2020). Direct selling is risky for businesses, but in can be quite profitable. direct sales can be made by establishing a successful e-commerce site. In this way, the profit margin can be increased. However, it is important for companies to bring together a community where customers can evaluate and communicate with other costumers rather than just a website where orders are taken (McMurtry and Hiam, 2019). In this framework a project that will provide significant benefits to the digitalization and economies of BRI (belt and road initiative) countries has been put forward which connects Asia, Africa and Europe and enables their economies to shift to the east in the western-dominated world. Many countries will be laid with optical cables and infrastructure will be created fort he internet and their joint business capacity will be increased through e- commerce (Karaca, 2020).

CONCLUSION

The main purpose of marketing is the realization of the sale transaction. No buyer or seller goes to do market without an expectation of gain. The marketing is the blueprint for how an organization identifies and reaches groups of sellers of buyers. When is bill is done correctly a transaction process will be carried out in which both parties will be profitable. The faster the speed of information generated in the marketplace the quicker and longer the relationship between the the buyer and seller will be (Curry, 1999).

With today's modern understanding of marketing, marketing is not only a sales process but rather a whole of activities that start before production, continue with the sales process and include after-sales (Eryılmaz et al., 2015).

Every business should create a scenario of how any product or service is purchased any consumed. This scenario should be based on the behavior of the individuals buying it not the party providing the service. Assuming that service workers are actors and buyers are viewers and listeners, physical resources are service equipment and these reveal service performance. This scenario means that the purchase and consumption transactions related to the service or product are investigated. Therefore, these methods guide the selection and training of service personnel, determining the roles and maintaining order in the service processes (İslamoğlu et al., 2011).

The purchasing stage constitutes the logistics of the product. It is managed by the person or department responsible for purchasing by evaluating the selection and alternatives. Includes details about payment, delivery, etc (Karabıyıkoğlu, 2018). In order to ensure the interests of the companies, it is important that all variables such as the 7 truths in the business environment supplier, quantity, time, quality, costs, delivery payment method are provided in the best conditions in order to provide the products, materials and services requested in the necessary operations of the purchasing unit (Erdal, 2018).

We lived in the age when first the power of water was discovered, then electricity. This new century is known as the digital age. At first, shells were used as money, later banknotes. Now digital currencies are used around the world. Switch from personal computers to localhosts. Now cloud-based systems have become widespread. In the new period, the new technologies equipped with blockchain technology will be used (Goodwin, 38: 2019)

In recent years, due to do widespread use of internet applications all over the world and the ease of access the purchasing address of industrial buyers has also been the web pages of suppliers as have the preferences of individual consumers. In addition to the long and complexity of the purchasing processes in industrial markets the digital purchasing method has started to be preferred more because of the reasons such as seeing the product variety on the web pages of the companies making price comparisons knowing the lead times and ease of payment. For this reason, product suppliers need to offer their products and services in line with the needs and demands of their customers who make up their chains. At this point, the necessity for suppliers to follow the technological developments in the digital transformation process and to evaluate and implement the technological opportunities brought by digitalization has emerged.

REFERENCES

Aksoy, M. (2013). *Satın Alma ve Diğer Departman İlişkileri*. http:// www.satinalmadergisi.com/2013/03/27/satinalma-ve-dige r-departman-iliskileri-2/

Ammer, D. S. (1968). *Materials Management*. Richard D. Irwin.

Bauer, R. A. (1960). Consumer Behavior As Risk Taking. In *Dynamic Marketing for A Changing World*. AMA.

Bonoma, T. V. (2008). *Satış Stratejileri*. Harvard Business Review.

Can, P. (2016). Endüstriyel Pazarlarda Algılanan Riskin Marka Duyarlılığına Etkisi Üzerine Bir Araştırma. *Uluslararası Alanya İşletme Fakültesi, 8*, 175-190.

Copley, T. P., & Callom, F. L. (1971). Industrial Search Behavior And Perceived Risk. *Annual Conference of the Association for Consumer Research*, 208- 231.

Curry, E. J. (1999). Uluslararası Pazarlama. BZD Yayın ve İletişim.

Deeter-Schmelz & Kennedy. (2002). An exploratory study of the internet as an industrial communication tool examining buyers perceptions. İnternational Marketing Management, 145-154.

Demir, H., & Gümüşoğlu, Ş. (2009). Üretim yönetimi (İşlemler yönetimi). Beta Basım A.

Erdal, M. (2018). *Tedarikçi araştırması ve örgütsel pazarlarda yeni istihbarat kaynakları: Akıllı mobil uygulamalar*. www. startupvadisi.com/...pazarlarda-yeni-istihbarat-kaynakla ri-akilli-mobil-uygulamalar

Eryılmaz, M., Kaygusuz, S., & Efil, İ. (2015). *İşletme Kuram ve Pratik*. Ekin Basım yayın.

Goodwin, T. (2019). Digital Darwinism/ Survival of the Fittest in the Age of Business Distruption. Dijital Darwinizm. Siyah Kitap.

İslamoğlu, A. H. (2009). Temel pazarlama bilgisi. Beta Basım.

İslamoğlu, A. H., Aydın, K., Candan, B., & Hacıefendioğlu, Ş. (2011). *Hizmet Pazarlaması*. Beta Basım A. Ş.

Karabıyıkoğlu, M. (2018). *Satıcının Kariyer Yolculuğu*. Optimist Yayın.

Karaca, E. K. (2020). *Çin'in Teknoloji Liderliği, Dijital Yönetim Kırılma & Yeni Bir Dünya Düzeni*. Altınbaş Üniversitesi Yayınları.

Karakış, İ. (2020). *Dijital Satın Alma Araştırması 2020 PwC EMEA*. https://www.pwc.com.tr/tr/Hizmetlerimiz/danismanlik/tedarik-zinciri-yonetimi/dijital-sat%C4%B1nalma-ara%C5%9Ft%C4%B1rmas%C4%B1-2020/dijital-satinalma-arastirmasi-2020.pdf

Kılvan, P., & Barutçu, S. (2018). Endüstriyel Satın Alımlardaki Tedarikçi İlişkilerinde Güven ve Operasyonel İş Birliğinin Etkisi: Denizli Tekstil Firmalarında Bir Araştırma. *IBANESS Congress Series*.

McMurty, J., & Hiam, A. (2019). Marketing for Dummies. Pazarlama for Dummies. Nobel Yaşam.

Mogee, M. E., & Bean, A. S. (1978). The Role of the Purchasing Agent. In T. V. Bonoma & G. Zaltman (Eds.), *Industrial Innovation, "in Organizational Buying Behavior* (pp. 126–137). American Marketing Association.

Mourtzis, D., Angelopoulos, J., & Panopoulos, N. (2020). *A Survey of Digital B2B Platforms and Marketplaces For Purchasing Industrial Product Service Systems: A Conceptual Framework*. Elsevier. doi:10.1016/j.procir.2020.05.246

Okay, A. (2013). *Kurum Kimliği*. Derin yayınları.

Öz, M. (2011). Lojistikte Yeni Yaklaşımlar. *Kahramanmaraş Sütçü İmam Üniversitesi İktisadi ve İdari Bilimler Fakültesi Dergi*, *1*(1), 141–155.

Özçelik, C., & Akyol, A. (2017). Endüstriyel alıcıların satın alma sürecinde yararlandıkları bilgi kaynakları ve KOBİ'ler üzerine bir araştırma. *Trakya Üniversitesi Sosyal Bilimler Dergisi*, *19*(1), 167–182.

Özdemir, Ş. (2006). *Endüstriyel Ürün Pazarlaması*. Seçkin Yayıncılık.

Peltekoğlu, B. F. (2019). *Kavram ve Kuramlarıyla Reklam*. Beta Basım Yayım. İkinci Basım.

Penpece, D. (2013). *Dijital İçerik Pazarlaması*. Karahan Kitabevi.

Saha, A. (2019). *UPS 2019 Industrial Buying Dynamics Study*. European Study. https://www.epressi.com/media/userfiles/136388/1562788200/whitepaper_ibd_v5_print.pdf

Tekin, V. N. & Şenol, L. (2011). *İşletme, Kavramlar. İlkeler. İşlevler.* Seçkin Yayıncılık.

Tektaş, Ö. Ö. (2008). Endüstriyel Pazarlarda Bilgi Kaynaklarına Verilen Önemin Kültürel Farklılıklar Açısından Kavramsal Olarak İncelenmesi. *H. Ü. İktisadi ve İdari Bilimler Fakültesi*, *26*(2), 239-255.

Tengilimoğlu, D. & Öztürk, Y. (2011). *İşletmelerde Halkla İlişkiler.* Seçkin Yayıncılık.

Upah, G. D. (1980). *Applying the Concept of Perceived Risk to Buying Influence in Industrial Firms*. Assocation for Consumer Research.

Webster, F. E. Jr. (1965). Modelling the Industrial Buying Process. *JMR, Journal of Marketing Research*, *II*(4), 370–376. doi:10.1177/002224376500200404

Yılmaz, K. M., & Erçiş, A. (2012). Endüstriyel Pazarlarda Kişisel ve Kişisel Olmayan Bilgi Kaynaklarının Marka Değeri Üzerindeki Etkilerinin İncelenmesi. *Uluslararası Alanya İşletme Fakültesi Dergisi*, *4*, 27–44.

Chapter 5
From Black to Green:
Eco-Friendly Learning With Blockchain Technology

Selin Göçen
Dicle University, Turkey

Gülsün Kurubacak
Anadolu University, Turkey

ABSTRACT

Blockchain technology, which is one of the powerful tools that can be used in the development of education systems, is among the technologies that are abstained from for environmentally friendly activities due to its high level of energy consumption and gas emissions. However, there is not enough data that blockchain technology harms the environment or consumes more energy than other technologies. It is thought that this technology, which is more flexible than many technologies and will strengthen education systems, can be used in the design of environmentally friendly learning environments. Based on this idea, the authors drew a general framework about eco-friendly learning, green technologies, and blockchain in the chapter and presented suggestions for the design of environmentally friendly learning environments in the context of use cases of blockchain technology in education.

DOI: 10.4018/978-1-6684-4153-4.ch005

Copyright © 2023, IGI Global. Copying or distributing in print or electronic forms without written permission of IGI Global is prohibited.

INTRODUCTION

The amount of solid waste produced, the carbon footprints left, the steps taken without calculating the depletion of vital resources can cause human beings to proceed step by step from green to black. According to the World Bank (2018) data, it is expected that by 2050, the global waste level will reach 3.40 billion tons and as the greenhouse gas emission equivalent, 2.6 billion tons. Apart from the increase in waste, many factors such as the climate crisis, the depletion of natural resources, epidemics, and economic crises make societies rethink the importance of their lives and ecological balance, enabling them to make sense of the connections between the problems experienced.

Arguably, it is vital to be aware that the future will bear the traces of the past and present. It can be stated that these problems that societies are experiencing right now and will feel more deeply in the future are not new. But they have started to increase with the increase in population and their needs, changes in social life, technological advances, incorrect planning and steps in the past. At this point, it may be helpful to remember the words of Nelson Mandela. With the phrase "Education is the most powerful weapon which you can use to change the world", it is seen that Nelson Mandela points out the most effective tool that societies should employ to solve the problems they experience.

By all accounts, to use education as an effective tool, restore the ecological balance, and move away from black to green, action plans that can be integrated into education systems are prepared with social collaborations and agreements. It seems that it would be beneficial to re-evaluate eco-friendly learning environments designed within the framework of action plans developed for reducing adverse environmental impacts, increasing environmental awareness and sustainability in the context of green technologies and especially blockchain. In this direction, brief explanations of eco-friendly learning, green technologies and blockchain technology and suggestions for eco-friendly learning environments were presented in the section within the context of blockchain technology use in education.

ECO-FRIENDLY LEARNING AND GREEN SCHOOLS

It can be said that the belief in the use of education as an effective tool is the basis of green or eco-friendly learning environments. Shaped on the principles of minimizing the harm environmental harm caused by human activities, increasing environmental awareness and integration with the environment, green and eco-friendly learning environments are high quality and sustainable learning ecosystems with high energy efficiency, where adverse environmental effects are minimized (Ontario Ministry of

Education, 2010). Although the concepts of green, eco-friendly and sustainability are used interchangeably from time to time, it can be said that sustainability is a more comprehensive concept. The concepts of green, environmentally friendly, and sustainability are brought together on common ground in this book chapter. Activities carried out to create and maintain green living spaces by catching a harmonious and productive life rhythm with nature, and learning spaces designed to create this awareness are defined as "eco-friendly".

Green or eco-friendly learning environments were piloted in Denmark after the Earth Summit held worldwide in Rio de Janeiro in 1992 and was initiated by the Foundation for Environmental Education (FEE) in 1994 with the support of the European Commission, then spread to 70 countries. This educational initiative, which started with the aim of environmental education, protecting and developing the environment under the leadership of young people, contributes to environmental awareness being an integral part of human life. Schools in this education initiative comply with many recommendations of international agreements, including the Kyoto Protocol United Nations Framework Convention on Climate Change (UNFCCC, 2016) to reduce greenhouse gas emissions, adopted in 1995 but ratified in 2005 (Li and Williams, 2006; Shakir, 2019). Eco-friendly regulations and training related to sustainable development are also carried out in educational institutions apart from these schools within the scope of ecological schools.

In educational institutions, with the education programs structured within sustainable development, it is tried to create eco-friendly learning environments that support raising environmental awareness, starting from early childhood and being applied at different education levels (Prabawani et al., 2017; Rashid et al., 2021). Apart from these actions carried out in educational institutions, activities are organized by various institutions to create environmental awareness for individuals of all age groups, starting from early childhood. Such efforts to create eco-friendly learning ecosystems and increase the environmental awareness of individuals are also related to combating the adverse effects of population growth and sustainability.

According to the population trends published by the United Nations Population Fund (UNFPA) in 2017, the world population may be over six billion by 2050. This rapid increase in the world population may cause pressure on natural resources and energy production and consumption. And its reflections can be observed in different areas related to sustainable development goals such as health, food, water, energy, climate crisis, security, education, housing, and employment. In this context, for a green future, local and global evaluations, new solutions, and action plans will be beneficial for societies to minimize the damages they have caused to the environment and may cause in the future.

The eco-friendly learning environment implementations towards a green future are beneficial in raising awareness and sensitivity on different issues such as creating ecological consciousness, carbon footprint reduction, energy efficiency, recycling, conservation of natural resources, and climate crisis. Despite eco-friendly implementations made in the learning environments, problems with waste production and energy consumption can be observed. In eco-friendly learning environments, papers are used in most assignments and processes. Despite the preference for eco-friendly design in face-to-face learning environments, there may be multidimensional environmental effects such as paper waste, ink traces, economic costs, energy use, impacts of paper production cycle on nature, and recycling process (Agarwal et al., 2013). Open and distance learning, with less classroom use and less need for heating, air conditioning, and lighting lessens the damage to nature by reducing the need for campus services. Nevertheless, considering the adverse environmental effects of printed books, learning materials, diplomas, qualification documents, intra-institutional and inter-institutional transaction flows, it may be beneficial to re-evaluate the design of eco-friendly learning environments in the context of both face-to-face education and open and distance learning. (Paykamyan, 2021). In order to minimize the damage to the environment, reduce the use of non-renewable resources, and minimize the footprints, technology support can be incorporated in learning processes and eco-friendly learning environments. However, at this point, it may be a wrong statement to say that technologies do not cause any harm to the environment. It may be beneficial to consider that all technologies that facilitate social life or increase individual life comfort also have adverse environmental effects such as energy consumption, greenhouse gas emissions, etc. At this point, it can be said that it would be useful to establish the profit-loss equation with more comprehensive analyzes.

GREEN TECHNOLOGIES AND GREEN ICT

Technological systems used in education are good energy consumers. However, energy efficiency is also essential in eco-friendly learning. For this reason, regulations that can minimize the energy consumption of the technologies to be employed to design education systems as eco-friendly are required (Woodard and Orr, 2015). It can be said that there is a need for efficient technology use to come to the fore within the framework of sustainability programs and for a system that works in favour of efficiency between consumed power, energy and efficiency.

In systems with fast data flow, such as education systems, both gas emissions and energy consumption are higher in new technologies employed due to increasing needs. In other words, the dependence on technology increases, the interaction of ICT with the natural environment deteriorates (Woodard and Orr, 2015; Bekaroo, Bohoree and Pattinson, 2016). While improving people's lives, ICT systems or technologies used also causes increased energy consumption and gas emissions. Computers used in many institutions cause carbon emissions. However, institutions also contribute to carbon emission reduction if they turn off the computers they use at night or reduce the necessity of travelling by working remotely. It is stated that ICT is responsible for two per cent of the world's carbon emissions, so even if the carbon footprint of all ICT functions is reduced by half, it will contribute to a pretty low emission reduction rate (Philipson, 2011; UNFCCC, 2016). It would be helpful to have better benefits from ICT usage and to employ technologies that help reduce carbon emissions for the wider community. At the same time, for ICT systems to continue to cooperate with sustainable development, they need to be able to control themselves and reduce the amount of energy they consume. The general term for technologies that reduce energy consumption and gas emissions is Green ICT. Used as an umbrella term, Green ICT covers the reduction of energy consumption in ICT applications, sustainable software development processes and life cycles that can be created (Mardamutu, Ponnusamy and Zaman, 2016; Zantout and Seyff, 2018).

Green ICT is defined as the study and application of designing, manufacturing, and using of computers, servers, and associated subsystems such as monitors, printers, storage devices, and networking and communications systems efficiently with minimal or no impact on the environment by San Murugesan (Agarwal and Nath, 2011; Agarwal et al., 2013). It can also be said that green ICT (Issa et al., 2015) which is directly related to sustainability, is a complex system that emphasizes recycling, consisting of four components such as hardware production, software, labour and application policies that support each other (Singh and Gond, 2017; Kasemsap, 2018) and programs that directly or indirectly address environmental sustainability (Hernandez and Ona, 2016).

Similar to green ICT, another term that symbolizes the positive relationship between technology and the environment is green technology. Green technology, which is considered one of the elements that can minimize the deterioration of environmental quality and provide a healthier environment, is the development and use of products, equipment, and systems that reduce the negative impact of human activities and conserve the environment and resources (Jasmi and Kamis, 2019). Green technologies and green ICT are expected to improve lives while minimizing the environmental damage of technologies and saving energy (Yang, Kuo and Yeh, 2011; Din, Haron and Ahmed, 2013; Hernandez and Ona, 2016; Kesswani and Jain, 2017). Strategies to reduce wasted energy and cost by institutions and individuals need to be integrated into ICT, and policies related to green ICT need to be developed (Deshpande and Unhelkar, 2011; Isaias et al., 2015; Kasemsap, 2018).

Green ICT is gaining importance due to many reasons, such as the widespread use of high-level technology, the effects of climate change, greenhouse gas emissions, and the socio-economic impact of the increasing population. By all accounts, technologies that can minimize resource sharing, energy consumption, or hardware need supports green ICT and sustainability. Overuse of many devices contributes to environmental damage (Garg, et al., 2010). With eco-friendly equipment, the power consumption of resources, electronic waste, energy consumption and gas emissions are reduced. With workload timers (Bhadra and Kundu, 2016; Shojafar, Cordeschi and Baccarelli, 2016) it is ensured that the systems work efficiently with controlled energy consumption. It would be beneficial to use timers and green equipment and to prefer supporting technologies and remote working systems for inter-institutional cooperation (Philipson, 2011) in reducing the environmental impact of extensive and intensive use of ICT. Energy consumption can be reduced by remote working and inter-institutional cooperation; recycling and sustainability can be supported. High energy consumption and adverse environmental effects, especially seen in systems with intensive technology use, can reach the maximum level due to increasing environmental needs.

Educational institutions are systems that use technology intensively and are in constant interaction with the environment. Educational institutions with high data flow rates are high energy consumers with intensive ICT use. In educational institutions, technology is used in many areas, such as computers used individually, teaching and learning processes, data storage units, student and personnel workflows, measurement and evaluation, heating and cooling systems. Using technologies can help make educational institutions more eco-friendly by reducing paper use, food waste, and transportation emissions and can reduce the environmental impact of educational institutions as a community (Paykamian, 2021). However, the technologies used by institutions are expected to be low in terms of cost, energy consumption and gas emissions and to provide high efficiency in meeting the needs (Woodard and Orr, 2015). For this reason, it would be beneficial for green ICT and green technologies to be adopted in educational institutions and to be preferred in design processes for a successful transition to a low-carbon future, even if they do not eliminate environmental damage (OECD, 2019). It may be useful to remember again that all technologies can harm the environment at a certain level, and the important thing is to minimize this damage or turn the damage into a benefit.

GREEN ICT AND BLOCKCHAIN TECHNOLOGIES

Although there are other technologies for societies to progress and achieve sustainable development goals, it would be beneficial to use blockchain technology that can be integrated into various systems and fields. Created to support bitcoin, the digital cryptocurrency that facilitates transactions without intermediaries and between unknown parties (Nakamoto, 2009), the blockchain is a distributed database system that enables the transfer of assets that may have value based on the consensus of the participants on the network (Yıldırım, 2018). In blockchain technology, which includes transactions recorded in decentralized encrypted ledgers, transactions are recorded in blocks; and time-stamped blocks link with previous blocks to form a chain (Kouhizadeh and Sarkis, 2018; Kouhizadeh, Saberi and Sarkis, 2021). Two types of blockchain technologies based on access control mechanisms, which are public blockchains with a direct transaction, and consortium or private enterprise blockchains with a transaction by invitation, can also be defined as a distributed ledger where data is extremely difficult to change and is cryptographically secure (Wang et al., 2019). Although blockchain technology became familiar to many people through cryptocurrencies, it has been integrated into many different fields (finance, service, food, education, etc.), and its usability for gas emissions and carbon pricing systems has attracted attention after the climate change crisis (Al Kawasmi, Arnautovic and Svetinovic, 2015; Khaqqi et al., 2018). Blockchain technology, which attracts the attention of energy companies and technology developers besides governments (Enescu et al., 2020; Janssen et al., 2020) stands out with its innovation potential, reliability and decentralization features (Woodside Augustine, ve Giberson, 2017; Andoni et al., 2019). With the increasingly interesting blockchain technology, many properties (identity, property, time, order) can be verified at once, and manipulation is prevented. Blockchain technology, which has progressed in four generations (Blockchain 1.0, 2.0, 3.0 and 4.0), has been used primarily in payment systems as Blockchain 1.0. This technology called cryptocurrency has become smart contracts as Blockchain 2.0. Blockchain 2.0, which is used in many areas related to finance such as property acquisition, settlement in payments, banking and securities, later started to be used in government, health, science, technology, culture and art. In this period called Blockchain 3.0, blockchain is associated with the regulation and management of applications in various fields. In the new era called Blockchain 4.0, it contributes to system integration in supply chains, financial management systems, asset management and workflows (Swan, 2015; Srivastava, 2018; Alladi, 2019; Bhaskar, Tiwari and Joshi, 2020). Although there are energy consumption and e-waste problems, it is also stated that blockchain technology can be used to support environmental protection (Kouhizadeh and Sarkis, 2018). In the green technology innovation based on blockchain, the security of the information is ensured, its authenticity can be confirmed with ease, and energy loads used by the systems in the transmission processes can be reduced with these. Blockchain technology can be used on issues that are very critical for societies and also in sustainability applications. It can be used for the intelligent management of water resources, which are essential resources for people (Thejaswini and Ranjitha, 2021). Systems that can follow the life cycles of waste can be developed for recycling systems, which is an essential issue in communal life. In the recycling process, blockchain technology can be used to follow the recycling processes transparently by classifying waste according to its different characteristics such as chemical qualities, colours, usage levels etc. (Gupta and Bedi, 2018; França et al., 2020). Blockchain technology eliminates third-party intermediaries and allows for direct verification and transactions. There is no definite statement on whether blockchain is among the green technologies due to the lack of an adequate analysis of its environmental and sustainability impacts, lack of enough reliable data on the current and future situations of energy consumption, and lack of sufficient consideration of alternative scenarios. However, it is foreseen that it can become an eco-friendly and green technology, as it is a technology that can minimize the damage caused by transportation and printing with its developable structure (European Environment Agency, 2020). Green technologies, green ICT and eco-friendly schools essentially focus on maintaining the ecological balance. It can be stated that green technologies are suitable for low carbon emission and energy efficiency characteristics. However, unlike green technologies, blockchain technology is said to have high carbon emissions and contribute to the climate crisis as a strong energy consumer. Despite these unfavourable views on blockchain technology over cryptocurrencies, it is also known that the technology can be developed and has a flexible structure (Blinder, 2018). In addition, it is stated that the opinions that blockchain technology has a bigger carbon footprint than other technologies, that its energy consumption is higher, or that its ecological impact is more is relative (Chenguel, 2021). There are unverified methodologies for predicting energy consumption and carbon emissions of blockchain; however, it is stated that these are only energy consumption and gas emission values based on estimated trends. (Giungato et al., 2017). In this direction, it is thought that the blockchain can be considered among the green technologies that support sustainability, taking into account the steps to harmonize it with green energy and green technologies.

BLOCKCHAIN TECHNOLOGIES IN EDUCATION AND ECO-FRIENDLY LEARNING

Institutions that want to benefit from the potential of technology in the integration of blockchain technology into education, which is a powerful technology that can be used in many different fields, may need to employ various applications or technologies. In other words, blockchain technology can also contribute to the integration of more applications and technologies to the institution where it is used. By all accounts, the integration of blockchain technology into education systems is mainly on the axis of verifying and sharing learning outcomes and certificate/diploma records (Chen et al., 2018). It can be said that this is related to the prominent security feature of blockchain. The convenience Blockchain provides in understanding who accesses the learners' data and how, can be evaluated as security in terms of education systems, protection, concealment and holistic storage of learner data (Gräther et al., 2018; Alammary et al., 2019). But more broadly, blockchain technology has great potential as it can be used in various fields such as to verify student documents for academic and administrative affairs, credit transfers, scholarship application processing, homework feedback, online resources, e-portfolio, intellectual property protection, and so on (Hussain and Cakir, 2020). Blockchain technology is used in many educational institutions for academic IDs, transcripts, certificates, and other records of learners (Alammary et al., 2019). The use of blockchain technology in certificate and qualification management processes and learner records facilitates fast, reliable and transparent data sharing between educational institutions. At the same time, it prevents the high level of paper waste that may occur, especially in institutions where the number of learners is high.

Blockchain technology can be used in different fields such as qualification, performance evaluation, skill training, learning environment design, fee and credit transfer, besides inter-institutional transcript, diploma or certificate verification processes. It is used for blockchain-based measurement and evaluation systems that can monitor learners' qualifications based on qualitative and quantitative parameters (Farah et al., 2018); blockchain-based learning environments that enable learners to gain skills, receive instant feedback and interact with instructors (Williams, 2019); analysis and reporting systems that support decision-making processes by storing data on all workflows in the education system reliably (Bdiwi et al., 2018), and fast and reliable credit and fee transfers (Alammary et al., 2019). Through blockchain technology, learners have an opportunity to store their lifelong learning data. Learners can securely store data associated with students' credentials (credits earned, skills and abilities gained, and courses they have attended, etc.), and allow them to see the institutions they want. Learners have the right to manage their own data while performing this process. By this way, learners can prove their identities/ academic credentials are accurate and have more control over what can be accessed by any one. It is sufficient for them to give the necessary authorization to the institutions they want to apply to. Thus, they do not need to prepare separate files for the institutions they want to apply for or for which they want to be accepted. At the same time, they do not deal with unnecessary paper waste, a waste of time and energy. This situation, which is important for both learners and institutions, is supported by the Digital Credentials Consortium (https://digitalcredentials.mit.edu). The Digital Credentials Consortium was launched an academic platform that is reliable, transferable, sharable and quickly updated. While this platform allows learners to securely store their identity information, it ensures that they do not spend additional effort to obtain, store or verify copies of proof documents (such as transcripts, certificates, diplomas) regarding their education from different institutions. In terms of institutions; contributes to the safer storage of papers prepared separately for each learner in the digital environment; it becomes easier to update faulty ones or check for frauds. It provides advantages to both institutions and learners in terms of energy and paper waste consumption and time management. When mentioning the storage of learners and institutions records, the storage space to be used should also be calculated. It can be said that there is a transition from printing to electronic, between printing and filing so much data and filing it on hard disks. In other words, it can be said that instead of paper waste, there is a transition to electronic waste. In this direction, considering the irreversible damages, blockchain-based storage services can be preferred (Cedeño, 2020).

Other usage areas of blockchain in educational institutions are about obtaining necessary parental permissions, and ethics committee permits for individuals, outside the institution (Gilda and Mehrotra, 2018) and protecting copyrights. Copyright protection is an important issue for both publishers (publishers, academic journals, universities, schools, etc.) and publication owners (learners, educators, researchers, etc.). Especially for digital resources, copyrights, and naturally, the publishing process can be a worrying issue. The fact that blockchain technology adds time and person stamps provides an advantage in terms of security of produced materials and protection of copyrights. At the same time, this technology can prevent the manipulation of materials and also can be said that this advantage of blockchain technology can also enable the production of more open education resources. With these qualities that support lifelong learning, and sustainability in education, it can be stated that blockchain technology will contribute to the transformation in education. It helps learners manage their learning processes, monitor their qualifications, quickly update their professional development and helps other permitted institutions see and specifically contribute to the continuation of the existence of disadvantaged learners in the system. Parental permissions and ethics committee permits are important for researchers. Parental and ethical committee permissions are important for researchers. Parental consent is required for studies with children, but it may be difficult for researchers outside the institution to reach parents directly. Ethics committee permissions are; important for all researchers. However, correspondence between institutions in ethics committee permissions; can be a worrying process for researchers in terms of time, energy, and cost. Following the permit processes, arranging the missing documents, making the desired revisions, and re-applying. Obtaining ethics committee permissions or parental permissions through a blockchain-based platform can make the process easier and healthier.

Apart from these usage areas, Woolf University is established as a blockchain university that does not have a physical presence and is based only on an application that will enable academics to introduce their expertise to potential learners (Broggi et al., 2018). Students use the application to select courses that suit their needs and interests. Blockchain Smart Contracts are used to regulate registrations and payments, thereby automating administrative processes. This application, which will contribute to reducing costs, is open to the participation of individual academic institutions, academicians, and students (Mikroyannidis et al., 2018). Considering its potential areas to be used and the Woolf University example, which does not have a physical presence, it can be concluded that Blockchain helps to reduce the adverse environmental effects of printing and contributes to providing transparency, speed, reliability, and flexibility to academic and administrative processes. Blockchain technology helps learners to introduce their competencies to different institutions. At the same time, it helps learners to plan their learning processes according to their own pace, abilities, past experiences or wishes. It provides the opportunity to monitor and record the changes that the learners want to make during their learning process (Sathler, 2020).

With smart contracts, certificates, and qualification records, learners can continue their lifelong learning processes with the digital identities given and easily access the institutions they want. Learners are required to obtain the approved documents from the institution they are affiliated with and submit them to the institution they want to apply to in the operation process of traditional education systems. This process, which the learner carries out to prove their proficiency, results in various adverse outcomes, such as paper waste, air pollution due to the transportation process, the amount of consumed energy, and costs. However, eco-friendly learning environments are expected to minimize the damage to the environment and to support sustainable development. In cases where blockchain technology is not used, these transactions are carried out in education systems. However, It may be helpful to consider the amount of time and energy spent, the amount of paper waste generated within and between institutions, and the amount of energy consumed for heating and cooling systems. The negative environmental effects of blockchain technology make the technology to be approached with caution. However, green, ecological or sustainable social living spaces should not be evaluated only with environmental effects; it should be considered in a multidimensional structure such as social equality, poverty, welfare, healthy and quality life. Education, which is one of the areas of social life, should not be evaluated only in the context of environmental outputs. It would be beneficial to comprehensively evaluate the benefit-harm balance to employ technologies in eco-friendly or green learning environments.

CONCLUSION

This chapter focuses on eco-friendly learning, green schools, green ICT, green technologies and blockchain. Examples of the use of blockchain technology in education are mentioned. The link that can be established between green technologies and green ICT and blockchain technology has been pointed out.

Climate crises, increase in greenhouse gas accumulation, depletion of natural resources, global warming, new problems faced by people, etc. has been the impetus to rethink what can be done individually or socially for a green future. In order to protect and re-create green living spaces, energy efficiency, time and cost management and sustainability have come to the fore. As producers and consumers, people had to follow the traces they left on nature more. ICT/technology, which is an integral part of social life, is widely used in education systems where people show the traces of production and consumption intensively. Such intensive use has also ensured that its environmental effects are not ignored. The intensive use of ICT and technologies has also caused people to choose the ones that cause the least damage to the environment. Green technologies and green ICT described in the chapter; has emerged and developed with the motivation of protecting the environment and green and ensuring sustainability. Low energy consumption, gas emission and cost; time-saving etc.

There are different opinions supporting that blockchain technology, which is the focus of the section, should be among the technologies that contribute to the climate crisis due to its energy consumption and gas waste to the environment, and there are different views that there is not enough data and that the technology can be among the green supportive and sustainable technologies due to its flexible structure. The benefits of using blockchain technology in eco-friendly learning environments are explained with reasons. It is explained why the damage to the environment can be less with blockchain-based eco-friendly learning environments and why sustainable development can be supported more. Blockchain technology is one of the powerful technologies that support the transition from black to green in education systems.

REFERENCES

Agarwal, S., Basu, K., & Nath, A. (2013). Green computing and green technology based teaching learning and administration in higher education institutions. *International Journal of Advanced Computer Research*, *3*(3), 295.

Agarwal, S., & Nath, A. (2011). Green computing - A new horizon of energy efficiency and electronic waste minimization: a global perspective. *2011 International Conference on Communication Systems and Network Technologies*, 688-693, 10.1109/CSNT.2011.148

Al Kawasmi, E., Arnautovic, E., & Svetinovic, D. (2015). Bitcoin-based decentralized carbon emissions trading infrastructure model. *Systems Engineering*, *18*(2), 115–130. doi:10.1002ys.21291

Alammary, A., Alhazmi, S., Almasri, M., & Gillani, S. (2019). Blockchain-based applications in education: A systematic review. *Applied Sciences*, *9*(12), 2400. doi:10.3390/app9122400

Alladi, T., Chamola, V., Parizi, R. M., & Choo, K. K. R. (2019). Blockchain applications for industry 4.0 and industrial IoT: A review. *IEEE Access : Practical Innovations, Open Solutions*, *7*, 176935–176951. doi:10.1109/ACCESS.2019.2956748

Andoni, M., Robu, V., Flynn, D., Abram, S., Geach, D., Jenkins, D., McCallum, P., & Peacock, A. (2019). Blockchain technology in the energy sector: A systematic review of challenges and opportunities. *Renewable & Sustainable Energy Reviews*, *100*, 143–174. doi:10.1016/j.rser.2018.10.014

Bdiwi, R., De Runz, C., Faiz, S., & Cherif, A. A. (2018, July). A blockchain based decentralized platform for ubiquitous learning environment. In *2018 IEEE 18th International Conference on Advanced Learning Technologies (ICALT)* (pp. 90-92). IEEE. 10.1109/ICALT.2018.00028

Bekaroo, G., Bokhoree, C., & Pattinson, C. (2016). Impacts of ICT on the natural ecosystem: A grassroot analysis for promoting socio-environmental sustainability. *Renewable & Sustainable Energy Reviews*, *57*, 1580–1595. doi:10.1016/j.rser.2015.12.147

Bhadra, S., & Kundu, A. (2016). Introducing eco friendly corporate system: A green approach. *International Journal of Green Computing*, *7*(1), 1–24. doi:10.4018/IJGC.2016010101

Bhaskar, P., Tiwari, C. K., & Joshi, A. (2020). Blockchain in education management: Present and future applications. *Interactive Technology and Smart Education.* Advance online publication. doi:10.1108/ITSE-07-2020-0102

Blinder, M. (2018). *Making cryptocurrency more environmentally sustainable.* https://hbr.org/2018/11/making-cryptocurrency-more-environmentally-sustainable

Broggi, J. D., Gallagher, M. A., Lilly, J., Duquette, J., Nimura, C., Pattenden, M., . . . Winkler, A. (2018). *Woolf - Building the first Blockchain University.* https://www.slideshare.net/eraser/woolf-building-the-first-blockchain-university-whitepaper

Cedeño, J. (2020). *How blockchain could impact education in 2020 and beyond*. https://www.gettingsmart.com/2020/02/23/how-blockchain-could-impact-education-in-2020-and-beyond/

Chen, G., Xu, B., Lu, M., & Chen, N. S. (2018). Exploring blockchain technology and its potential applications for education. *Smart Learning Environments*, *5*(1), 1–10. doi:10.118640561-017-0050-x

Chenguel, M. (2021). Blockchain and ecological impact: Between reality and accusation? *International Conference on Global Economic Revolutions: The Era of the Digital Economy: ICGER 2021*.

Deshpande, Y., & Unhelkar, B. (2011). Information systems for a green organisation. In B. Unhelkar (Ed.), *Handbook of research on green ICT: Technology, business and social perspectives* (pp. 116–130). IGI Global. doi:10.4018/978-1-61692-834-6.ch008

Din, N., Haron, S., & Ahmad, H. (2013). The level of awareness on the green ICT concept and self directed learning among Malaysian Facebook users. *Procedia: Social and Behavioral Sciences*, *85*, 464–473. doi:10.1016/j.sbspro.2013.08.375

Enescu, F. M., Bizon, N., Onu, A., Răboacă, M. S., Thounthong, P., Mazare, A. G., & Șerban, G. (2020). Implementing blockchain technology in irrigation systems that integrate photovoltaic energy generation systems. *Sustainability (Basel)*, *12*(4), 1540. doi:10.3390u12041540

European Environment Agency. (2020). *Blockchain and the environment*. https://www.eea.europa.eu/publications/blockchain-and-the-environment

Farah, J. C., Vozniuk, A., Rodríguez-Triana, M. J., & Gillet, D. (2018, July). A blueprint for a blockchain-based architecture to power a distributed network of tamper-evident learning trace repositories. In *2018 IEEE 18th International Conference on Advanced Learning Technologies (ICALT)* (pp. 218-222). IEEE. 10.1109/ICALT.2018.00059

França, A. S. L., Neto, J. A., Gonçalves, R. F., & Almeida, C. M. V. B. (2020). Proposing the use of blockchain to improve the solid waste management in small municipalities. *Journal of Cleaner Production*, *244*, 118529. doi:10.1016/j.jclepro.2019.118529

Garg, M., Gupta, S., Goh, M., & Desouza, R. (2010). Sustaining the green information technology movement. In N. Bajgoric (Ed.), *Always-on enterprise information systems for business continuance: Technologies for reliable and scalable operations* (pp. 218–230). IGI Global. doi:10.4018/978-1-60566-723-2.ch013

Gilda, S., & Mehrotra, M. (2018, January). Blockchain for student data privacy and consent. In *2018 International Conference on Computer Communication and Informatics (ICCCI)* (pp. 1-5). IEEE. 10.1109/ICCCI.2018.8441445

Giungato, P., Rana, R., Tarabella, A., & Tricase, C. (2017). Current trends in sustainability of bitcoins and related blockchain technology. *Sustainability, 9*(12), 2214. doi:10.3390/su9122214

Gräther, W., Kolvenbach, S., Ruland, R., Schütte, J., Torres, C., & Wendland, F. (2018). Blockchain for education: Lifelong learning passport. In W. Prinz & P. Hoschka (Eds.), *Proceedings of the 1st ERCIM Blockchain Workshop 2018, Reports of the European Society for Socially Embedded Technologies*. doi:10.18420/blockchain2018_07

Gupta, N., & Bedi, P. (2018). E-waste management using blockchain based smart contracts. In *2018 International Conference on Advances in Computing, Communications and Informatics (ICACCI)* (pp. 915-921). IEEE. 10.1109/ICACCI.2018.8554912

Hernandez, A. A., & Ona, S. E. (2016). Green it adoption: Lessons from the philippines business process outsourcing industry. *International Journal of Social Ecology and Sustainable Development*, *7*(1), 1–34. doi:10.4018/IJSESD.2016010101

Hussain, I., & Cakir, O. (2020). Blockchain technology in higher education: prospects, issues, and challenges. In R. Sharma, H. Yildirim, & G. Kurubacak (Eds.), *Blockchain technology applications in education* (pp. 97–112). IGI Global. doi:10.4018/978-1-5225-9478-9.ch005

Isaias, P., Issa, T., Chang, V., & Issa, T. (2015). Outlining the issues of cloud computing and sustainability opportunities and risks in European organizations: A SEM study. *Journal of Electronic Commerce in Organizations*, *13*(4), 1–25. doi:10.4018/JECO.2015100101

Issa, T., Tolani, G., Chang, V., & Issa, T. (2015). Awareness of sustainability, green IT, and cloud computing in Indian organisations. In Green services engineering, optimization, and modeling in the technological age (pp. 269–287). IGI Global. doi:10.4018/978-1-4666-8447-8.ch011

Janssen, M., Weerakkody, V., Ismagilova, E., Sivarajah, U., & Irani, Z. (2020). A framework for analysing blockchain technology adoption: Integrating institutional, market and technical factors. *International Journal of Information Management*, *50*, 302–309. doi:10.1016/j.ijinfomgt.2019.08.012

Jasmi, N., & Kamis, A. (2019). Importance of green technology, education for sustainable development (ESD) and environmental education for students and society. *Journal of Engineering Research and Application*, *9*(2), 56–59.

Kasemsap, K. (2018). Cloud computing, green computing, and green ict. In K. Munir (Ed.), *Cloud Computing Technologies for Green Enterprises* (pp. 28–50). IGI Global. doi:10.4018/978-1-5225-3038-1.ch002

Kaza, S., Yao, L., Bhada-Tata, P., & Van Woerden, F. (2018). *What a waste 2.0: a global snapshot of solid waste management to 2050*. World Bank Publications. doi:10.1596/978-1-4648-1329-0

Kesswani, N., & Jain, S. K. (2017). Schematic classification model of green computing approaches. In Nature-inspired computing: concepts, methodologies, tools, and applications (pp. 1643-1650). IGI Global. doi:10.4018/978-1-5225-0788-8.ch063

Khaqqi, K. N., Sikorski, J. J., Hadinoto, K., & Kraft, M. (2018). Incorporating seller/buyer reputation-based system in blockchain-enabled emission trading application. *Applied Energy*, *209*, 8–19. doi:10.1016/j.apenergy.2017.10.070

Kouhizadeh, M., Saberi, S., & Sarkis, J. (2021). Blockchain technology and the sustainable supply chain: Theoretically exploring adoption barriers. *International Journal of Production Economics*, *231*, 107831. Advance online publication. doi:10.1016/j.ijpe.2020.107831

Kouhizadeh, M., & Sarkis, J. (2018). Blockchain practices, potentials, and perspectives in greening supply chains. *Sustainability (Basel)*, *10*(10), 3652. doi:10.3390u10103652

Li, Z., & Williams, M. (Eds.). (2006). *Environmental and geographical education for sustainability: cultural contexts*. Nova Publishers.

Mardamutu, K., Ponnusamy, V., & Zaman, N. (2016). Green energy in data centers. In Biologically-inspired energy harvesting through wireless sensor technologies (pp. 234–249). IGI Global. doi:10.4018/978-1-4666-9792-8.ch012

Mikroyannidis, A., Domingue, J., Bachler, M., & Quick, K. (2018). Smart blockchain badges for data science education. *2018 IEEE Frontiers in Education Conference (FIE)*, 1-5. 10.1109/FIE.2018.8659012

Nakamoto, S. (2008). Bitcoin: A peer-to-peer electronic cash system. *Decentralized Business Review*, 21260.

Ontario Ministry of Education. (2010). *Green schools resource guide: a practical resource for planning and building green schools in Ontario.* ZAS Architechs Inc. http://www.edu.gov.on.ca/eng/policyfunding/GreenSchools_Guide.pdf

Paykamian, B. (2021). *Could ed tech and virtual learning help schools 'go green'?* https://www.govtech.com/education/k-12/could-ed-tech-and-virtual-learning-help-schools-go-green.html

Philipson, G. (2011). A Comprehensive and Practical Green ICT Framework. In B. Unhelkar (Ed.), *Handbook of Research on Green ICT: Technology, Business and Social Perspectives* (pp. 131–145). IGI Global. doi:10.4018/978-1-61692-834-6.ch009

Prabawani, B., Hanika, I. M., Pradhanawati, A., & Budiatmo, A. (2017). Primary schools eco-friendly education in the frame of education for sustainable development. *International Journal of Environmental and Science Education, 12*(4), 607–616.

Rasyid, S., & Ardi, M. (2021). Development of environmentally friendly school model ın biak numfor regency of papua province. *Turkish Journal of Computer and Mathematics Education, 12*(13), 4096–4109.

Sathler, L. (2020). For an open innovation platform dedicated to education: a blockchain approach. In R. Sharma, H. Yildirim, & G. Kurubacak (Eds.), *Blockchain technology applications in education* (pp. 153–169). IGI Global. doi:10.4018/978-1-5225-9478-9.ch008

Shakir, Q. (2019). *Green schools (Eco-school).* https://www.researchgate.net/publication/331481308_Green_SchoolsEco-School

Shojafar, M., Cordeschi, N., & Baccarelli, E. (2016). Resource scheduling for energy-aware reconfigurable Internet data centers. In Q. Hassan (Ed.), *Innovative research and applications in next-generation high performance computing* (pp. 21–46). IGI Global. doi:10.4018/978-1-5225-0287-6.ch002

Singh, S., & Gond, S. (2017). Green computing and its impact. In *Nature-inspired computing: Concepts, methodologies, tools, and applications* (pp. 1628–1642). IGI Global. doi:10.4018/978-1-5225-0788-8.ch062

Srivastava, A., Bhattacharya, P., Singh, A., & Mathur, A. (2018). A systematic review on evolution of blockchain generations. *International Journal of Information Technology and Electrical Engineering, 7*(6), 1–8.

Swan, M. (2015). *Blockchain: Blueprint for a new economy*. O'Reilly Media, Inc.

The Organisation for Economic Co-operation and Development. (2019). *Financing climate futures: blockchain technologies as a digital enabler for sustainable infrastructure. in financing climate futures: rethinking infrastructure initiative*. https://www.oecd.org/finance/Blockchain-technologies-as-a-digital-enabler-for-sustainable-infrastructure-key-findings.pdf

Thejaswini, S., & Ranjitha, K. R. (2021). Blockchain for management of natural resources using energy trading as a platform. In *International Conference on Communication, Computing and Electronics Systems: Proceedings of ICCCES 2020* (*Vol. 733*, p. 475). Springer Nature. 10.1007/978-981-33-4909-4_36

United Nations Framework Convention on Climate Change. (2016). *ICT Sector Helping to Tackle Climate Change*, https://www.unfccc.int/news/ict-sector-helping-to-tackle-climate-change

United Nations Population Fund. (2017). *World population trends*. https://www.unfpa.org/world-population-trends

Wang, Y., Singgih, M., Wang, J., & Rit, M. (2019). Making sense of blockchain technology: How will it transform supply chains? *International Journal of Production Economics*, *211*, 221–236. doi:10.1016/j.ijpe.2019.02.002

Williams, P. (2019). Does competency-based education with blockchain signal a new mission for universities? *Journal of Higher Education Policy and Management*, *41*(1), 104–117. doi:10.1080/1360080X.2018.1520491

Woodard, H. C., & Orr, R. L. (2015). Green technology for green schools. In *Marketing the green School: Form, function, and the future* (pp. 106–117). IGI Global. doi:10.4018/978-1-4666-6312-1.ch007

Woodside, J. M., Augustine, F. K. Jr, & Giberson, W. (2017). Blockchain technology adoption status and strategies. *Journal of International Technology and Information Management*, *26*(2), 65–93. doi:10.58729/1941-6679.1300

Yang, M., Kuo, C., & Yeh, Y. (2011). Dynamic rightsizing with quality-controlled algorithms in virtualization environments. *International Journal of Grid and High Performance Computing*, *3*(2), 29–43. doi:10.4018/jghpc.2011040103

Yıldırım, H. (2018). Açık ve uzaktan öğrenmede blokzincir teknolojisinin kullanımı. *AUAd*, *4*(3), 142–153.

Zantout, H., & Seyff, N. (2018). *Green ict for sustainable development.* https://www.goodfellowpublishers.com/free_files/Chapter%206-600bcc6f6d9ccd4d6521451487b23bfe.pdf

ADDITIONAL READING

Fernando, Y., Rozuar, N. H. M., & Mergeresa, F. (2021). The blockchain-enabled technology and carbon performance: Insights from early adopters. *Technology in Society, 64*, 101507. Advance online publication. doi:10.1016/j.techsoc.2020.101507

Ganesh, K., & Anbuudayasankar, S. P. (2013). *International and interdisciplinary studies in green computing*. IGI Global., doi:10.4018/978-1-4666-2646-1

Guustaaf, E., Rahardja, U., Aini, Q., Maharani, H. W., & Santoso, N. A. (2021). Blockchain-based Education Project. *APTISI Transactions on Management, 5*(1), 46–61. doi:10.33050/atm.v5i1.1433

Kamani, K., Kathiriya, D., Virparia, P., & Parsania, P. (2011). Digital Green ICT: Enabling Eco-Efficiency and Eco-Innovation. In B. Unhelkar (Ed.), *Handbook of Research on Green ICT: Technology, Business and Social Perspectives* (pp. 282–289). IGI Global. doi:10.4018/978-1-61692-834-6.ch019

Li, D. (2021). Green technology innovation path based on blockchain algorithm. *Sustainable Computing: Informatics and Systems, 31*, 10. doi:10.1016/j.suscom.2021.100587

Passarelli, M., Cariola, A., & Bongiorno, G. (2022). Trends in blockchain technologies: a bibliometric analysis. In *Blockchain Technology Applications in Businesses and Organizations* (pp. 208–238). IGI Global. doi:10.4018/978-1-7998-8014-1.ch010

Polas, M. R. H., Saha, R. K., & Ahamed, B. (2022). Leveraging green iot and blockchain technology in the era of transformative digitalization: A green energy usage perspective. In *Handbook of Research on Social Impacts of E-Payment and Blockchain Technology* (pp. 115–135). IGI Global. doi:10.4018/978-1-7998-9035-5.ch008

Steiu, M.-F. (2020). Blockchain in education: Opportunities, applications, and challenges. *First Monday, 25*(9). Advance online publication. doi:10.5210/fm.v25i9.10654

Yuvaraj, M. (2015). Green libraries on cloud computing platform. In Encyclopedia of information science and technology (3rd ed., pp. 3901–3911). Hershey, PA: IGI Global. doi:10.4018/978-1-4666-5888-2.ch384

KEY TERMS AND DEFINITIONS

Blockchain Technology: Blockchain is a digital, decentralized, and distributed ledger technology that allows data to be recorded and shared securely.

Carbon Footprint: Carbon footprint is the trace of the damage caused by the life cycle to the environment, measured in units of carbon dioxide.

Digital Identity: An online or networked identity.

Green ICT: Green ICT is about minimizing the negative environmental effects of Information and Communication Technologies and maximizing the positive effects.

Green Schools: Eco-friendly learning environments that support global sustainability.

Green Technology: Green tech is a comprehensive research area that aims to reduce the damage done directly or indirectly to the environment and to protect the environment.

Sustainability: Being aware of the fact that the past and the present shape the future, it can achieve a harmonious and efficient life rhythm with nature.

Technology: The application of techniques, skills, methods, and processes to the aims of life or put into practical use to solve problems.

Chapter 6
Blockchain:
An Exploratory Review of Applications in Marketing

Mohan Menon
University of North Georgia, USA

Ashraf Mady
University of North Georgia, USA

ABSTRACT

This chapter provides an exploration to the enormous potential of blockchain and its strategic benefits. The application of blockchain has the potential to revolutionize existing business models. Blockchain is a peer-to-peer model that can speed up processes resulting in robust tracking and reducing costs of transactions. It is essential for business professionals to understand the significance of the implications of this far-reaching technology on business activities. Consequently, it is imperative that business leaders comprehend aspects of the technology to be integrated in strategic initiatives. The purpose of this chapter is to inform readers about the strategic benefits of blockchain technology. The chapter explains blockchain in the context of business strategies while highlighting its applications. This chapter is structured to explain the idea of blockchain technology, expound on the importance of the technology's business applications, and illustrate how professionals can integrate this technology in business.

DOI: 10.4018/978-1-6684-4153-4.ch006

Copyright © 2023, IGI Global. Copying or distributing in print or electronic forms without written permission of IGI Global is prohibited.

INTRODUCTION

The realization of the bidirectional impactful relationship between business and technology is transforming the professional world. Achievements in technologies continue to revolutionize business. In the same time, business needs continue to accelerate and influence technological advancements. This portends exceptional future that will link the two in ways that were unimaginable thus far. Historically, technology has always played a significant role in a business enterprise. In the past, Information Technology (IT) implementation focused on efficiency and automation to drive cost reduction. In recent years, the digitization of information has changed the role of IT in the organization. IT has become a core driver of organizations' operations, innovations, and competitiveness (Allam, 2018; Bharadwaj, El Sawy, Pavlou, & Venkatraman, 2013). In order to build and advance business capabilities, another technology known as blockchain has come into prominence in the recent decade. According to the World Economic Forum, by 2025 10% of the world's GDP may be on the blockchain system (Carson, Romanelli, Walsh, & Zhumaev, 2018). In addition, in recent years, the business world have become acquainted with terms such as cryptocurrency, Bitcoin, blockchain, etc. According to Deloitte, blockchain is at an inflection point with companies moving away from simply understanding it to building practical applications (Deloitte, 2018). Blockchain applications have attracted the interest of investors across a wide range of industries such as banking, marketing, insurance, communications, entertainment, transportation, healthcare, education, and public services (Friedlmaier, Tumasjan, & Welpe, 2018). This indicates that blockchain is here to stay (Menon & Mady, 2021).

BLOCKCHAIN EXPLAINED

Blockchain is the underpinning technology for cryptocurrencies such as Bitcoin (Nakamoto, 2008). However, the technology has diverse applications far beyond Bitcoin (Zheng, Xie, Dai, Chen, & Wang, 2017). The genesis of blockchain type of technology might be traced to an article, titled, "How to Time-Stamp a Digital Document (Haber & Stornetta, 1990). They proposed the concept of blockchain – a system consisting of a chain of blocks connected by cryptography and timestamps that cannot be modified (Deshpande, 2019). The Bitcoin White Paper titled, "Bitcoin: A Peer-to-Peer Electronic Cash System," was published in 2008 (Nakamoto, 2008) provided the modern foundations for the technology. Blockchain was envisioned as a way to send payments digitally, securely, and anonymously between parties without needing third-party verification.

Initially, blockchain was defined as a "digital, decentralized and distributed ledger in which transactions are logged and added in chronological order with the goal of creating permanent and tamperproof records" (Treiblmaier, 2018). Similarly, the authors of Blockchain Revolution defined blockchain as "an incorruptible digital ledger of economic transactions that can be programmed to record not just financial transactions but virtually everything of value" (Tapscott, 2018). It is a trusted decentralized peer-to-peer (P2P) network of shared ledgers thereby assuaging the need for certain third-party dependent activities (Linn & Koo, 2016). Blockchain systems do not require a central authority and interparty conflicts can be resolved based on the collective opinion of all parties involved. The technology facilitates the distribution of auditable ledgers in an immutable (unchangeable), transparent, and secure way (Reyna, Martín, Chen, Soler, & Díaz, 2018).

The conceptual foundations of blockchain lies in the old public ledgers that were used in medieval societies to record notable events – transfer of assets, births, marriages, deaths, judicial rulings, election results, etc. However, unlike the older (human) master of the town ledger, cryptography and distributive networks are used to achieve "a secure, transparent, immutable repository." Blockchain can be envisioned as a sequence of blocks (Figure 1), which holds a complete list of transaction records like a conventional public ledger. Each block points to the immediate previous block, called the parent block, via an encrypted reference (Zheng et al., 2017). Thus, it is a distributive system or shared ledger of transactions, each of which "depends on a logical relationship to all its predecessors" with the first block called the genesis block. This type of linked data is recorded and re-recorded thousands of times but not copied. Once the blocks are published, they are linked to other blocks in the chain and the entirety of the chain is automatically verified for accuracy by miners or human checkers (DePatie, 2017). This system removes the need of trusted intermediary such as a broker, clearinghouse, or a bank. Deloitte Insights calls blockchain "democratized trust." (Deloitte, 2016).

Figure 1. Blockchain data structure

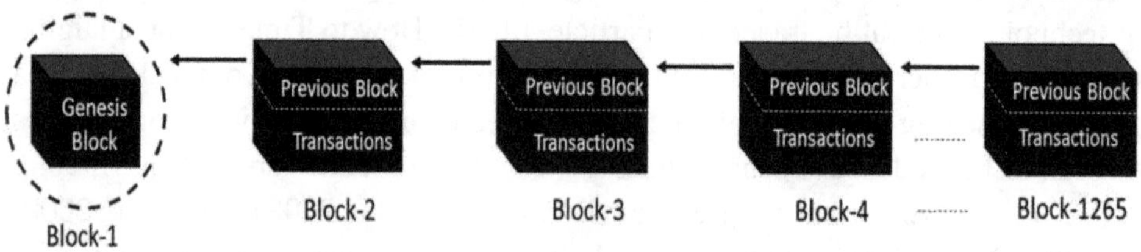

The blocks contain an incessantly growing list of entries or records consisting of the encryption of new data, such as a financial payment, merged with the encryption of the prior entry. These blocks periodically are time stamped and made available to those in the system (DePatie, 2017). Such sequential merging of encrypted transactions creates robust forms of security because any retroactive alteration or change will require changes in all subsequent entries.

Researchers have categorized the blockchain system into public, private, and consortium blockchains (Xu et al., 2016; Zheng et al., 2017). Either way, the data is spread over, stored in various physical/virtual locations, and continuously reconciled on all the databases within that network. This allows the records of transactions to be public and verifiable rather than in a hidden central location. Thus, breaching, and degrading data is nearly impossible since each computer (node) in the network has a copy of the ledger, not controlled by any one entity, as in a centralized system, and thus less prone to failure.

The following diagram (Figure 2) depicts a basic blockchain transaction in which customer A wants to buy widgets from seller B. The system collects and stores all such transactions during a certain period. The requested transactions are sent (broadcast) to all participating computers in the specific blockchain network.

Figure 2. The basic blockchain process

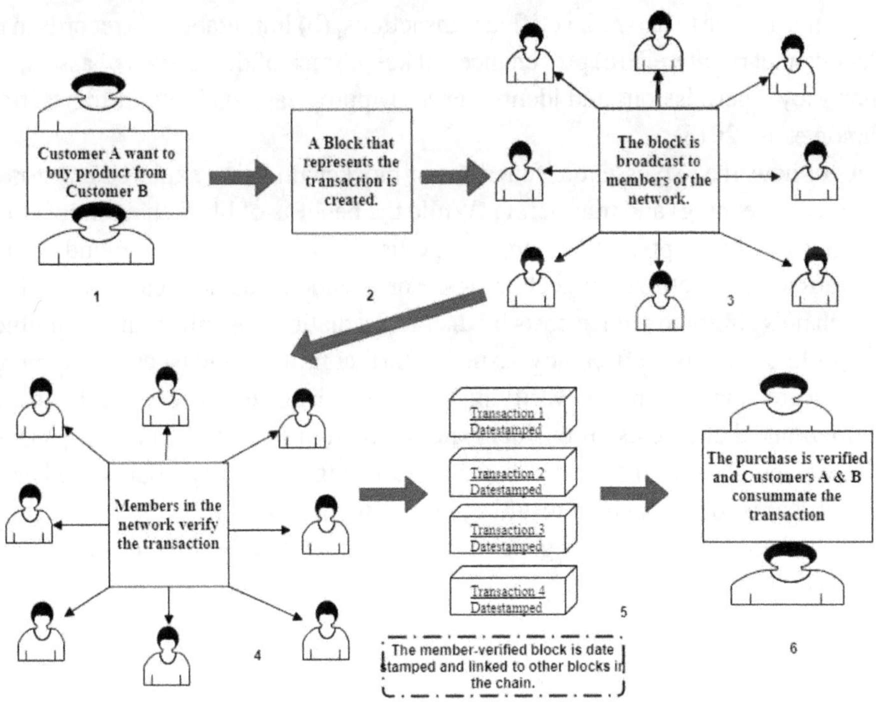

When the participants/miners receive the request, it checks, i.e., validates the transaction against a set of rules/protocols set by the creators of the specific blockchain network. The protocol establishes the way new entries are initiated, edited, validated, recorded, and finally distributed. Cryptology, the "keeper of trust," is used to certify the integrity of data/transactions/blocks. These verified transactions are assembled in packages (blocks) and added/linked in a certain manner and the block is locked within the system. When other computers in this specific network validate if the lock on the block is correct, it becomes part of the chain of blocks that can be seen but not altered.

BENEFITS OF BLOCKCHAIN INNOVATIONS

The business aspect represents the "what" and "why," while the technology aspect explains the "how" of a blockchain system. As with many other technologies, one does not need to know the inner workings of blockchain to be able to use it effectively. A survey for McKinsey found that financial, technology, media, and public sectors have better opportunities with respect to feasibility and impact of blockchain utilization. With decentralization, transparency, distributive nature of the ledgers, and cryptographic security being the cornerstones of blockchain technology, some of the core properties of blockchain include: (a) consensus – all parties must agree to network verified transactions; (b) immutability – records in the ledger cannot be altered; (c) provenance – it keeps track of the records of assets; and (d) privacy – permissions and identity ensure appropriate visibility of transactions (Carson et al., 2018).

As of now, the two essential functions of blockchain being explored are storage (i.e., record keeping) and transacting. While the benefits of blockchain innovations may go to early adopters, there are compelling strategic reasons, beyond storage and transacting to utilize the technology. For instance, business can benefit from blockchain's ability to reduce costs by disintermediation, tokenization of tangibles/intangibles, increase efficiency using smart contracts and secure payments, authenticate transactions by verifying customer data, etc. These might inspire providers and their clients to consider experimentation with the technology. As with all new technologies, sustained efforts in unearthing its full potential could lead to optimal scaling of blockchain resulting in transformation of business strategies, as we know them today. A few of general business applications of the technology are presented (Table 1).

Table 1. Sampling of blockchain applications by industry sectors

Industry Sectors	Application Briefs and Examples of Companies Involved
Banking/Finance	Whether it is being used to secure transactions or to pro-actively compete against the newer Fintechs, nearly all banks are experimenting with the technology.
Media Applications including IP rights.	Smart contracts can protect copyrights and automate the sale of creative endeavors online, thus eliminating the risk of file copying and re-distribution. Ex. TRST.im, Blockparti, BitProof, etc.
Advertising	Provides transparency in cases of data inflation and discrepancy in advertising platforms; disintermediation of ad buying and selling; ad verification, content management, etc. Ex. Rebel AI, Lucidity, etc.
Supply Chain Auditing and other SCM applications	Blockchains provide an easy way to certify origins of products. Blockchain timestamping of dates and locations would correspond to product numbers. Ex. IBM, Skuchain, Vavc, etc.
Education	A more diversified, democratized, decentralized education system needs to maintain reputation, trust in certifications and proof of learning. Ex. Blockcerts, Appii, etc.
Internet of Things	IoT technologies are by nature distributed and thus blockchain will play are role in how devices communicate with each other. In addition, smart contracts can automate remote systems through a combination of software, sensors, and network facilities. Ex. Filament, Chimera-inc.io, etc.
Healthcare	Blockchain systems can transform the space by making patients the center of the healthcare ecosystem and providing increased privacy, security, and interoperability of patient data. Ex. BitHealth.
Stock Trading	When p2p trade confirmations become instant, intermediaries like auditors and custodians are removed from the process. Ex. equityBits, Spritzle, BitShares, etc.
Sharing Economy	By enabling p2p payments, blockchain opens door to direct interactions between parties. Ex. BTC Jam, Codius, BitnPlay, La'zooz, etc.
Prediction Markets	Prediction markets pay out according to event outcomes. Blockchains can bring the 'wisdom of the crowd' technology that can find new applications in the future. Ex. Augur.
Voting/Election Systems	Votes can be counted, followed, secured, and correlated while maintaining privacy of the voters. Ex. Follow My Vote.
File Storage	Decentralizing file storage on the Internet helps prevent hacking and lost files. Ex. Ascribe, ArtPlus, Stampery, Storj.lo, Peernova, etc.
Identity Management	Distributed ledgers offer enhanced methods to prove who a customer is and her/his status. Having a secure, digitized identity will be important as online services continue to evolve. Examples could include civil registries, voting records, identity fraud, etc. Ex. The Real McCoy, BlockVerify, Uniquid, Onename, etc.
Static and Dynamic Registries	Static registry for storing older, non-changing reference information such as patents, land titles, origin for produce, etc. Dynamic registry for updating transactions or assets when they are exchanged on the network. Ex. Factom.

Building Trust

As indicated earlier, blockchain allows for a democratized, transparent, and empowering way for marketers to interact with their consumers. The promise of a consumer-owned and controlled identity and personal information within the blockchain helps them limit the type and amount of data being shared (Prescher, 2018). Customers are more likely to transact with businesses with verifiable reputation and this has been a major limitation for newer and smaller businesses online. Thus, these businesses have to rely on third party platforms such as Amazon or Facebook to be proxies of trust. This proxy trust system has its own drawbacks. Blockchain presents a solution to such limitations by enabling trust and credibility of transactions since one of its foremost characteristics is trust-enabling (Seebacher & Schüritz, 2017). In addition, blockchain can mandate proof-of-identity to commit any transactions or information change (Iansiti & Lakhani, 2017). Everledger, an independent technology company, utilizes the use of blockchain to promote trust, enable transparency and allow tracking the history of assets. Stolen credit or debit cards domestically or across borders could be automatically rendered unusable within a blockchain with other location-verifying blocks in the chain. Similarly, chargeback fraud activities could be slowed in the process (Vaccaro, 2019).

Blockchain can also enhance consumer trust and engagement by leveraging new channels across the digital business space (Treiblmaier, 2018). Given that many business activities involve settlements with third-party companies for services rendered, smart contracts may be one of the most transformative application of blockchain (Iansiti & Lakhani, 2017). A smart contract, for instance, can automate the process of sending a payment to a supplier once the system confirms shipment is delivered and allows disintermediation between transacting parties to protect from malicious or accidental exceptions (Christidis & Devetsikiotis, 2016). Smart contracts can also be used to validate payments, international trade arrangements, loans, and voting (Cong & He, 2019); in healthcare and clinical trials (Nugent, Upton, & Cimpoesu, 2016); Internet of Things (Christidis & Devetsikiotis, 2016), or other applications such as gaming (Bartoletti & Pompianu, 2017).

Cost Savings

Blockchain's ability to influence cost savings is likely to be driven by operational efficiencies, including disintermediation of transactions (Menon & Mady, 2021). Some of these savings result from the reduction of time, effort, and human involvement in keeping records and the need for reconciliation between parties. While there is likely to be some start up, implementation and training costs, savings from efficiencies and reduction in paperwork is projected to effect cost savings gradually. Blockchain can bring about efficiencies in the verifying and transference of information and payments/assets between sellers and buyers. Another example of cost savings can be realized in the behavioral targeting strategies. In the current system, display ads follow the customer from site to site even after they have purchased the item. A blockchain-enabled implementation can provide tracking and transparency to halt ad-buying networks to programmatically bid for space for the item already purchased.

Disintermediation in pricing platforms benefit both business and consumers with lower prices as the number entities siphoning off a percentage from the total sum are eliminated. Usually, businesses (hence consumers) are forced to pay transaction fees/discount points to credit card companies and online payment portals thereby increasing the cost of sales, resulting in higher prices, lowering profits, and losing customers (Business Insider – India, 2018; PricewaterhouseCoopers, 2017). Blockchain system offers "almost" zero transaction cost because it can eliminate intermediaries – much to the chagrin of Visa, MasterCard, and others including their issuing banks (Dalton, 2018). Whether the credit card vendors and their banks will allow this scenario is yet to be determined. Businesses could motivate consumers to share information directly with them as reduced cost of transactions lead to lowering prices (Harvey et al., 2018). User tailored deals and personalized pricing could become commonplace in online transactions.

Similarly, in cross-border payment situations, typical money transfer/remittance cost can be up to 20% of total amount, blockchain could reduce the costs significantly while guaranteeing transparency, verification, and speedy real-time processing (Reiff, 2020).

Virtual Assets or Tokenization

Tokenization is the conversion of non-digital assets and product to virtual assets (tokens) for transaction and recording purposes. Although most physical assets and intangibles cannot be virtualized, i.e., converted to digital forms, blockchain allows their conversion into digital equivalents or tokens for transactions purposes. In the real estate market, for instance, if one owns a house worth 200,000, USD it can be converted into 200,000 tokens, each worth 0.000005% share of the property. Therefore, when a homebuyer obtains one token through the blockchain system, she/he has 0.000005% ownership of the house. Similarly, if somebody buys 100,000 tokens representing the asset, she/he owns half of the house or owns the asset completely when all 200,000 tokens are bought. In addition, because the purchases take place in a blockchain where transactions are transparent and immutable, the token owner is ensured of the ownership, and nobody can erase it even if the deal is not yet registered with any county authority. The current NFT (Non-Fungible Token) phenomenon illustrates this idea well with artworks being tokenized but are not fungible – meaning that it is not easily replaced/duplicated (Murphy, 2021). With graphic designer Beeple (Mike Winkelmann) selling his "Everydays: The First 5000 Days" artwork for 6.5 USD million, this phenomenon attracted much attention. Beyond artworks, other items are being traded in the NFT space such as sports trading cards, Twitter cofounder Jack Dorsey's first tweet, Nyan Cat GIF, etc. Some companies/brands have also tried to ride the bandwagon – Charmin NF toilet paper, Pizza Hut's NF pizza and the like.

Another example is the use of "vAtoms" within the "BLOCKv" platform that enables bridging the gap between physical and virtual worlds. In other words, vAtoms represent digital tokens for physical objects that can be transacted and can be linked to customers' e-mails, phone numbers, added to digital banners, inserted into a video game, etc. Similarly, gold, diamonds, lettuce, haircuts, medical services, etc. can be tokenized and traded within the blockchain. Therefore, tokenization is a valuable tool to transact intangible, fungible, and non-fungible goods/services with all the benefits of blockchain such as transparency, trackability, immutability, etc.

Blockchain can also help reduce the incidence of counterfeit products that violate IP rights by facilitating end-to-end tracking of the product from the origins (Galvez, Mejuto, & Simal-Gandara, 2018). Sourcing transactions between all parties in the supply chain can be recorded by the blockchain system, which provides both reliability and integrity to the chain. Since parties have the access to the same version of the ledger, the transactions are transparent and the tracking easier. This is especially critical for life-threatening nature of supplies of counterfeit auto parts, medicines, medical devices, food supply adulteration, etc. all of which can be brought into compliance using blockchain technology.

Blockchain can also help reduce the incidence of IP rights violations by by involving consumers (Galvez et al., 2018). For instance, LVMH Moet Hennessy Louis Vuitton, the Prada Group, and Compagnie Financiere Richemont are allied with the Aura Blockchain Consortium to utilize and promote a single blockchain solution for all luxury brands worldwide. Bulgari, Cartier, and Hublot have also been using this system. The way it works is when consumers buy the authentic brand, they are provided access to the system that lists the history of the product, origin, components, environmental info, proof of ownership, warranty, and care information (Paton, 2021).

BLOCKCHAIN APPLICATIONS IN LOGISTICS AND DISTRIBUTION

As mentioned previously, blockchain technologies provide promising solutions to the current limitations in food supply chain management (Tan, Yan, Chen, & Liu,

2018). It allows autonomous supply chain by collecting real-time data and automating triggered actions based on predetermined conditions. With evolving consumer behaviors that require business to provide information on the sourcing of products, sustainability and investigate contamination (thus recalls) of a variety of produce and meats, there is a need to track all items to its source. Implementing blockchain systems in supply chain/logistics yield three distinct benefits – verifying products, reducing paperwork, and the possibility of end-to-end tracking (Dobrovnik, Herold, Fürst, & Kummer, 2018). For instance, Walmart has been working with IBM on a blockchain enabled food supply chain solution (Kamath, 2018; Tiwari, 2016). In this case, Walmart can track pork that was sourced from China – and keep track of where the meat comes from, where it was processed and later stored (Marr, 2018).

Reports indicate that China has implemented the world's first agricultural commodity blockchain. In this case, banks such as ABN AMRO, ING, Société Générale, Louis Dreyfus Co., and the Shandong Bohi Industry Company teamed up to develop and complete the "first full agricultural commodity transaction using a blockchain platform" (ABN-AMRO, 2018). The system reduced administrative time spent on process documents and data. In the test scenario, the blockchain platform was used in the following manner:

1. Shandong Bohi of China, the buyer entered into a smart contract on the platform with the seller, Louis Dreyfus Co.
2. Shandong Bohi requested a smart letter of credit.
3. Banks involved review, confirm, and issue the letter of credit on the blockchain platform.
4. Transaction agents and inspectors issue documents such as bill of lading, certificates of quality, etc. Using the platform, the documents can be issued faster with lower risk of fraud.
5. Louis Dreyfus Co. presents the documents to the banks electronically in less time.
6. Auto checking the documents on the platform to help reduce discrepancies thereby increasing efficiency.
7. The banks sign off on the documents and the title to the assets is transferred while the payments to the seller were made simultaneously (ABN-AMRO, 2018).

The actors in the value chain can provide/share information concerning their goods making consumers feel safe about buying the product. Information about sustainability and other critical aspect of the product/process can be provided through blockchain.

Similarly, DeBeers Jewelers uses the technology to track diamonds from the point they are mined to the final sale to customers. This helps the company to avoid selling "blood diamonds" and certify the authenticity of product to reassure customers. Another mining company, BHP Billiton, uses blockchain to track the mining and supply processes and record data throughout the mining process with its vendors (Marr, 2018). It is designed to increase efficiency internally, but also allows the company to have more effective communication with its partners.

Anheuser-Busch (AB) earlier piloted the technology for their global shipping activities (Zhao, 2018). AB was part of a blockchain consortium using the Accenturedeveloped system that included other participants such as APL Logistics, Kuehne+Nagel and others. Later in 2019, the company joined the IBM blockchain initiative called, "Trust Your Supplier" network along with others such as GlaxoSmithKline, Lenovo, Nokia, and Vodafone (Morrison, 2019). The system allows AB to track and verify documents such as ISO certifications, bank account information, tax certifications and those related to insurance throughout the lifecycle of each supplier. In addition, by doing so on a real-time basis, it reduces costs associated with these activities (Morrison, 2019).

Thus, blockchain in SCM/Logistics allow for crucial data storage, retrieval, tracking, and verifying across the value chain with little chance of alterations, given that the original documentation is accessible by all parties in the chain. Therefore, the bill of lading, for example, cannot be tampered without the consensus of all the parties involved. An added benefit is the resulting reduction in documentation/paperwork.

ADVERTISEMENTS AND PROMOTIONS

Marketers that implement advertising within their own websites to promote the contents, products, increase memberships, drive subscriptions, or direct sales can benefit from blockchain technologies by deploying smart contracts for actions/responses and verify consumer engagement. The smart contract tech within the blockchain system also enables marketers to direct users' activities based on their behaviors (viewership patterns, website engagement, etc.) without the need to use intrusive ads or spam. At the same time, it is possible for the advertiser and website owner to transact business without an intermediary such as Google. Toward this end, BitClave has developed a blockchain system that allows users of the site and clicks to be validated in a secure manner without a need for an intermediary.

Google Trends reported a peak in searches for blockchain advertising during 2018 (Guzenko, 2019). During the same year, marquee brands such as Unilever, Kimberly-Clark, Kellogg's, and others joined the consortium for ad-buying blockchain to disintermediate ad buying and enable them to audit spending. More recently, companies such as Lucidity have begun to develop advertising supply chain transparency to minimize fraud and discrepancies. In this system, marketers can create a sole source of recorded & verified information regarding ad impressions, transactions, and related data, thereby creating blockchain-enabled decentralized system to ensure accountability at every step of the buying process.

Given the influence of advertising/promotions in both consumer and business lives, use of blockchain in the digital advertising space is critical in delivery and billing verification. For instance, it has been difficult to determine how money is being spent on various advertising platforms especially in an increasingly programmatic environment. Programmatic advertising was expected to reach close to 100 USD billion, about 70% of total digital spend in recent years (Dayal, 2020) and by 2022, digital ad spend is set to rise to 427.26 USD billion. Given the enormous sums of money involved, transparency in transactions is of paramount importance (Guzenko, 2019). In an automated programmatic environment with split-second space bidding scenarios, there are bad players trying exploiting the system to the tune of seven USD billion to 23 USD billion a year in fraudulent impressions, clicks and conversions that drain marketing budgets (Bailey, 2020). Studies forecast that click fraud could increase to 44 USD billion by 2022 (Juniper Research, 2017). With multiple parties to a digital ad buy, blockchain can effectively verify that all transactions are transparent and fulfilled (English, 2019).

Blockchain technologies can mitigate some of the risks by allowing the ad industry to function in a collaborative and open network where one can authenticate the clicks with comprehensive and verifiable information about the parties involved in the eco-system. For instance, using blockchain's immutability principles, the adChain platform provides a protocol that allows advertisers to benefit from campaign auditing and real-time impression tracking (Goldin, Soleimani, & Young, 2017). This is an example of using immutability to restrict attempts of pay-per-click providers to profit from fraudulent traffic and clicks.

As marketing practitioners comprehend the value of the technology and it becomes embedded in various spheres of marketing, additional applications are likely to be developed and refined. For instance, blockchain might solve the problem of tracking celebrity/influencer ROI and performance by using smart contracts that will facilitate payments upon completion of specific actions or delivering contracted results/metrics (Deshpande, 2019). Both Social media.market and indaHash are enablers in this space.

Similarly, blockchain can be utilized in realizing the required results in loyalty programs. Survey reports suggest an average consumer is affiliated with 14.8 different loyalty programs but only active in 6 of them (Morgan, 2020). Consumers find it difficult to keep track and utilize them before expiration dates and this is a lose–lose situation for both consumers and brands. Companies such as OrionCoin, Loyalcoin, and Sandblock have tried to solve this problem by using blockchain. For instance, OrionCoin developed loyal points system called ORC that is finite in supply unlike the typical loyalty points awarded by credit card companies. In the case of ORC, the built-in transparency enables consumers to retain the value of their points rather than have them diluted by merchants looking for quarterly profits (Single Grain Blog, 2018).

The decentralized blockchain system can facilitate collaboration among multiple brands allowing customers to redeem loyalty points with any brands within the network. Since the loyalty points are deposited in real time, customers can redeem them on the go (Deshpande, 2019). For example, virtual trusted agreements enabled by blockchain can allow customers to validate rewards programs, e-mail, or newsletters subscriptions. Smart contracts also allow the automation of payments when customers sign up for these reward programs or newsletters. Coupons, another part of the loyalty inducing strategy, have seen paltry redemption rates in the past decades even with slightly better rates for digital coupons (Bickerton, 2018; Zhang, Ma, Popkowski Leszczyc, & Zhuang, 2020). Rouge, a blockchain service provider in this space, allows brands to offer unique coupons and track them through dates coded into them, so they expire automatically online. Similarly, these blockchain-enabled coupons can only be used/shared once and cannot be duplicated.

Heineken and the Los Angeles Football Club developed another example of engagement/loyalty program experiment. The two brands, intending to build relationships with fans in a unique way, used BlockV, a system that bridges the gap between physical and digital goods. In this case, fans were able to win prizes (caps, meeting with players, tickets, beer, etc.) in the form of vAtoms, the digital tokens. Heineken also offered a vAtom empty pint glass (virtual) and when fans shared it the glass would fill up 25% and after 4 shares, the glass would be full, and they could redeem it for the real thing at participating retailers. Coke used a similar tactic to share and redeem virtual bottles of coke. In both cases, the sharing was traced, authenticated, and limited to only one receiver each time.

MEDIA APPLICATIONS

Blockchain technology can enable new opportunities in media and publishing businesses. For instance, the IP rights that belong to a person, band, songwriter, or author are difficult to guarantee under the traditional system. Content creators do not yet have a common standard to utilize, protect and police violations. Blockchain might offer the infrastructure to do just that and a way to establish attribution that cannot be falsified. This might take care of fake news, content piracy and so on. "Blockchain-registered publications provide uncontestable records about a publication's attribution and origins" (Dicker, 2018). A good example of this can be found in Bitcoin Magazine, an early blockchain publication. According to Dicker (2018), the magazine timestamps the entirety of its articles and the associated metadata to a blockchain. This allows the magazine to create a transparent and publicly verifiable record of the article creation and attribution information. When the article is published, its metadata is placed within a blockchain transaction. The transaction block is attached to a blockchain and is stored. The timestamp and attribution information now reside in the blockchain and provides proof of its existence and ownership (Dicker, 2018). Another platform, the Basic Attention Token (BAT), created by Brenden Eich, allows more efficient digital advertising "by creating a new token that can be exchanged between publishers, advertisers, and consumers" within a blockchain network (Delisle, 2018).

RADICAL DISRUPTION OF TOURISM BUSINESS MODELS

Blockchain can lead to a more radical disruption of the current business models in tourism. Major companies of the sector like TUI and Eurowings have already expressed their intentions to test and adopt blockchain technology in their booking, reservation, and payment systems (Antoniadis, Spinthiropoulos, & Kontsas, 2020). Blockchain-enabled systems allow hotels access to real-time data to have better control over occupancy and apply instant special offers when needed (Marr, 2018). Winding Tree, a global travel agent platform, announced its blockchain-enabled platform could bypass intermediaries such as Expedia and Booking.com. In 2019, Hahn Air issued its first airline ticket via Winding Tree blockchain platform. Blockchain platforms also offer small businesses strategic opportunities and connect directly with customers without having to pay commissions to intermediaries. It will be interesting to see if the intermediaries such as Airbnb or Uber might be rendered ineffective when homeowners/car owners are able to use blockchain networks to transact business (Schiller, 2018).

In the last year and a half, travel and tourism have been tremendously affected by the pandemic. A total of 217 countries imposed some form of international and domestic travel restrictions, world airline traffic fell by 55% compared to 2019, and the job losses in the sector reached 100 million in 2020 (Lee, 2020). In terms of revenue, the estimates are that COVID-19 cost the industry over a trillion dollars (Madden, 2021). However, with vaccinations on the rise around the world, there is hope that travel might rebound to profitable levels shortly. To achieve this goal, countries and companies are exploring ways to make travel safer by experimenting with various forms of vaccine passports such as Travel Pass (IATA), AOKPass

(Air France), Digital Green Certificate (EU), Green Pass (Israel), etc. Some of these including AOKpass, IBM's Digital Health Pass, and GuardTime pass are blockchain-enabled to add another layer of security and make them more secure and tamper-proof (Patel & Drozdiak, 2021).

SOCIAL RESPONSIBILITY

Studies have addressed some of the positive implications of blockchain technologies on socially responsible companies as related to the promotion of their brands' quality and sustainability (Ertemel, 2018). The system, when implemented appropriately, allows consumers to instantly verify the extent of the company's environmental, social responsibility claims, and the brands' contributions to charities (Deshmukh, 2020). Companies such as Ecovadis produce sustainability ratings to help procurement teams monitor the corporate social responsibility – environmental, social, and governance (CSR-ESG) practices within the chain. Verifying such claims can enhance the brand's image in the marketplace while reducing the risks to the reputation of the brand/company since all the compliance information is accessible and verified in real-time allowing issues to be identified before they become public (Morrison, 2019). In addition, the system can be utilized to authenticate other product/marketing claims such as being organic, halal, kosher, etc. Therefore, working within the blockchain system allows businesses to signal their brands' positive attributes to consumers, thereby, projecting better public image.

GENERAL BLOCKCHAIN APPLICATIONS IN MARKETING

Other general blockchain applications in marketing related industries are provided in this section. Blockchain technology can be deployed to allow research firms record customers' activities and associate these activities with micropayments. These payments can go directly to customers based on their online activities such as a click of a button, answering a survey, watching an ad, reading a report, etc. The technology allows information to be protected from modifications and to be shared directly between users in a secure manner. This presents an opportunity to exchange relevant and accurate information allowing the marketing of products, services, and programs to be more relevant to each user. There are a myriad of other ways blockchain might revolutionize marketing – including affiliate marketing, social commerce, data and privacy protection, digital marketing security, etc. Suffice to say that any innovation that results in cost savings and transparency of transactions will have effective use cases in marketing now and in the future (Tönnissen & Teuteberg,

2020). Companies implement blockchain to strengthen their market positions. Google is using blockchain in its cloud ventures to differentiate itself, Walmart is using it to track produce to its source, Maersk to access its cargo data, British Airways to manage flight data, UPS for its customs brokerage business, FedEx to solve customer disputes, etc. (Krauth, 2018). Walmart also has filed a patent application to create its own digital currency to target the unbanked by making transactions cheaper and faster (Ennis, 2019). While this is not an exhaustive list, it does provide a glimpse into some of the bigger players and their utilization of this technology.

BLOCKCHAIN SERVICE PROVIDERS

This brief section highlights providers in this space for reference. There are several smaller blockchain service providers trying to onboard their SME clients into the ecosystem. While many of them have been mentioned throughout the text and in Table 1 as examples in specific areas, bigger companies such as IBM, Microsoft, Amazon, and Alibaba offer full-service in this space. Many of them have been securing patents, investing heavily in this space, and creating large employee pools for related projects. One big player, Facebook, seems to have run into some difficulty as some of the initial backers of the Libra project (ex. PayPal, Visa, MasterCard, etc.) have severed ties (Ingram, 2018).

IBM markets its multi-cloud flexibility and ease of use for almost any industry or organization. They provide learning tools, use case studies, white papers, and bring together many members/partners to provide end-to-end solutions for clients (Androulaki et al., 2018). Their Hyperledger system can be used to support supply chain traceability, enhance financial services, improve customer service, secure financial transactions, maintain accurate public records, and to protect citizen information. Microsoft launched its Azure Blockchain Service in 2015 and later introduced the Enterprise Smart Contracts, providing users the infrastructure to build their own blockchain services. Recently, the company began collaborating with others to offer an "open source, blockchain-based identity system that allows people, products, apps, and services to interoperate across blockchain nodes, networks, cloud providers, and organizations (Patrizio, 2018). SAP's Cloud Platform Blockchain can be accessed from any device and provides a blockchain service platform for supporting IoT within a single ecosystem. Similarly, Deloitte, the consulting major, also offers a service to companies to build their own private blockchain network customized for an industry or for use within an organization (Deloitte, 2018a; Patrizio, 2018).

CONCLUSION

While the debate continues about cryptocurrencies, there is an agreement about blockchain's potential to revolutionize every aspect of business. In this exploratory chapter, the authors attempted to respond to two questions pertaining to (1) the foundational concepts of blockchain; and (2) the business applications of the technology in various industries. The idea is to target business professionals and introduce the strategic business benefits of blockchain. In addition, this exploratory work provide a sampling of business use situations; and enlighten blockchain concepts could be used in various business domains. The hope is that, armed with this topical knowledge and expertise, business professional can capture opportunities utilizing the benefits of blockchain.

The explored strategies of blockchain implementations supports its ability to empower existing or newer business models if deployed to its full potential. However, at this point, blockchain implementation has not migrated much beyond early adopters. A survey of over 1,000 senior executives in large companies from several different advanced economies indicates that while the majority of respondents see "compelling business case" for using the technology, only 34% said that their companies initiated any sort of blockchain technology implementation (Deloitte, 2018b). Therefore, it is safe to say that business professionals are at varying stages of utilization of the full potential of the technology. While first instances of blockchain application was in the finance sector, other applications in business have not been far behind. Suffice to say, the technology is here to stay but as with every paradigm-changing innovation, the speed of adoption is influenced by growing pains, lack of awareness/understanding, pushback, and other forces that are likely to (at least initially) hinder its full implementation in industries and the lives of consumers (Jao, 2019). Also, anticipated cost of implementation, disintermediation worries, and loss of employment in certain sectors are likely to lead the resistance to change. Intermediaries, such as banks and credit card companies, are likely to pushback competitively or legislatively thereby slowing the pace of blockchain implementation (Casey, 2017). Resistance also can be expected from lobbies or those in power that might have stake in maintaining status quo.

Non-technology companies are likely to be slower adopters given its perceived complexity. Research firm, Gartner found that a vast number of "blockchain projects" are used just for recording data via the decentralized ledger while the other features were being ignored (Gartner, 2019). Blockchain technology and its marketing value propositions are innovative and, as with other technological breakthroughs, it will be a while before blockchain implementation passes the 50% industry standard benchmark for adoption. Yet, it might be adopted faster as a B2B tool than as a B2C tool. The 2019 CIO Agenda Survey of more than 3,000 executives showed that only 11% of the CIOs have deployed or are in the short-term planning phase with blockchain implementation since there is yet no consensus "on key components such as product concept, feature set, and core application requirements" (Gartner, 2019; Roe, 2019). Yet, as per another report, future spending on blockchain applications looks promising and predicted to go beyond 12.4 USD billion by 2022 (Soohoo, 2019).

Blockchain technology use examples presented in this chapter are experimental in nature with enormous potential. Value realized from the technology in one area are likely to drive usage in other areas as well. In terms of the immediate future of blockchain, one of the areas that is ripe for transformation is the identification management systems. This is different from identity management systems that focuses on access authorization and authentication. Identification management systems allow parties to confirm who they are during business transactions. Managing a customer or company's identity online has implications for their online presence, their confidential information, and their reputations. Blockchain's independent and transparent verification process can be used to authenticate individuals and companies. Such developments and proper implementation in the near future are likely to be welcomed by both merchants and their banks.

REFERENCES

ABN-AMRO. (2018, January 22). *First agricultural commodity trade through blockchain completed—ABN AMRO Bank.* https://www.abnamro.com/en/news/first-agricultural-commodity-trade-through-blockchain-completed%20%20

Allam, Z. (2018). *On Smart Contracts and Organisational Performance: A Review of Smart Contracts Through the Blockchain Technology.* doi:10.1515/rebs-2018-0079

Androulaki, E., Barger, A., Bortnikov, V., Cachin, C., Christidis, K., De Caro, A., Enyeart, D., Ferris, C., Laventman, G., Manevich, Y., Muralidharan, S., Murthy, C., Nguyen, B., Sethi, M., Singh, G., Smith, K., Sorniotti, A., Stathakopoulou, C., Vukolić, M., ... Yellick, J. (2018). Hyperledger fabric: A distributed operating system for permissioned blockchains. *Proceedings of the Thirteenth EuroSys Conference*, 1–15. 10.1145/3190508.3190538

Antoniadis, I., Spinthiropoulos, K., & Kontsas, S. (2020). Blockchain Applications in Tourism and Tourism Marketing: A Short Review. In A. Kavoura, E. Kefallonitis, & P. Theodoridis (Eds.), *Strategic Innovative Marketing and Tourism* (pp. 375–384). Springer International Publishing. doi:10.1007/978-3-030-36126-6_41

Bartoletti, M., & Pompianu, L. (2017). *An empirical analysis of smart contracts: Platforms, applications, and design patterns*. doi:10.25300/MISQ/2013/37.2.14

Bharadwaj, A., Sawy, O., Pavlou, P., & Venkatraman, N. (2013). Visions and Voices on Emerging Challenges in Digital Business Strategy. *Management Information Systems Quarterly*, *37*(2), 633–661. doi:10.25300/MISQ/2013/37.2.14

Bickerton, J. (2018, January 8). Digital Coupon Redemption Rates. *RAVEN5*. https://www.raven5.com/blog/digital-coupon-redemption-rates/

Business Insider. (2017, March 20). *The Blockchain in banking report: The future of blockchain solutions and technologies*. Business Insider. https://www.businessinsider.com/blockchain-in-banking-2017-3

Business Insider - India. (2018, August 21). *Here's how blockchain reduces the overall costs associated with trade*. Business Insider. https://www.businessinsider.in/heres-how-blockchain-reduces-the-overall-costs-associated-with-trade/articleshow/65484887.cms

Carson, B., Romanelli, G., Walsh, P., & Zhumaev, A. (2018, June 19). *Blockchain beyond the hype: What is the strategic business value?* https://www.mckinsey.com/business-functions/mckinsey-digital/our-insights/blockchain-beyond-the-hype-what-is-the-strategic-business-value

Casey, K. (2017, April 27). *Blockchain: The Pro's and Cons of a Technology that Will Affect our Future*. Medium. https://medium.com/totvslabs/blockchain-the-pros-and-con-s-of-a-technology-that-will-affect-our-future-f67037da7d64

Christianson, S. J. (2019, April 29). *The Blockchain Game: A great new tool for your classroom*. Blockchain Pulse: IBM Blockchain Blog. https://www.ibm.com/blogs/blockchain/2019/04/the-blockchain-game-a-great-new-tool-for-your-classroom/

Christidis, K., & Devetsikiotis, M. (2016). Blockchains and Smart Contracts for the Internet of Things. *IEEE Access : Practical Innovations, Open Solutions*, *4*, 2292–2303. doi:10.1109/ACCESS.2016.2566339

Chuen, D. (2015). *Handbook of Digital Currency: Bitcoin*. Innovation, Financial Instruments, and Big Data.

Cong, L. W., & He, Z. (2019). Blockchain Disruption and Smart Contracts. *Review of Financial Studies*, *32*(5), 1754–1797. doi:10.1093/rfs/hhz007

Cvitanović, P. L. (2018). New Technologies in Marketing as Competitive Advantage. *Proceedings of the ENTRENOVA - ENTerprise REsearch InNOVAtion Conference, 4*(1), 266–274.

Dalton, M. (2018, November 9). *Feeless Cryptocurrency: How Do Blockchains Achieve Free Transactions?* https://www.bitrates.com/news/p/feeless-cryptocurrency-how-do-blockchains-achieve-free-transactions

Dayal, P. (2020, April 9). *Why Blockchain Occupies a Unique Seat at the Advertising Table*. https://www.martechadvisor.com/articles/blockchain/why-blockchain-occupies-a-unique-seat-at-the-advertising-table/

Delisle, B. (2018, January 30). An Introduction to Basic Attention Token (BAT)—Blockchain-Based Digital Advertising. *CryptoSlate*. https://cryptoslate.com/basic-attention-token/

Deloitte. (2016, February 24). *Blockchain: Democratized trust*. Deloitte Insights. /content/www/us/en/insights/focus/tech-trends/2016/blockchain-applications-and-trust-in-a-global-economy.html

Deloitte. (2018a). *Deloitte and Chronicled Alliance*. Deloitte United States. https://www2.deloitte.com/us/en/pages/about-deloitte/solutions/deloitte-and-chronicled-alliance.html

Deloitte. (2018b). *Global Blockchain Survey 2018*. Deloitte United Kingdom. https://www2.deloitte.com/uk/en/pages/innovation/articles/global-blockchain-survey-2018.html

DePatie, J. (2017). *New Kid on the Blockchain – Exchange*. http://scienceandentertainmentexchange.org/article/new-kid-on-the-blockchain/

Deshmukh, S. (2020, September 27). *3 ways blockchain can accelerate sustainable development*. World Economic Forum. https://www.weforum.org/agenda/2020/09/3-ways-blockchain-can-contribute-to-sustainable-development/

Deshpande, I. (2019, July 25). *What Is Blockchain in Marketing? Strategies, Best Practices, Benefits and Examples*. https://www.martechadvisor.com/articles/blockchain/what-is-blockchain-how-to-use-it-in-marketing/

Dhaou, S., Zalan, T., & Toufaily, E. (2017). Blockchain Technology Adoption: Implications and Challenges. *ICEB 2017 Proceedings*. https://aisel.aisnet.org/iceb2017/39

Dicker, J. (2018, May 1). *3 Steps Toward Accepting Blockchain's Place in Media*. https://www.adweek.com/performance-marketing/3-steps-toward-accepting-blockchains-place-in-media/

Dimitrov, B. (2020, April 9). *Why Fighting COVID-19 Needs More Than Blockchain To Succeed*. Forbes. https://www.forbes.com/sites/biserdimitrov/2020/04/09/why-fighting-covid-19-needs-more-than-blockchain-to-succeed/

Dobrovnik, M., Herold, D. M., Fürst, E., & Kummer, S. (2018). Blockchain for and in Logistics: What to Adopt and Where to Start. *Logistics*, *2*(3), 18. doi:10.3390/logistics2030018

Eickhoff, M., Muntermann, J., & Weinrich, T. (2017). *What do FinTechs actually do? A Taxonomy of FinTech Business Models*. Academic Press.

English, T. (2019, January 7). *What is the Future of Blockchain and Advertising?* https://interestingengineering.com/what-is-the-future-of-blockchain-and-advertising

Ennis, D. (2019, August 6). *Walmart may tap unbanked market with cryptocurrency plan*. Marketing Dive. https://www.marketingdive.com/news/walmart-crypto-patent-banking-credit-card-fees/560318/

Ertemel, A. V. (2018). Implications of Blockchain Technology on Marketing. *Journal of International Trade. Logistics and Law*, *4*(2), 35–44.

Friedlmaier, M., Tumasjan, A., & Welpe, I. M. (2018). Disrupting Industries with Blockchain: The Industry, Venture Capital Funding, and Regional Distribution of Blockchain Ventures. *Proceedings of the 51st Hawaii International Conference on System Sciences*, 10. 10.24251/HICSS.2018.445

Galvez, J. F., Mejuto, J. C., & Simal-Gandara, J. (2018). Future challenges on the use of blockchain for food traceability analysis. *Trends in Analytical Chemistry*, *107*, 222–232. doi:10.1016/j.trac.2018.08.011

Gartner. (2019, June 12). *Gartner Reveals Seven Mistakes to Avoid in Blockchain Projects*. Gartner. https://www.gartner.com/en/newsroom/press-releases/2019-06-12-gartner-reveals-seven-mistakes-to-avoid-in-blockchain

Goldin, M., Soleimani, A., & Young, J. (2017). *The AdChain Registry*. Academic Press.

Guzenko, I. (2019, September 17). *Council Post: How Blockchain Is Affecting the Marketing and Advertising Industry*. Forbes. https://www.forbes.com/sites/forbestechcouncil/2019/09/27/how-blockchain-is-affecting-the-marketing-and-advertising-industry/

Haber, S., & Stornetta, W. S. (1991). How to timestamp a digital document. *Journal of Cryptology*, *3*(2), 99–111. doi:10.1007/BF00196791

Harvey, C. R., Moorman, C., & Toledo, M. (2018, October 1). How Blockchain Can Help Marketers Build Better Relationships with Their Customers. *Harvard Business Review*. https://hbr.org/2018/10/how-blockchain-can-help-marketers-build-better-relationships-with-their-customers

Iansiti, M., & Lakhani, K. R. (2017, January 1). The Truth About Blockchain. *Harvard Business Review*. https://hbr.org/2017/01/the-truth-about-blockchain

Ingram, D. (2018, May 9). Facebook shakes up management, launches blockchain division. *Reuters*. https://www.reuters.com/article/us-facebook-moves-idUSKBN1IA05S

Jao, N. (2019, July 16). *Lack of understanding is blockchain's biggest hurdle in Asia: Survey*. TechNode. https://technode.com/2019/07/16/lack-of-understanding-is-blockchains-biggest-hurdle-in-asia-survey/

Juniper Research. (2017, September 26). *Ad Fraud to Cost Advertisers $19 billion in 2018, Representing 9% of Total Digital Advertising Spend*. Juniper Research. https://www.juniperresearch.com/press/press-releases/ad-fraud-to-cost-advertisers-19-billion-in-2018

Kaaru, S. (2020, March 21). *Toyota launches blockchain lab for verification, supply chain and more*. CoinGeek. https://coingeek.com/toyota-launches-blockchain-lab-for-verification-supply-chain-and-more/

Kamath, R. (2018). Food Traceability on Blockchain: Walmart's Pork and Mango Pilots with IBM. *The Journal of the British Blockchain Association*, *1*(1), 1–12. doi:10.31585/jbba-1-1-(10)2018

Krauth, O. (2018, February 9). *5 companies using blockchain to drive their supply chain*. TechRepublic. https://www.techrepublic.com/article/5-companies-using-blockchain-to-drive-their-supply-chain/

Lee, Y. N. (2020, May 6). *5 charts show which travel sectors were worst hit by the coronavirus*. CNBC. https://www.cnbc.com/2020/05/06/coronavirus-pandemics-impact-on-travel-tourism-in-5-charts.html

Linn, L. A., & Koo, M. B. (2016). *Blockchain For Health Data and Its Potential Use in Health IT and Health Care Related Research*. Academic Press.

Madden, D. (2021, January 14). *The Covid-19 Pandemic Has Cost the Global Tourism Industry $935 Billion*. Forbes. https://www.forbes.com/sites/duncanmadden/2021/01/14/the-covid-19-pandemic-has-cost-the-global-tourism-industry-935-billion/

Marr, B. (2018, March 23). *How Blockchain Will Transform the Supply Chain and Logistics Industry*. Forbes. https://www.forbes.com/sites/bernardmarr/2018/03/23/how-blockchain-will-transform-the-supply-chain-and-logistics-industry/

Menon, M., & Mady, A. (2021). Blockchain: A non-technical primer for marketing academics. *Marketing Education Review*, 1–19.

Morgan, B. (2020, May 7). *50 Stats That Show the Importance of Good Loyalty Programs, Even During A Crisis*. Forbes. https://www.forbes.com/sites/blakemorgan/2020/05/07/50-stats-that-show-the-importance-of-good-loyalty-programs-even-during-a-crisis/

Morrison, O. (2019, August 6). *Anheuser-Busch InBev joins new IBM blockchain*. Foodnavigator.Com. https://www.foodnavigator.com/Article/2019/08/06/Anheuser-Busch-InBev-joins-new-IBM-blockchain

Murphy, J. (2021, March 8). *What Is an NFT? Inside the Next Billion-Dollar Crypto Sensation*. Entrepreneur. https://www.entrepreneur.com/article/366648

Nakamoto, S. (2008, October 31). *Bitcoin: A Peer-to-Peer Electronic Cash System*. https://nakamotoinstitute.org/bitcoin/

Nugent, T., Upton, D., & Cimpoesu, M. (2016). Improving data transparency in clinical trials using blockchain smart contracts. *F1000 Research*, *5*, 2541. doi:10.12688/f1000research.9756.1 PMID:28357041

Patel, T., & Drozdiak, N. (2021, March 24). *Building a Covid Travel Passport Is a Serious Tech Challenge*. BloombergQuint. https://www.bloombergquint.com/businessweek/covid-19-vaccine-passports-face-security-concerns-technical-glitches

Pathak, S. (2016, December 15). How blockchain might be useful in marketing and advertising. *Digiday*. https://digiday.com/marketing/blockchain-tech-might-useful-marketing/

Paton, E. (2021, April 20). LVMH, Richemont and Prada unite behind a blockchain consortium. *The New York Times*. https://www.nytimes.com/2021/04/20/business/lvmh-richemont-prada-blockchain.html

Patrizio, A. (2018, March 27). *Top 10 Blockchain as a Service Providers*. Datamation. https://www.datamation.com/data-center/top-10-blockchain-as-a-service-providers/

Prescher, C. (2018, March 22). *Using Blockchain to Build Connections with Consumers*. Chief Marketer. https://www.chiefmarketer.com/using-blockchain-build-connections-consumers/

PricewaterhouseCoopers. (2017, February). *Blockchain: A new tool to cut costs*. PwC. https://www.pwc.com/m1/en/media-centre/articles/blockchain-new-tool-to-cut-costs.html

PwC Global Power & Utilities. (2015). *PwC-Blockchain Ppportunity for Energy Producers and Consumers.pdf*. https://www.pwc.com/gx/en/industries/assets/pwc-blockchain-opportunity-for-energy-producers-and-consumers.pdf

Reiff, N. (2020, February 25). *Forget Bitcoin: Blockchain is the Future*. Investopedia. https://www.investopedia.com/tech/forget-bitcoin-blockchain-future/

Reyna, A., Martín, C., Chen, J., Soler, E., & Díaz, M. (2018). On blockchain and its integration with IoT. Challenges and opportunities. *Future Generation Computer Systems*, *88*, 173–190. doi:10.1016/j.future.2018.05.046

Roe, D. (2019, June 21). *What Blockchain Companies Need to Do to Succeed*. CMSWire.Com. https://www.cmswire.com/information-management/what-blockchain-companies-need-to-do-to-succeed/

Rosic, A. (2016, September 18). *What is Blockchain Technology? A Step-by-Step Guide for Beginners*. Blockgeeks. https://blockgeeks.com/guides/what-is-blockchain-technology/

Schiller, B. (2018, February 5). *On This Blockchain-Based Version of Airbnb, There's No Middleman*. Fast Company. https://www.fastcompany.com/40524021/on-this-blockchain-based-version-of-airbnb-theres-no-middleman

Seebacher, S., & Schüritz, R. (2017). Blockchain Technology as an Enabler of Service Systems: A Structured Literature Review. *Proceedings of the International Conference on Exploring Services Science*, 12–23. 10.1007/978-3-319-56925-3_2

Sharma, R. (2020, May 2). Toyota Set to Explore the Application of Blockchain Technology. *Business Blockchain HQ*. https://businessblockchainhq.com/business-blockchain-news/toyota-set-to-explore-the-application-of-blockchain-technology

Single Grain Blog. (2018, March 13). *Case Studies—Companies Doing Blockchain Marketing Well*. Single Grain. https://www.singlegrain.com/blockchain/blockchain-marketing-case-studies/

Soohoo, S. (2019, April). *Worldwide Blockchain 2019-2022 Forecast: Market Opportunity by Use Case—1H18 Update*. https://www.reportlinker.com/p05770124/Worldwide-Blockchain-Forecast-Market-Opportunity-by-Use-Case-1H18-Update.html

Tan, B., Yan, J., Chen, S., & Liu, X. (2018). The Impact of Blockchain on Food Supply Chain: The Case of Walmart. In M. Qiu (Ed.), *Smart Blockchain* (pp. 167–177). Springer International Publishing. doi:10.1007/978-3-030-05764-0_18

Tapscott, D., & Tapscott, A. (2018). *Blockchain Revolution*. Portfolio/Penguin. https://blockchain-revolution.com/

Tiwari, T. (2016, December 6). *Profit Alert: Walmart is adopting the blockchain right now...* Palm Beach Research Group. https://www.palmbeachgroup.com/palm-beach-daily/profit-alert-walmart-is-adopting-the-blockchain-right-now/

Tönnissen, S., & Teuteberg, F. (2020). Analysing the impact of blockchain-technology for operations and supply chain management: An explanatory model drawn from multiple case studies. *International Journal of Information Management*, *52*, 101953. doi:10.1016/j.ijinfomgt.2019.05.009

Treiblmaier, H. (2018). The impact of the blockchain on the supply chain: A theory-based research framework and a call for action. *Supply Chain Management*, *23*(6), 545–559. doi:10.1108/SCM-01-2018-0029

Vaccaro, S. (2019, April 2). *The Role of Blockchain in Fighting Chargeback and Ecommerce Fraud*. Chargeback. https://chargeback.com/the-role-of-blockchain-in-fighting-chargeback-and-ecommerce-fraud/

Xu, X., Pautasso, C., Zhu, L., Gramoli, V., Ponomarev, A., Tran, A. B., & Chen, S. (2016). The Blockchain as a Software Connector. *2016 13th Working IEEE/IFIP Conference on Software Architecture (WICSA)*, 182–191. 10.1109/WICSA.2016.21

Yermack, D. (2019). Blockchain technology's potential in the financial system. *Blockchain Technology*, 20.

Zhang, Z., Ma, M., Popkowski Leszczyc, P. T. L., & Zhuang, H. (2020). The influence of coupon duration on consumers' redemption behavior and brand profitability. *European Journal of Operational Research*, *281*(1), 114–128. doi:10.1016/j.ejor.2019.08.029

Zhao, W. (2018, March 15). *Anheuser-Busch Owner Pilots Blockchain for Shipping*. CoinDesk. https://www.coindesk.com/anheuser-busch-owner-pilots-blockchain-shipping

Zheng, Z., Xie, S., Dai, H.-N., Chen, X., & Wang, H. (2017, June 25). An Overview of Blockchain Technology: Architecture, Consensus, and Future Trends. *IEEE International Congress on Big Data (BigData Congress)*. 10.1109/BigDataCongress.2017.85

Section 3
Education Implications

Chapter 7
Augmenting Learner Support Services With the Use of Blockchain Technology

S. K. Pulist
https://orcid.org/0000-0003-3720-216X
Indira Gandhi National Open University, India

ABSTRACT

The potential of implementation of blockchain technology in education to manage different internal processes seems quite obvious. The chapter discusses different areas of the open and distance learning system in general and learner support services in particular in which use of BCT can be a promising pay-off. The chapter focuses on the introduction of blockchain: how it started and recent developments, open and distance learning system in India, learner support services in open and distance learning, BCT in India, strategies of the government of India to augment blockchain, application of blockchain in open and distance learning in general and learner support services in particular, challenges in implementation of blockchain in open and distance learning and learner support in the country, and its implications.

INTRODUCTION

Blockchain Technology is here to revolutionise all walks of human life. It has already become an important part of financial transactions and digital negotiable instruments replacing the hard currency. The virtual currencies are the most apt example of use of Blockchain Technology (BCT). The recent addition to this continuum is the Non-Fungible Token (NFT) which is another product of the BCT. It would not be

DOI: 10.4018/978-1-6684-4153-4.ch007

Copyright © 2023, IGI Global. Copying or distributing in print or electronic forms without written permission of IGI Global is prohibited.

out of place to name it as monitoring and tracking technology. It can be used in any area that requires monitoring, tracking, authentication and verification of activities done by the users in a distributed manner. At the core, Blockchain is a system that facilitates recording of important information in a way that makes it almost impossible to change it in any way afterwards. The chain of building blocks keeps it complete and growing further. Each of the new blocks builds upon the earlier block and record information as an add-on to the existing information. The distributed structure of the technology enables appendage of the new information and record of the transaction to the ledger of all the existing partners. This new block will have information about the sender, receiver and other relevant important information (metadata) that needs to be shared over and above the existing information.

As the name suggests Blockchain is a database that stores information in a distributed manner in the form of blocks of information. These blocks of information are linked to each other in a linear way in a specified pattern. Thus, these blocks become a permanent repository of that chunk of information with historical chain log. Whenever a transaction/information is added in the form of a new block, the partners of that chain confirm the validity of the transaction/ addition of the information. On confirmation by all the partners, it becomes a permanent constituent of the cluster and is appended to the existing chain as the last block in that order. Any type of information can be stored in the information blocks. However, since the information stored in the blocks is not editable, the information not requiring any change in its basic structure only is stored in the blocks. Subsequent updates of the information are linearly and chronologically linked to earlier blocks of information so as to put all the information blocks to scrutiny and validation. Each of the blocks in the chain contains important information in the form of a hash of the previous block. "A hash function takes an input of variable length and produces an output of fixed length. This way, hashing within the blockchain (i.e., hash chain) makes it very difficult to change previous blocks, thus ensuring immutability" (Steiu, 2020). The Proof of Work is another important component of the process that provides a testimony for adding a block to the chain. Figure 1 presents the cryptographic linking of blocks in the blockchain.

Figure 1. Cryptographic linking in blockchain
Source: Thomas (2020)

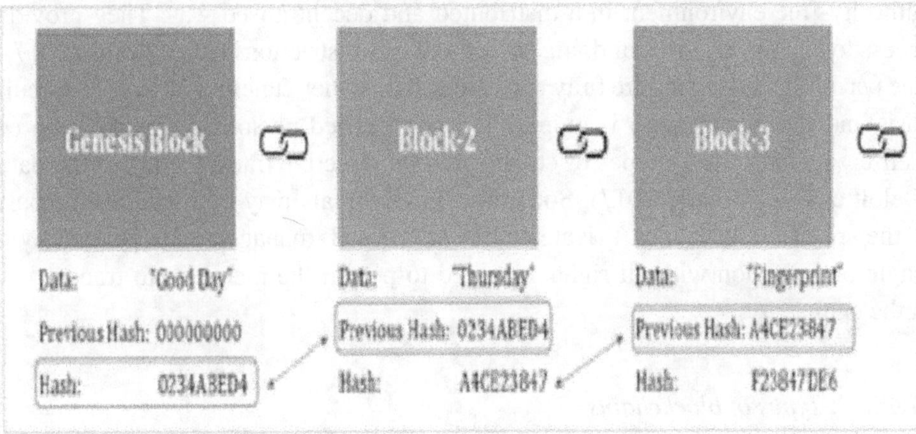

The first block of information in Blockchain is called genesis block. The subsequent blocks are connected to each other in chronological order in such a way that it is easy to trace the original information in the genesis block. As Mahankali (2019) puts it in simple terms, 'Blockchain is a growing list of records called blocks which are linked using cryptography'. Depending upon the number of blocks and number of partnering users, the size of the chain keeps on expanding. The uniqueness of the blockchains is that they are appendable but not editable. The mechanism of digital blockchain was invented by Staurt Haber and W. Scott Stornetta way back in 1991. This was developed out of the necessity of providing a robust tamper-proof mechanism that could store timestamps also in the same uneditable way (Conway, 2021). However, Satoshi Nakamoto first used BCT for devising a digital cryptocurrency in the form of Bitcoin. Thus, the cryptocurrency Bitcoin created in 2009 became the major tangible product of blockchain application. Inversely, the evolution of cryptocurrencies is attributed to the birth of BCT (Yuan & Wang, 2018). The Woolf University established by the academics of Oxford is claimed to be the first Blockchain University without any physical appearance (Broggi *et al.*, 2018). The university platform is open to the learners and teachers along with other educational institutions to join and offer their academic programmes to the global learners. Other renowned names of the educational institutions which have stepped into the use of BCT in some of their processes are: MIT, University of Nicosia, and UK Open University.

The Blockchain can be a private, a permissioned or a permissionless platform depending upon the requirements (Deloitte & Assocham, 2017). Figure 2 presents the types of blockchains (Varghese et al., 2018). The permissionless chains are public blockchains which permit anyone to participate and contribute. These types of blockchains profess the open philosophy in real sense, since, they operate in an authority-free environment in a distributed and decentralised way. They provide access to anyone desirous of doing so without any restrictions (Zheng et al., 2017). The permissioned chains are fully restricted chains, they function on a close-circuit basis and the membership is granted to pre-specified persons on verification of their credentials. These types of chains provide selective transparency to the data (Deloitte & Assocham, 2017). Sometimes, individuals may be invited to be part of the specific blockchain. Private chains are owned, managed and operated by a single organisation with all rights reserved to permit the members to transaction in the chain.

Figure 2. Types of blockchains

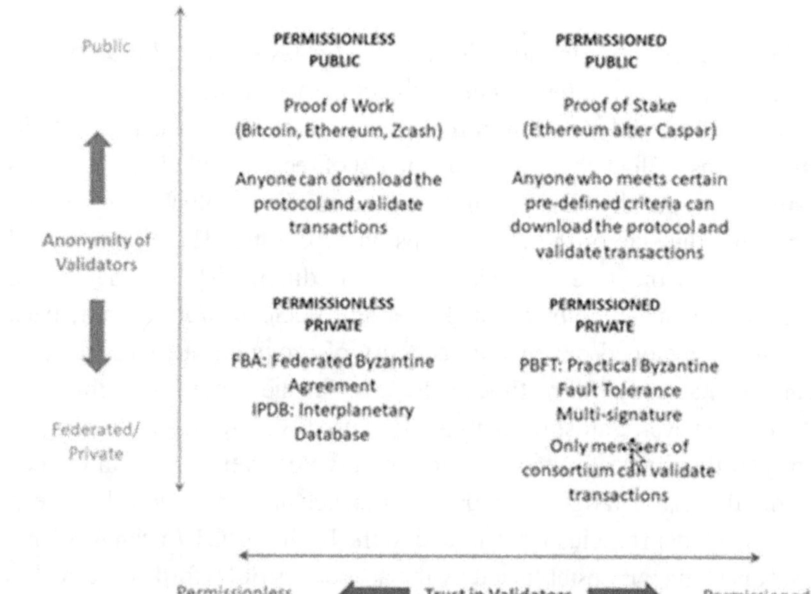

The researchers are keen on making researches on how to make valuable use of BCT other than the safe and secure financial transactions. The application of BCT in specialised areas like medical and health, and digital currencies has been found very useful. The technology can be used to transfer the sensitive data across platforms and users in a distributed environment. Dutta (2021) has identified some new areas of use of BCT as: supply chain monitoring, loyalty rewards regulation, creation of digital identities, secure data sharing, copyright and patent protection, digital voting, and title transferring in real-estates.

Governments across the globe are increasingly taking interest in BCT. However, quite a few have come up with a regulatory framework for it and recognise it for the overall economic ecosystem of the country. The concern is that how to make it usable for socio-economic development of the people minimising potential risks and putting a check over its negative areas. India has taken strong steps for development of BCT ecosystem in the country paving the way for its substantial use in different areas. After India, China and USA are the biggest aspirants for use of this technology (Sharma, Yildirim, & Kurubacak, 2020). The US Government has implemented the Education Blockchain Initiative (EBI) to promote use of BCT in education. Some of other countries who have taken keen interest in BCT are: Vietnam, Estonia, UAE, Brazil, Chile, Canada, Singapore, and Switzerland. As per the report of the World Economic Forum predictions, the total spending on the BCT world over is expected to be 10% of the global GDP by the year 2025 (WEF, 2018).

The chapter focuses on introduction of Blockchain - how it started and recent developments, open and distance learning system in India, learner support services in open and distance learning, BCT in India, strategies of government of India to augment blockchain, application of blockchain in open and distance learning in general and learner support services in particular, challenges in implementation of blockchain in open and distance learning and learner support in the country, and its implications.

DEFINING BLOCKCHAIN TECHNOLOGY

The developers have tried to define the BCT in different ways depending upon their structure, usage, and operational specificities. Swan (2015a) defined Blockchain Technology as “the secure decentralised computing ledger that underlies Bitcoin and cryptocurrencies, and more profoundly is a next-generation global scale decentralized infrastructure and mechanism for securely updating distributed computing notes with ongoing consensus truth States”. Borge et al. (2017) considers Blockchain in the form of consensus algorithms which are used to authenticate different activities and credentials such as work done, stake claimed, importance held, activity completed, deposits made, time elapsed, etc. While the Blockchain can be regarded as an infrastructure technology, it has different uses in different sectors (Bozkurt & Ucar, 2020). It is transformational in nature and functions well in non-linear and decentralized learning ecospaces. The record of information and the block made of this record, are the two important components of Blockchain. These blocks are linked to each other in a specific pattern and structure to maintain the chain. Zheng et al. (2017) define Blockchain as "an immutable, decentralized database — a chain of 'blocks' which stores information such as transactions' dates, times, amounts, and/or participants (participants on the blockchain are usually not personally identifiable)". Mikroyannidis, Third, Domingue, Bachelor, and Quick (2020) define characteristics of the BCT as: "unique timestamp, cryptographic hash links, cryptographic puzzle mechanism, and dedicated participants' network". Blockchain follows "secure by design paradigm which makes it a unique system that makes business transactions transparent and trusted in a consortium environment without any reliance on a centralized or third-party authority" (GoI, 2021).

Bozkurt and Ucar (2020) view BCT as a bridging infrastructure among formal, non-formal and informal learning processes. Since, it is a chain of blocks that follows a distributed mechanism, all the participating nodes individually keep a full copy of the chain in the form of a ledger, and keep on verifying and updating the existing and the new information for authenticity. The information is secure even if any of the participating nodes has compromised with the information (Cedeno, 2020). As part of a distributed leger system, the Blockchain has three important components to function: Applications, Protocol, and Networks and Infrastructure (Lemoie & Soares, 2020). The application is the user interface that communicates with the distributed ledger system. The software and other process controlling the transactions in the distributed ledger, are called the protocol and network. The hardware part of the ledger system such as computers and servers are the infrastructure component. The four major technological components of Blockchain as identified by Chen, Xu, Lu, & Chem (2018) are: currency, decentralization, immutability, and traceability. Swan (2015b) views the development of BCT in three phases on a continuum i.e. Blockchain 1.0 - encrypted currency, Blockchain 2.0 - intelligent/ smart contracts, and Blockchain 3.0 - widely used applications based on BCT.

Hash Tree and Smart Contracts

Blockchain technology has defined the constructs of 'trust' and 'authentication' in a new perspective. The BCT makes use of 'Merkle Tree' data structure which is also named 'Hash Tree' as devised by Ralph Merkel in the year 1971 (Filipova, 2018). In this structure, the blocks of data are "grouped in pairs and the hash of each of these blocks is stored in a parent node. This grouping of hash codes continues till the root node" that makes the information blocks tamperproof (Deloitte & Assocham, 2017). The 'Smart Contract' is another important component of AI based Blockchain ecosystem. The Blockchain makes use of auto-executable simple programmes called 'smart contracts'. These programmes run and execute an agreement as soon as the predetermined conditions of the agreement are met. There is no limit on specifying the number of conditions that should be met before taking an action. However, a specific protocol needs to be decided beforehand as to how the data/updated information would be executed/updated on the Blockchain. This process eliminates the requirement of a third-party intervention and all the participants of the Blockchain are notified about the execution of the agreement without loss of time due to human intervention. Any type of workflow can be automated that requires certain conditions to be met before execution of an action. The relevant Blockchain is updated with the information on completion of such an action (IBM, n.d.). Clark, Bakshi, and Braine (2016) defined the smart contracts as "automatable and enforceable agreements that are controlled by contract code and are activated by externally owned accounts". The smart contacts could be in the form of best contract, skill awareness contract, job posting contract or job store contract, etc. (Clark, 2020).

BLOCKCHAIN POLICY IN INDIA

The governments across the world are yet to come up with a progressive framework on BCT which would promote its fool proof use along with regulating it for best and positive use while maintaining accountability at different levels. In order to boost research in application of BCT, Government of India has provided support to a project which is multi-institutional in nature. The project "Distributed Centre of Excellence in Blockchain Technology" has been launched in collaboration with Centre for Development of Advanced Computing (C-DAC), Institute for Development and Research in Banking Technology (IDRBT) and Veermata Jijabai Technological Institute (VJTI) that will be executing the research initiatives on BCT. The major areas of research as proposed in the National Strategic Plan are: 'standards & interoperability, scalability & performance, consensus mechanisms, security & privacy, key management, secure smart contracts and detection of vulnerabilities in Blockchain Technology based solutions' (MeitY, 2021). The Strategic Document provides a thorough SWOC analysis of national ecosystem vis-a-vis implementation of BCT in India.

The MeitY (2021) has envisioned the implementation of BCT in India through a three-component schema vis. "development of a unified blockchain framework, blockchain services, and blockchain applications for different domains". Development of an indigenous Blockchain platform is among the objectives of Government of India. Since, a single platform will not be able to cater to the requirements of different fields, multiple blockchain platforms would be required to be developed. One of the platforms would be used for hosting education chain. The standard APIs will be deployed for different domains using National Blockchain API. The National Level Blockchain Framework (NLBF) is proposed with three types of participants i.e. "a) Confident user of Technology (Application Developers), b) Provider or Operator of

Technology (Infrastructure & Services, BaaS), and c) Complete Technology Stack Builder (IP Creator)" (MeitY, 2021).

The National Strategy on Blockchain as propounded by India further aims 'to create trusted digital platforms through shared Blockchain infrastructure; promoting research and development, innovation, technology and application development; and facilitating state of the art, transparent, secure and trusted digital service delivery to citizens and businesses, thus making India a global leader in Blockchain Technology' (GOI, 2021). The niche areas where the BCT can be deployed on purpose are: trade finance, cryptocurrency, crowd-funding, insurance banking, sharing economy, legal, real estate, Internet of Things, healthcare, IT cloud storage, and supply chain in addition to education, governance, cyber security, media, and power sector. The educational institutions and the agencies providing financial assistance to the learners can be the potential users for BCT for providing achievement credentials and scholarship details. The government of India plans to introduce course on BCT at UG and PG levels. The research students would be encouraged to undertake research project involving this technology. The identification of data to be kept off-chain and on-chain also is proposed before implementation of BCT in any domain.

BLOCKCHAIN INITIATIVES IN INDIA

The Blockchain technology is still evolving and its usage in different fields is being explored, though, it "has the potential to transform multiple industries and make processes more democratic, secure, transparent, and efficient" (Deloitte & Assocham, 2017). In the Indian context, the government has already initiated different steps in this direction. Since, BCT is considered to be a disruptive innovation, monitoring is a must to protect the interest of the users and other stakeholders. There is all the likelihood that miscreants and early adopters may put this technology to an abuse and encroach upon the privacy of the masses. This apprehension might motivate people and Institutions to go slow with adoption of BCT in serious businesses requiring privacy. Government of India has identified that the application of BCT can be helpful in areas such as "healthcare, Governance, cyber security, automobiles, media, travel, logistics & hospitality, education, legal, energy, smart cities, etc." (MeitY, 2021). The Blockchain platforms would be created as a national resource and would be offered to the users as Blockchain as a Service (BaaS). The integration of BCT with Artificial Intelligence and other emerging technologies is also in the offing. A regulatory framework covering different aspects such as "Infrastructure, Research, Technology Stack, Testing & Certification, Capacity Building, etc." also is under development (GOI, 2021).

The Government of India has established a centre of excellence in Blockchain technology that is mandated to provide Blockchain technology-based solutions to meet and augment the governance requirements. The Centre operates "as a coordinated, interoperable blockchain ecosystem around the nation, allowing all partners to benefit from shared learning, experiences and resources" (GOI, n.d.). The Government of India (2021) has already released a document on National Strategy on Blockchain that focuses on technological and administrative aspects of Blockchain implementation. The document provides an insight to the strategies to develop the Blockchain ecosystem in the Indian context. A blockchain framework at the national level has been envisaged in the document. As many as 44 domains have been identified by the Government of India in which potential use of BCT can be made (GOI, 2021). As per the forecast made in the above document, "the Blockchain would be used as a foundational technology for 30% of the global customer base". As per Global Crypto Adoption Index by Chainalysis, India holds second position in adoption of cryptographic initiatives just next to Vietnam (Chainalysis, 2021). National eGovernance Division, Govt. of India, C-DAC, NIC, NIELIT, NITI Ayog, Gujarat Narmada Valley Fertilizers & Chemicals Ltd, Govt. of Telangana, Govt. of Tamil Nadu, Reserve Bank of India, Mahindra and Mahindra, IBM, State Bank of India are working coordination for development of Blockchain infrastructure at different levels. Amazon, Microsoft, IBM, Oracle are already in the business of providing Blockchain infrastructure to the institutions for their use.

The use of the BCT in the field of education is still at the primary stage. It is going to play a wider role in education in the times to come. While other uses of this technology in education are a matter of exploration and discovery, quite a few institutions have started making its use for academic credential storage. The educational institutions can use it for "enhancing opportunities for lifelong learning, creating greater efficiencies for educators through smart contracts, and offering students ownership of their academic records" among others (Maryville University, 2021). The employers also can keep a track of the performance of employees and any skills acquired during the job period can well be chained with the current educational and experience credential of the employee. This would enhance authenticity of the experience and post-studies skills acquired by an employee for further elevation within and outside the organisation also. This way, the skill development programmes, internships, projects, etc. undertaken by an employee can be made part of his/her credentials very easily. In the Indian context, different national level services like eSign, ePramman, DigiLocker, etc. will be integrated with the application of this technology.

DigiLocker

The DigiLocker is one of the Government of India services made available through Ministry of Electronic and Information Technology that facilitates opening of an account by any of the citizens of the country. The personal documents, certificates, credentials and other important documents such as vehicle registration certificate, driving license, academic certificates, etc. can be stored on the cloud through DigiLocker which is considered to be very secure and protected. These documents can be submitted online to any agency, employer or educational institution for verification purposes by creating a tag and link to the portal of the concerned institution. The system functions in a completely safe environment. The smart contracts on the DigiLocker are automatically triggered for execution with the help of tags and permissions accorded by the user. Another service connected to the DigiLocker is the National Academic Repository (NAD). As the name suggests, the repository facilitates the academic institutions to push the grade cards, degree certificates and other testimonials of their students to this central repository. The students are able to access this repository with the help of DigiLocker credentials and can submit and tag any of these documents online to an application to be submitted to any institution, employer or service agency. As per the circular issued by University Grants Commission of India, these documents will be accepted for all the purposes for which a printed copy of a certificate is required (UGC, 2022). Online acceptance of applications for providing different services has become ubiquitous now-a-days and this process requires scanning of the printed certificates and get them uploaded while filling in the online application forms. With this circular, the mandatory requirement of submission of scanned copies of certificates would be eliminated. Instead, the applicants would be able to tag the digital certificates available on the NAD to the application as an authentic document.

The Academic {Blockchain} Document (ABCD) of Central Board of Secondary Education is a platform that houses academic documents of the students in a secure and tamperproof manner. Any changes to the mark sheets are recorded in the history. The certificates are accessible to the students, higher education institutions and employers in addition to government departments (CBSE, 2021). Drugs Logistics is an initiative by Government of Karnataka in association with Government of India to manage the supply chain of drugs in the state. The mechanism prevents any threats pertaining to faking of critical data regarding a medicine such as ingredients, date of expiry, quality, etc. (NIC, n.d.). Certificate Chain is another Blockchain based initiative by Government of Karnataka. The certificates issued by different government departments and educational institutions re stored on this Blockchain to be accessed by the users in a secure and verifiable manner. It gives access to these documents to the students, government departments, educational institutions and non-governmental agencies for verification of the documents submitted by the applicants. Some other important services in which the BCT is being used are: Public Distribution System, Blood Bank, Goa Goods and Services Tax, Goa Excise, and State excise supply chain (eAbgari) (NIC, n.d.).

eSign

The eSign is an online service based on Public Key Infrastructure and is part of Digital India Mission of Government of India. It can be integrated with any online portal with the help of an API in order to facilitate Aadhar based authentication. The authentication process uses the crypto token-based certification in which the "primary key of the subscriber shall be stored in the Hardware Security Module (HSM)" (CCA, 2021). The authentication for eSign can be done online or offline with the help of Aadhar number. The authentication of documents through eSign is considered as secure and reliable. With the help of this service, the users are able to put their eSignatures on any document in the digital environment. The eSignatures are valid even in legal documents (GOI, 2021). The government agencies, private organisations and individual users are making use of this service.

ePramman

The ePramman number / Adhaar number is the unique identity number provided to the citizens of India. It works on a standard eAuthentication Framework. The number is used for authentication of individual's identity while applying for any of the services provided by the government agencies based on identification authentication (GOI, 2021). It eliminates the requirement of multiple identities for availing different government services. "Aadhaar is a strategic policy tool for social and financial inclusion, public sector delivery reforms, managing fiscal budgets, increase convenience, and promote hassle-free people-centric governance" (UIAI, n.d.). The Aadhar infrastructure is scalable that currently can handle 100 million authentications a day.

Academic Bank of Credits

The Blockchain technology can be utilized to track the progress of the students who register for short term courses with other institutions while pursuing their main degree in one institution. The credits earned by the students this way, can be redeemed and reckoned for awarding the degree by the host institution. According to Langard (2019), such a decentralised system works on the "Ethereum Blockchain Certified Data Taken" system to access the certified data. Automation of issue of final diploma and degree certificates with the help of smart contracts is another option using the BCT. The educational institutions are now focusing on its use for facilitating accumulation of credits ensuring authenticity of the achievements made by the learners in classroom, MOOCs and other online courses, and skill development programmes including internship, hands-on projects, etc. The University Grants Commission of India has already established the Academic Bank of Credits to regulate the credits earned by the learners in a staggered way from different institutions (UGC, 2021). It provides for consolidation of these credits with one institution which can grant a certificate reckoning such accumulation of credits.

OPEN AND DISTANCE LEARNING SYSTEM

The open and distance learning (ODL) system has taken a tangible leap from being an alternative system of education parallel to the conventional face to face education system. The ODL system can be considered to be part of the mainstream education system in India. As per the report of the All India Survey on Higher Education (AISHE), the number of students seeking higher education in India during the year 2019-20 is estimated to be 38.5 million. As many as 11.1% of the learners sought admission to different programmes offered by the ODL system (AISHE, 2020). Thus, ODL system is the major contributor to gross enrolment ratio (GER) which was estimated to be 27.1 in the year 2019-20. This could be attributed to different propellers such as: quality of programmes, credibility of the programmes, limited number of reputed face-to-face institutions, regulatory framework supporting the ODL system, among others.

The recent disruptions in face-to-face activities of the education system have forced the world to rethink and relook the ODL system as a dynamic mainstream educational intervention strategic system to cater to the changing learning requirements of the educational aspirants in daunting situations. This journey has been from instruction-centered to learner-centered to learner-personalised and now to situation-specific-learner centered education system. Each new development in ICT has added a new dimension to the ODL processes. These developments have further expanded the scope of ODL system in meeting the requirements of the aspirants and dwindling conventional boundaries e.g. digital learning has helped the learners to do away with the conventional printed self-learning material. Different digital resources are available to the learners on a click. Keagan (1996) views distance education as the constitutional activity that advances the transfer of knowledge and learning between the teacher, self-learning material and the learners not necessarily meeting at the same time or the same place. Over the period, the ODL system has been accepting the technological developments taking place from time to time though its assimilation has been at a moderate pace. These technologies have changed the very characteristics and nature of the open and distance learning with the passage of time. This has given birth to online and blended learning. The ODL system has been able to experiment with these new technologies to integrate and enrich the learning experiences of the learners. The admissions, instruction /learning delivery, support services, and evaluation are the core components of the system. The extensive use of ICT and its assimilation, have facilitated the system to switch over to an online mechanism of performing the above processes.

The ODL institutions get a limited time for receiving the admission forms from the fresher learners and verify them with the original certificates they have submitted. It is a herculean task and cannot be completed with the required level of precision in a given time. Hussain and Cakir (2020) see development of quality learning material, instructional delivery, learner interaction, assessment, grading and scoring, result declaration, and certification as major challenges for open and distance learning system in the current scenario. The BCT joining hands together with other technologies, can prove to be a resource and capability multiplier for open and distance education system. The BCT can help the open and distance learning institutions in making available the affordable, equitable and quality educational eContent to the learners at the global level (McGreal, 2021). The digital certificates and credential issued by the educational institutions through BCT perhaps would be a great contribution in easing out the admission process.

LEARNER SUPPORT SERVICES IN ODL

The role of learner support services in the open and distance learning system is of utmost importance. As observed by Chattopadhyay (2014), learner support services are not limited to counselling, formative assessment and term end examinations. The scope of these services commences with providing relevant information as part of pre-counselling to the prospective learners so as to enable them to choose relevant programmes from a variety of programmes offered by an ODL institution. The academic counselling becomes crucial to leverage the learning acquired by the learner after having gone through the self-learning material. Since, the learner is at a distance from the parent institution, the support services rise to the occasion to provide hand-holding to the learners at all steps. The support services can broadly be divided into administrative services and academic services. Ciftci (2020) feels that development in ICT and more and more educational applications have led to change in the learning needs of the learners. The cultural change due to this, has further led to changes in learning habits of the aspirant.

Administrative Support

The Administrative support services play a crucial role in providing all sorts of supportive help to the learners. To Tait (2003), "advising, assessment, giving feedback to the individual on non-academic aptitudes and skills, practical help to promote study, making out a case for funding, writing a reference, promoting changes within the institution to benefit students and organizing student support" are some of the administrative support services provided to the learners by an ODL institution. Chattopadhyay (2014), counts pre-admission services, information services, and post–admission services under administrative service. Broadly speaking, the academic trajectory of a learner and his/her learning experiences can be seen in the form of a broad canvas. So many associated hand-holding activities and institutional support take place beyond the process of direct teaching and learning in the academic life of the learner making his/her learning effective and engaging. All these services can be considered as learner support in case of open and distance learning.

Academic support

All the teaching and learning activities including academic counselling, tutorial, practicals, teleconferencing, interactive radio counselling, and doubt classes are part of the academic support extended to the learners by the institution. Feedback on the work submitted by the learners as part of the evaluation process and comments on the performance are also crucial part of the teaching and learning activities. The teacher-learner interaction is helpful to the learners in guiding them and setting their direction leading to motivation and enthusiasm for studies. Sharing online resources with the learners and using digital content as flipped classroom add to the riches of the teaching and learning process.

BLOCKCHAIN IN OPEN AND DISTANCE LEARNING

The use of Blockchain technology is yet in the infancy stage in the field of education, and the technology has not been able to pick up as in some other areas of use. It is believed that the distance education would soon witness the transformative impact of application of BCT to handle different processes (Sharma, 2018). This technology would provide a secure environment through data encryption to the academic institutional that which could be easily tampered, manipulated or breached in the wake of cut-throat competition for admission to institutions of repute and employment purposes.

The educational institutions are yet in the process of assessing the specific advantages of use of BCT in their activities and processes. The Gartner Survey conducted in the year 2019 revealed that the use of BCT by the higher education institutions in US was to the tune of 2% only. Another 18% of the institutions were planning to implement the technology in next 2 years (Gartner, 2019). For the time being, the usage is limited to storage of the student data only. Chen et al. (2018) sees a great potential in use of BCT for performance assessment of the learners by tracking the path of progression in sync with the teaching and learning activities designed by the teacher in addition to formative and summative assessment. Grech and Camilleri (2017) visualise different uses of Blockchain in open and distance learning as: permanent trust certificates, multistage accreditation verification, automatic recognition and credit transfer, lifelong learning passport, monitoring of intellectual property, tuition payments, funding learners, and learner authentication in educational organisations. The implementation of blockchain in education will enhance the motivation level of the learners when they own the data connected with their profile, learning, and credentials.

Though, the researchers do believe that if used, it can play a bigger role in education, since it has the potential to revolutionise the teaching and learning. As clark (2020) puts it, overall spending on these technologies in education is going to increase substantially over the coming years at the global level. NISG (2019) predicted that India would be among the leading countries who would be using BCT for different purposes by 2025. Technology is going to disrupt educational activities at a massive level (Grech & Camilleri, 2017). The BCT can offer a learner-centred model of education that can offer a framework to the learning that can be controlled by them in different ways to manage their learning and the ways the learning activities should happen on the platform (Mikroyannidis, Third, Domingue, Bachelor, & Quick, 2020). This opens up new possibilities of its use in the area of education. The Open Source University (https://os.university/) has already provided a common platform to the educational institutions, students and employers where they can register for their purpose. It is called the "smart ledger for personal and organisational development" This facility is provided to the users as a service. The chapter will focus on the application of BCT for augmenting learner support services for distance learners.

USE OF BLOCKCHAIN FOR LEARNER SUPPORT SERVICES

The learner support services are the core component of the open and distance education system. The academic and administrative support is the mainstay for the distance learners that is provided through learner support centres. The application of Blockchain will lead to standardisation of the protocols through which different institutions will store the student information and share on the Blockchain platform to be used by different stakeholders. The driving force for the BCT to stay is its transparency, traceability, verifiability, scalability and unchangeability. The distributed ownership is one of the core benefits of the use of this technology.

Standardisation of support services while keeping the space for customisation of learning activities, communities built upon the BCT, and academic culture would facilitate the ODL system in providing required support services to the learners. The pedagogical models designed with the help of Blockchain can be used for the learners who newly joined the system. Blockchain technology is an important tool in the hands of the academic counsellor who can assist the learners in making their educational journey a joyful learning experience. The education system can be benefited with this Technology in maintaining grade-sheets/transcripts, performance badges, personal records of the learners including payment details, tracking daily attendance, and academic performance through different assessment tools, efficient data storage, and security. The technology provides a state of permanence to the information generated in all these areas.

POTENTIAL USES OF BCT IN SUPPORT SERVICES

The educationists have now started exploring the possibilities of using the semantic web and BCT to further customise learning for every learner. Specifically with the online resources and technologies being used by the education system, more so the open and distance learning system, BCT would enable the educational institutions to create a scaffolding system to provide academic and administrative support services to its learners. The pandemic situation has charged the mind-set of the institutions and cleared the roadblocks to accept and integrate technology wholeheartedly in all the pedagogical processes including the learner support services.

Much talked about uses of Blockchain technology in education are: decentralisation, transparency, security consistency. Some of the benefits as identified by Cedeno (2020) are: easy maintenance of student records and final degree certificates, badges, file storage, publishing, lower costs, and rewards and digital tokens for computer nodes. The University of Melbourne has already started using BCT for academic/administrative activities to help the learners. Now the students can use the digital verified copies of their certificates with the prospective employers and other organisations as an immutable manner (Moore, 2019).

Personalised Leaner Support

The implementation of Blockchain technology in providing personalised support services to the learners will add value to the individual and collective efforts of the educational institutions in providing unique teaching and learning experiences. The use of technology not only will help the teachers to understand the learners' learning trajectory but also help the learners to understand and analyse their own progress and setbacks in a better way. The learning patterns of the learners can be analysed and understood in a better way by a teacher with the help of information stored in the blocks and customise the learning methodology to shoot his/her learning requirements and learning styles (Tapscott & Tapscott, 2017). Automaticity of process on meeting certain conditions by keeping all lessons in a linear or nonlinear personalised way, will help the teachers in tracking learner progress. The better understanding of the learner's learning style and needs will lead the teachers with the probability of providing customised academic support, and thus, provide optimum benefit to them. Thus, use of Blockchain technology can improve adaptive learning and enable the teacher to know the plight of a stagnant learner and information about a learner likely to drop out. It can help in identifying popular courses based on the performance and influx of the learners, learning patterns of the learners, and reading-content difficulty for the learners.

The open and distance learning system transforms a learner in to a lifelong learner in a real sense. Unlike face-to-face mode, this system makes the learner learn special skills and techniques which develop in him/her the sense of self-motivation, self-study, self-reliance, and above all self-propulsion. As a life-long learner, his/her journey of learning new things never stops. Blockchain technology fully meets the requirements of lifelong learners. The personalised progress-path tagged with achievement reward tokens can play an important role for them as a personal trainer and educator. These progress-driven reward tokens help the learners in sustaining their motivation

Learning Outcomes

The use of blockchain Technology can help the ODL institutions in different ways by enhancing the quality of services and deliverables for the learners thereby matching the learning outcomes with their requirements and expectations. Use of BCT in pedagogical process has the potential to enhance the impact of activities across different key dimensions (Tapscott & Kaplan, 2019). The programmes running through the Blockchain can facilitate the teacher and the learner alike to know the extent of learning that is happening on the trajectory and further customise the delivery mechanism, learning resources and methodology suiting to the individualised requirements of the learner. Thus, adaptive learning can be facilitated during the process.

In Blockchain framework, the results / assessment records are stored in the form of blocks on the chain and the teacher feedback is always available to the learner alone with the assessment component. The words of appreciation and tokens awarded by the teacher on the work accomplished by the learner, stay on the block forever. The assessment of work presented by the learner and constructive comments given by the teacher will help the learner in referring back and forth while focusing on the required areas and prepare well for the term end examinations. The updated information about the extent of learning they have acquired, will also motivate the distance learners to focus on specific areas requiring attention. This will help them in getting intrinsic motivation to improve performance. The intelligent contracts can keep a watch over the performance of the learners and direct them for their efforts towards specific areas of curriculum. This would improve the counselling system quantitatively as well as qualitatively. The counsellors can match their efforts and teaching skills with the grasping power of the learners so as to adjust the gap between them appropriately. This would be an opportunity for the academic counsellor also to evaluate his/her counselling patterns that can help in redesigning instructional strategy.

Seeking feedback of learners, teachers, stakeholders, employees, etc. on the design of the curriculum, learning outcomes and expectations of the employers, is as important. A state of integration and equilibrium can be achieved by the ODL institution by making necessary adjustments at different points in the teaching and learning process which can also be adjusted with the help of implementation of BCT in Education.

Assessment

The application of Blockchain in ODL would facilitate project-based learning, collaborative learning, group assignments, and discussion forums as part of assessment tools. Participation by the learners in collaborative projects would in turn develop collaborative learning skills among them. The learning can happen in a distributed and decentralised way with the help of learning blocks and exchanges dedicated to the personal and professional development of the learners. The learners in groups can facilitate mutual learning and play an important role in peer assessment. From teacher assessment point of view, the contributions made by different learners will be saved in the form of blocks on the chain. The teacher will be able to easily assess the work done by individual learners and give grades. This will prevent the learners from attraction of copying content from other fellow learners or other educational resources available in the public domain. All this will be available in a verifiable format for rest of the time since all the blocks will be time stamped at the time of creation.

The formative assessment in ODL is a tool applied as *'assessment for learning'* of the learners. The performance made by the learners during mid-term/class tests and quizzes helps the learner to focus on weak performance areas. The assessed scripts and comments offered by the teacher become a reference point for the learner to focus on. The subsequent performance of the learners builds heavily on their learnings from the formative assessment. The BCT can help the learners and teacher alike in tracking the progress subsequent to feedback on documents submitted for formative assessment. The revision of the work done for preparing these documents will later help the learner in preparing for final assessment at the end of the term.

Management of Learner Payments

The Blockchain technology can help the institutions and the learners to manage the payment of tuition fee. Thayer (2018) emphasises that the learner fee payment records can be safely maintained with the help of BCT that will be shared among the learner and educational institution in real-time in a secure way. The fee can be divided based on components of learning. As soon as one component is completed by the learner, the system can demand payment for the next module/segment from the learner. Personalised monitoring and tracking system for fee payment with the help of smart contracts can trigger reminders to the learner for payment of fee based on fulfilment of certain criteria whenever payment is due. The Maryville University (2021) has already launched courses on the *'pay-as-you-go'* basis to make the programmes with high tuition fee, affordable and accessible.

The payment of learner fees through Blockchain powered cryptocurrencies will help in lessening the arduous work of maintaining all transaction records in physical form. The King's College in New York has already taken a lead in 2014 and tested its use as a case for accepting programme fee from the students in Bitcoin (Moore, 2019). The scholarship and financial assistance process can be automated minimising human intervention to make the system efficient and self-driven based on meeting certain specified conditions. The financial institutions providing funding to the learners may be able to track the payment of fee to the educational institution on behalf of the learner and seek status of learner's academic achievements directly.

Curriculum Monitoring

At national level, regulatory agencies such as UGC, AICTE, MCI, etc. can ask the educational institutions to store their curriculum on the Blockchain platform. In the absence of a unified system, monitoring of the curriculum is difficult. While other educational institutions can match their curriculum with it in case of a need, the regulatory agencies also will be able to monitor the update, revision, and modification in the curriculum of these educational institutions, and give suggestions from time to time. This will help in standardising the curriculum at class and discipline levels. The prospective learners can find out an institution of interest based on better curriculum for their study and, thus, shortlist the institutions where they would like to further their studies. Currently, the UGC has already suggested a curriculum for undergraduate classes based on choice-based credit system (CBCS). The educational institutions have the liberty to deviate from this curriculum up to 30% so as to contextualise it and incorporate some of the peculiar topics depending upon the local context. As per the practice in vogue, the learners have to present an authenticated copy of the syllabus to the new institution for deciding on the admissibility of the credits.

Institutional Collaboration

The application of Blockchain technology in education at global level can promote inter-institutional research collaborations. Teachers and learners from multiple educational institutions can come together and make use of the expertise available at different institutions for research purposes (Jain, 2019). The work done by different scholars can be stored and preserved linearly and chronologically on the cloud. Since, the Blockchain based system maintains peer-to-peer connectivity without a need or intervention of a third party, institutional intervention can be minimised to a great extent except in cases where it is necessary. The mechanism will facilitate the learners at lower levels e.g. UG and PG levels also to collaborate with peers at other institutions for preparing inter-institutional projects and undergoing internships programmes. The work done by the learners at different institutions under such collaborations can be transferred seamlessly in an authentic and verifiable manner. While cases of plagiarism can be curbed with this practice to a great extent, the system will encourage the learners to think beyond the institutional boundaries and participate in community driven projects and activities.

Credit Transfer

Blockchain technology can help the learners in undergoing different courses from different institutions of their choice and seek credit transfer to any one institution he/she desires to get the degree from. This will facilitate the learner to travel in multiple learning spaces for earning specific course credits. Thus, Blockchain will promote and systematise the vertical and horizontal mobility of the learners from one institution to another. The mobility for learners within the institution also will be seamless and eliminate the constraints of waiting for testimonials and awards to get promoted from one programme to another. At the advanced level of use of BCT, the intelligent contracts can facilitate transfer of academic credits from one institution to the other as an automatic process on fulfilment of certain conditions for such a transfer without forcing the learner to complete all the credits from one institution in a continuum (Bittencourt, Goodert, Sharma, & Bortolozzy, 2020). The mechanism of transfer of credits can be used by the learners to complete some specialised courses from renowned professional institutions, and thus, enhance credibility of his/her achievements.

ePortfolio

Use of Blockchain technology for creating ePortfolios of learners can prove to be much helpful for their emphatic social appearance and professional visibility. The learners can use the mechanism to store their personal data gathered from the classroom or brought from external sources. The activities and other verified credentials earned by the learners can be stored by them in Blockchain in wallets and shared with their peers, teachers, and other learning facilitators (Steiu, 2020). Clark, Bakshi, and Braine (2016) give the 'portfolio' of the learner the name of 'Learning Passport' that contains all the information about credentials, certificates and badges earned by the learner on the semantic web. This is the Blockchain to which the credentials and badges are added in the form of blocks one by one. This source is owned by the learner which can be easily shared with any other employer or educational institution. This helps the learner not only in showcasing his/her academic work and achievements, but also in bringing the acquired credentials of learning and experiences at one place thereby presenting the lifelong academic achievements as exhibits and evidence. This way, the reskilling and up-skilling credentials of the learners also can be easily tracked by their prospective employers.

Publishing of Learners' Work

The application of Blockchain technology can help the ODL institutions in getting the quality work done by the learners published. The projects, internship reports, assignments, research reports prepared by the learners of senior classes can benefit and guide the junior learners of similar disciplines. The documents can become valuable open educational resources (OERs) in the hands of the educational institutions. This type of good work is available in every academic session from the learners. The use of the BCT will eliminate the necessity of printing the documents in tangible form. This will also benefit the graduating learners to showcase their quality work to the educational institutions they would like to join next or the prospective employers planning to hire them for jobs. In Indian context, 'ShodhGanga' (https://shodhganga.

inflibnet.ac.in/) – a Government of India initiative implemented through UGC - is already available to the higher education institutions for preserving the research reports and theses for the benefit of new researchers in different disciplines. While this type of publication activity will recognise the good work done by the learners, it will help in preventing the tendency of plagiarism.

The use of OERs fits perfectly into the philosophy of open and distance learning. The use of BCT can help in standardising the quality benchmarks in support services with enhanced access and openness. The most relevant and popular resources on specific topics by the renowned authors, can be shared by the teachers with the help of BCT. This will be a win-win situation for the teachers, the learners and the authors of the content since they will be able to directly know the number of users making use of the resources created by them. This can ensure a widespread access to free and cost-effective resources for learning in a secure way.

Credentials and Placement Service

The Blockchain technology can transform the way the degrees and certificates are issued by the educational institutions to the successful learners. Verbert, Sharples and Klobucar (2016) view the digital credentials of the learners as the "single secure record of educational attainment accessible and distributed across many institutions". Thayer (2018) propagates the use of blockchain for student record keeping including credentials management. Tapscott and Kaplan (2019) consider use of Blockchain as trust anchor among the learners, educational institutions and employers since the credentials of the students once authenticated by an institution cannot be altered in any way and can be fully trusted for subsequent use. Thus, mapping of the skills and knowledge acquired by the learner and required by the employer can be done authentically. This will enhance the scope of campus placement of the learners bu renowned business houses. At the same time, the learner can have full control over who can access his/her credentials and who cannot. Thus, the provisions made under the BCT will help the learners in seeking ownership of their personal data with the institution.

Since the certificates issued with the help of Blockchain will be in digital format, the learners will not be required to pay the institutions for issuing the transcripts every time these are required by another educational institution or employer since the credentials in digital format can be used time and again without any hassle. This helps the learners to freely and selectively share the testimonials authenticated by the institution in support of their qualifications they have earned, with the potential employers and other third-party agencies needing such authentication to serve the learner. The issue of virtual transcripts will, thus, facilitate inter-institutional horizontal and vertical mobility of learners. Devine (2015) predicts use of Blockchain beyond normal activities and asserts that the educational timeline and individual learning history of the learner can be used for predicting future potential of the applicants for specific jobs by the employers. The educational institutions have already started experimenting with issue of digital transcripts and degrees. As the first ever educational institution in the Asia-pacific region, the University of Melbourne has issued its Bachelor of Science certificate credentials based on the BCT (Newswire, 2017).

CHALLENGES

Implementation of new technologies in education have always posed certain challenges and they needed to be overcome before reaping the fruits of their integration in education. Blockchain technology is not the exception to it. The strengths of BCT warranting its use in distance education themselves are the challenges also in its wider implementation by implication. There are different impediments in fully exploiting the benefits of the technological application of Blockchain posed before the educational institutions. Therefore, the stakeholders need to resolve those issues before the Blockchain becomes a solution itself for making the learning experience of the learners a satisfying and fulfilling endeavour. Different researchers have flagged different issues and challenges keeping in view the type of use of BCT in different areas. Since, majority of them are technical constraints and feasibility issues, they are equally application to the area of distance education also. Hussain and Cakir (2020) highlighted the major challenges in implementation of BCT in higher education as: challenge of acceptance, scalability, security of data, usability and interoperability, governance and operational rights, operational cost, and management of personally identifiable information, among others. Security, scalability, adoption rate, and cost are some of the challenges identified by Maryville University (2021). A few of the crucial challenges are discussed in the forthcoming paragraphs.

Privacy and security

Application of a technology is governed by a legal framework in order to protect all the stakeholders from any untoward incident. The provisions of BCT need to fit into the existing statutory provisions and legal framework. Therefore, the implementing ODL institutions need to adhere to the national/international legal framework governing the privacy and security of personal data of the learners while ensuring the same level of security for their institutional data as well. There are different activities such as evaluation, assessment and certificate, the data related to which cannot be thrown open to the virtual world. There would always be a requirement of checks and balances to preserve the breach of data. Any lapse on the part of the institution may lead to cybercrime since personally identifiable information (PII) may put the learners at risk of exploitation and misuse of the information. Classification of the institutional data into off-chain and on-chain data would require substantial thinking. By putting appropriate restrictions on data access with multilevel permission system, activating latest firewalls and strictly adhering to the legal framework, the issue of privacy and security can be addressed by the institutions.

Scalability

Alammary et al. (2910) visualise a serious challenge in scalability of the usage of Blockchain in education due to huge number of blockchain transactions making it extremely slow over the internet. In case of distance education and online learning institutions, enrolment of students is huge. The institutions with massive number of learners may need to develop IT infrastructure in that proportion only. A huge number of blocks in a chain would make it time consuming since a transaction needs to be completed from all the concerned nodes. Keeping a track of the data of learners who are moving from one institution to another, for the last admitting institution will add to the load on Blockchain since all these transactions would, in the process, require peer to peer validation. Zheng, Xie, Dai, Chen, and Wang (2018) noted that Bitcoin which is based on the BCT can handle only three to seven transactions in a second which would be inadequate in case of transactions connected with learners in bulk. In order to ensure scalability of operations, the institutions need to ensure deployment of state of the art ICT infrastructure. The educational institutions can collaborate with an ICT firm for providing required software with perpetual license so that issue of frequent upgrade is addressed appropriately.

Nascent Stage of Technology

The advent of Blockchain technology is recent and it is yet to make its impact in different areas (Alammary, Alhazmi, Almasri & Gillani, 2019). The education system, being not sure of sustainability of this technology, may not be able to adopt and adapt it so quickly in the pedagogical and learner support processes. As Thayer (2018) puts it, the technology is yet in its nascent stage and is yet to mature. Lots of changes and improvisations would be required before it becomes stable and as reliable. He further claims that its failure rate is very high and 90% of the projects fail before they become viable. At this stage, the people and organisations with pilot mind-set only will try to analyse the risk and move in this direction. Therefore, the BCT is attractive but not as mature as to gain full trust for its implementation (Kandaswamy & Furlonger, 2018). Steiu (2020) insists that instead of implementing BCT as a solution for all problems, its application should be need-based and specific to the problems which can be resolved by it. The institutions can try the technology to begin with on pilot basis in selected operations and slowly expand the scope based on requirements and level of comfort.

Lack of Trust

Blockchain Technology is comparatively new in the field. Though, it is being tested in different areas including financial transactions, the educational institutions are a bit reluctant to adopt this technology for the core and confidential areas of their processes. As pointed out by Steiu (2020), the institutions find it difficult to trust a comparatively new technology for sharing student data and other secure data with third parties. This might be due to lack of appropriate information, knowledge, and skills to handle the data on the system in a secure way. However, convincing them to implement and use it, will enhance the ease of verifying the credentials of their learners and employing them for the specific jobs for which they are certified and validated. Once the employers start raising the demand for instant verification of credential of the learners, the institutions will be forced to implement the BCT in their operations to expedite the system of credential verification. In the Indian context, government has already requested all the universities to accept the certificates and credentials shared by the students through National Academic Depository as authentic (UGC, 2022). The support provided by the government organisations will go a long way in building the trust of the educational institutions in the new technologies such as blockchain.

CONCLUSION

The potential of implementation of Blockchain technology in education to manage different internal processes seems quite obvious. The chapter has discussed different areas of ODL system in general and learner support services in particular in which the use of BCT can be a promising pay-off. It can help in automatising different activities and promoting personalised scaffolding service to the learners while tracking their progress and providing instant feedback/help in the required knowledge domains. This personalised hand-holding of the learners may help in reducing the drop-out rate in ODL institutions substantially. The support of the state functionaries will go a long way in boosting its proliferation. The policy makers, government, innovators, educational institutions, implementers and users need to come together to launch Blockchain based services, test them, and gain confidence in their use for confidential and restricted access activities in order to make them more sustainable and successful. In the past also, different technologies in education have come up to the expectations of the stakeholders in making the process of learning more engaging, rationalising, cost efficient, confidence building and trustworthy, and providing security and privacy to the sensitive, confidential and personal data. The ODL institutions need to deliberate on the sensitive data to be kept off-chain and on-chain to secure the interests of all concerned stakeholders. As the technology matures leading to more and more applications in different sectors, its challenges and opportunities of use will increase.

IMPLICATIONS

In order to implement the Blockchain technology, Jain (2019) invites the educational institutions to promote understanding about the technology, experiment its usage for different activities, and become part of the movement. Unless the educational institutions at massive scale from across the boarders become part, the movement cannot achieve the required pace where all the stakeholders start benefitting. The adoption of BCT in educational processes will facilitate improvement in different key dimensions (Tapscott & Kaplan, 2019). Its usage will transform different aspects of education such as: student record keeping, streamlining verification process, automation of learning continuum, mobility of students, and authentication by employers in favour of the students.

REFERENCES

AISHE. (2020). *All India Survey on Higher Education 2019-20*. New Delhi: Ministry of Education, Government of India. Retrieved from https://www.education.gov.in/sites/upload_files/mhrd/files/s tatistics-new/aishe_eng.pdf

Alammary, A., Alhazmi, S., Almasri, M., & Gillani, S. (2019). Blockchain-based applications in education: A systematic review. *Applied Sciences (Basel, Switzerland)*, *9*(12), 2400. doi:10.3390/app9122400

Barr, A. (2014). Artificial Intelligence in eLearning. E-learning *Guild Online Forum on New Instructional Design for New Instructional Technology*. Retrieved from https://www.elearningguild.com/showfile.cfm?id=5284

Bittencourt, D. F., Goodert, A. R., Sharma, R. C., & Bortolozzi, F. (2020). Framework blockchain education: Rupture in higher education. In R. C. Sharma, H. Yildirim, & G. Kurubacak (Eds.), *Blockchain Technology Applications in Education*. IGI Global. doi:10.4018/978-1-5225-9478-9.ch004

Borge, M., Kokoris-Kogias, E., Jovanovic, P., Gasser, L., Gailly, N., & Ford, B. (2017). Proof-of-personhood: Redemocratizing permissionless cryptocurrencies. In *2017 IEEE European Symposium on Security and Privacy Workshops (EuroS & PW)* (pp. 23-26). IEEE. 10.1109/EuroSPW.2017.46

Bozkurt, A., & Ucar, H. (2020). Blockchain technology as a bridging infrastructure among informal, non-formal, and informal learning processes. In R. C. Sharma, H. Yildirim, & G. Kurubacak (Eds.), *Blockchain Technology Applications in Education*. IGI Global. doi:10.4018/978-1-5225-9478-9.ch001

Broggi, J. D., Gallagher, M. A., Lilly, J., Duquette, J., Nimura, C., Pattenden, M., Richter, F., Arbide, L. S. M. N., Avin, S., Kelley, K., & Lidova, M. (2018). *Woolf - Building the first Blockchain University*. https://woolf.university/assets/doc/whitepaper.pdf

Buzkurt, A., & Ucar, H. (2020). Blockchain Technology as a bridging infrastructure among formal, non-formal, and informal learning processes. In R. C. Sharma, H. Yildirim, & G. Kurubacak (Eds.), *Blockchain Technology Applications in Education*. IGI Global. doi:10.4018/978-1-5225-9478-9.ch001

CBSE. (2021). *Academic {BlockChain} Documents (ABCD)*. New Delhi: Central Board of Secondary Education. Retrieved from https://cbse.certchain.nic.in/aboutcc

CCA. (2021). *e-Authentication guidelines for eSign- Online Electronic Signature Service: Report of the Committee on New PKI based Digital/Electronic Signature Schemes*. Controller of Certifying Authorities, MeitY, Govt. of India. Retrieved from https://cca.gov.in/sites/files/pdf/esign/CCA-EAUTH.pdf

Cedeno, J. (2020, February 23). *How Blockchain Could Impact Education in 2020 and Beyond*. Retrieved from https://www.gettingsmart.com/2020/02/23/how-blockchain-could-impact-education-in-2020-and-beyond/

Chainalysis. (2021). *The 2021 Global Crypto Adoption Index: Worldwide Adoption Jumps Over 880% With P2P Platforms Driving Cryptocurrency Usage in Emerging Markets*. Retrieved from https://blog.chainalysis.com/reports/2021-global-crypto-adoption-index/

Chattopadhyay, S. (2014). Learner Support Services in Open Distance Learning System: Case Study on IGNOU. *Conference Proceedings 'Distance Learning and Reciprocal Library System: Is Public Library Network the better Option!'*. Retrieved from http://eprints.rclis.org/29071/

Chen, G., Xu, B., Lu, M., & Chem, N. S. (2018). Exploring Blockchain technology and its potential applications for education. *Smart Learning Environments*, *5*(1), 10. doi:10.118640561-017-0050-x

Ciftci, G. T. (2020). Change of the learning cycle after Blockchain: Changing trust society. In R. C. Sharma, H. Yildirim, & G. Kurubacak (Eds.), *Blockchain Technology Applications in Education*. IGI Global. doi:10.4018/978-1-5225-9478-9.ch003

ClarkC. D.BaksiV. A.BraineL. (2016). Smart contract templates: Foundations design landscape and research directions. arXiv:1608.00771

Clark, D. (2020, February 20). *Forecasted expenditure on advanced EdTech worldwide 2025*. Retrieved from https://www.statista.com/statistics/1085930/edtech-expenditure-forecast/

Conway, L. (2021, November 4). *Blockchain Definition: What you need to know*. Retrieved from https://www.investopedia.com/terms/b/blockchain.asp

Deloitte & Assocham. (2017). *Blockchain technology in India: Opportunities and challenges*. Retrieved from https://www2.deloitte.com/content/dam/Deloitte/in/Documents/strategy/in-strategy-innovation-blockchain-technology-india-opportunities-challenges-noexp.pdf

Dutta, A. (2021, July 25). *Real-World applications of Blockchain Technology* [Blog]. Retrieved from https://www.analyticsinsight.net/real-world-applications-of-blockchain-technologies/

Filipova, N. (2018). Blockchain - an opportunity for developing new business models. Business Management. *D. A. Tsenov Academy of Economics*, *20*(2), 75–92.

Gartner. (2019, December 12). *Gartner Survey Reveals Blockchain Adoption Combined with IoT Adoption Is Booming in the U.S.* Retrieved 27 December, 2021 from https://www.gartner.com/en/newsroom/press-releases/2019-12-12-gartner-survey-reveals-blockchain-adoption-combined-with-iot-adoption-is-booming-in-the-us

Government of India. (2021). *National Strategy on Blockchain*. New Delhi: MeitY, Government of India. Retrieved from https://www.meity.gov.in/writereaddata/files/NationalStrategyBCT_%20Jan2021_final.pdf

Government of India. (n.d.). *Centre of Excellence in Blockchain Technology*. Retrieved from https://blockchain.gov.in/aboutus.html

Grech, A., & Camilleri, A. F. (2017). *Blockchain in education. European Commission JRC*. https://www.gartner.com/en/newsroom/press-releases/2019-12-12-gartner-survey-reveals-blockchain-adoption-combined-with-iot-adoption-is-booming-in-the-us doi:10.2760/60649

Hussain, I., & Cakir, O. (2020). Blockchain technology in higher education: Prospects, issues and challenges. In R. C. Sharma, H. Yildirim, & G. Kurubacak (Eds.), *Blockchain Technology Applications in Education*. IGI Global. doi:10.4018/978-1-5225-9478-9.ch005

IBM. (n.d.). *What are smart contracts on Blockchain?* Retrieved November, 2, 2021 from https://www.ibm.com/in-en/topics/smart-contracts#:~:text=Smart%20contracts%20are%20simply%20programs,intermediary's%20involvement%20or%20time%20loss

Jain, A. K. (2019). *Blockchain Goes to School: Cognizant 20-20 Insights*. Retrieved 27 December, 2021 from https://www.cognizant.com/whitepapers/blockchain-goes-to-school-codex3775.pdf

Kandaswamy, R., & Furlonger, D. (2018). *Blockchain-based transformation: A Gartner Trend Insight Report*. Gartner Research. Retrieved from https://www.gartner.com/en/doc/3869696-blockchain-based-transformation-a-gartner-trend-insight-report

Keagan, D. (1996). *Foundations of Distance Education* (3rd ed.). Routledge.

Langard, V. (2019, 23 May). *B C Diploma raises €1.2 million to become the global standard for certifying diplomas and certificates online*. Retrieved from https://medium.com/bcdiploma/bcdiploma-raises-1-2m-58cdba9b3b85

Lemoie, K., & Soares, L. (2020). *Connected Impact: Unlocking Education and Workforce Opportunity Through Blockchain*. American Council on Education.

Maryville University. (2021). *How Blockchain is used in education* [Blog]. Retrieved October 21, 2021 from https://online.maryville.edu/blog/blockchain-in-education/#:~:text=Blockchain%20gives%20students%20ownership%20of,what%20an%20employer%20can%20access

McGreal, R. (2021, January 27). *What are the advantages of Blockchain as an educational resource? What is Blockchain and how could it improve access to Education?* World Economic Forum. Retrieved 27 December, 2021 from http://people.uis.edu/rschr1/?p=58137

Meit, Y. (2021). *National strategy on blockchain: Towards enabling trusted digital platforms*. Ministry of Electronics & Information Technology, Government of India.

Mikroyannidis, A., Third, A., Domingue, J., Bachelor, M., & Quick, K. (2020). Blockchain applications in lifelong learning and the role of the semantic blockchain. In R. C. Sharma, H. Yildirim, & G. Kurubacak (Eds.), *Blockchain Technology Applications in Education*. IGI Global. doi:10.4018/978-1-5225-9478-9.ch002

Moore, S. (2019, October 16). *4 Ways Blockchain will transform higher education*. Retrieved from https://www.gartner.com/smarterwithgartner/4-ways-blockchain-will-transform-higher-education#:~:text=Improve

Mosinegutu, V. (2021, July 23). *What is Web 3.0 and how will it change the internet?* Retrieved from https://academy.coinzilla.com/what-is-web-3-0/

Newswire. (2017, October 8). *Melbourne first in Asia-Pacific to issue recipient-owned blockchain records*. Retrieved from https://www.newswire.com/news/university-of-melbourne-first-in-asia-pacific-to-issue-recipient-owned-19980513

NIC. (n.d.). *Drugs Logistics*. Government of Karnataka. Retrieved from https://blockchain.gov.in/drugslogistics.html

NISG. (2019). *National Strategy for Blockchain*. National Institute for Smart Government.

Pradhan, D. (2018, July 18). *Blockchain This Week: Two Educational Institutions Adopt Blockchain, Oracle to Launch Blockchain Cloud Service in India, and more* [Blog]. Retrieved from https://inc42.com/buzz/blockchain-this-week-two-indian-education-institutions-adopt-blockchain-and-more/

Sharma, R. C., Yildirim, H., & Kurubacak, G. (2020). Preface. In R. C. Sharma, H. Yildirim, & G. Kurubacak (Eds.), *Blockchain Technology Applications in Education*. IGI Global. doi:10.4018/978-1-5225-9478-9

Steiu, M.-F. (2020). Blockchain in education: Opportunities, applications, and challenges. *First Monday*, *25*(9). Retrieved from https://journals.uic.edu/ojs/index.php/fm/article/download/10654/9726 doi:10.5210/fm.v25i9.10654

Swam, M. (2015a). Blockchain thinking: The brain as a decentralised autonomous corporation. *IEEE Technology and Society Magazine*, *34*(4), 41–52. doi:10.1109/MTS.2015.2494358

Swam, M. (2015b). *Blockchain: Blueprint for a new economy*. O'Reilly.

Tait, A. (2003). Management of services to students. In S. Panda (Ed.), *Planning and management in open and distance education*. Kogan Page.

Tapscott, D., & Kaplan, A. (2019). *Blockchain revolution in education and lifelong learning: Preparing for disruption, leading the transformation*. Blockchain Research Institute and IBM Institute for Business Value. Retrieved from https://www.ibm.com/downloads/cas/93DDVAKE

Tapscott, D., & Tapscott, A. (2017). The blockchain revolution and her education. *EDUCAUSE Review*, *52*(2), 11–24.

Thayer, T. L. (2018). *4 Promising and ambitious blockchain initiatives for higher education*. Gartner Research. Retrieved from https://www.gartner.com/en/documents/3893582/4-promising-and-ambitious-blockchain-initiatives-for-hig

Thomas, S. (2020, June 29). *Fundamentals of Blockchain, Data Driven Investor*. Retrieved 11 April 2022 from https://medium.datadriveninvestor.com/fundamentals-of-blockchain-180de46b472f

UGC. (2021, July 29). *The University Grants Commission (Establishment and Operation of Academic Bank of Credits in Higher Education) Regulations' 2021*. Retrieved from https://www.ugc.ac.in/pdfnews/1767992_Notice-Academic-Bank-of-Credits.pdf

UGC. (2022, January 3). *Acceptance of Degree, Mark-sheet and other educational documents presented through DigiLocker platform – reg*. New Delhi: University Grants Commission. Retrieved from https://www.ugc.ac.in/pdfnews/0790412_DigiLocker-Platform_001.pdf

UIAI. (n.d.). *What is Aadhar?* New Delhi: Unique Identification Authority of India, Government of India. Retrieved from https://uidai.gov.in/my-aadhaar/about-your-aadhaar.html

Varghese, B., Villari, M., Rana, O., James, P., Shah, T., Fazio, M., & Ranjan, R. (2018). Realizing Edge Marketplaces: Challenges and Opportunities. *IEEE Cloud Computing.*, *5*(6), 9–20. doi:10.1109/MCC.2018.064181115

K. Verbert, M. Sharples, & T. Klobucar (Eds.). (2016). Adaptive and adaptable learning. In *Proceedings of the Conference on Technology Enhanced Learning*. Springer. 10.1007/978-3-319-45153-4_7

World Economic Forum. (2018). *Building Block(chain)s for a better planet*. The World Economic Forum Report 2018. Retrieved from https://www.weforum.org/reports?year=2017#filter

Yuan, Y. & Wang, F.–Y. (2018). Blockchain and cryptocurrencies: Model, techniques, and applications, *IEEE Transactions on Systems, Man, and Cybernetics: Systems*, *48*(9), 1,421–1,428. doi:10.1109/TSMC.2018.2854904

Zheng, Z., Xie, S., Dai, H., Chen, X., & Wang, H. (2017). An overview of blockchain technology: Architecture, consensus, and future trends, *2017 IEEE International Congress on Big Data (BigData Congress)*. 10.1109/BigDataCongress.2017.85

Zheng, Z., Xie, S., Dai, H. N., Chen, X., & Wang, H. (2018). Blockchain challenges and opportunities: A survey. *International Journal of Web and Grid Services*, *14*(4), 352–375. doi:10.1504/IJWGS.2018.095647

APPENDIX: ABBREVIATIONS

ABC		Academic Bank of Credits
ABCD		Academic {Blockchain} Document
AICTE		All India Council for Technical Education
AISHE		All India Survey on Higher Education
ASSOCHAM		The Associated Chambers of Commerce and Industry of India
BaaS		Blockchain as a Service
BTC	:	Blockchain Technology
CBSE		Central Board of Secondary Education
CCA		Controller of Certifying Authorities
C-DAC		Centre for Development of Advanced Computing
EBI		Education Blockchain Initiative
ePramman		Electronic Certification
eSign		Electronic signatures
GDP		Gross Domestic Product
GER		Gross Enrolment Ratio
GOI		Government of India
IDRBT		Institute for Development and Research in Banking Technology
MCI		Medical Council of India
MeitY		Ministry of Electronics & Information Technology, Government of India
MIT		Massachusetts Institute of Technology
NAD		National Academic Repository
NFT		Non-Fungible Token
NIC		National Informatics Centre
NIELIT		National Institute of Electronics & Information Technology
NISG		National Institute for Smart Government, Government of India
NITI Ayog		A Planning Commission like body in India
NLBF		National Level Blockchain Framework
ODL		Open and Distance Learning
OER		Open Educational Resource
PG		Postgraduate
UG		Undergraduate
UGC		University Grants Commission of India
UIAI		Unique Identification Authority of India
VJTI		Veermata Jijabai Technological Institute
WEF		World Economic Forum

Chapter 8
Get Ready for Blockchain Technology:
A Probe Into Its Potential for Indian Schools

Arnab Kundu
https://orcid.org/0000-0002-7169-7189
Bankura University, India

ABSTRACT

The fourth industrial revolution is peeping through blockchain technology since the invention of the steam engine, electricity, and information technology. It has come up with a lot of expectations and is now considered a 'disruptive technology' stirring the fabric of society to its roots. This desk research concentrates on evaluating its potential for Indian schools. How could blockchain contribute to effective schooling in India? What could be the possible framework to implement this technology for progressive changes in schools? A thorough review of the existing literature and the Indian context reveals that blockchain is promising for Indian schools providing enhanced digital accuracy, openness, cost-saving operations, and a heightened constructivist learning paradigm. Based on these assumptions, the author has proposed a framework for blockchain implementation (FBI) in school operations which could be a substantial contribution to the 'glocal' management of blockchain technology in schools.

DOI: 10.4018/978-1-6684-4153-4.ch008

Copyright © 2023, IGI Global. Copying or distributing in print or electronic forms without written permission of IGI Global is prohibited.

INTRODUCTION

The famous playwright and polemicist George Bernard Shaw (n.d.) once said progress is impossible without change, and those who cannot change their minds cannot change anything. Change is one of the major characteristics of technology through the process of invention, innovation, and diffusion. Blockchain (henceforth BC) is the name of such a technological innovation that allows decentralized and distributed records of digital transactions. As the name suggests a 'blockchain' is a chain of blocks that contain information (Casino, Dasaklis & Patsakis, 2019). It is the core technology used to create the cryptocurrency, Bitcoin, through the maintenance of immutable distributed ledgers in thousands of nodes (Nakamoto, 2008).

The first implementation of BC took place in 2009 in the context of Bitcoin as a digital currency and now its connotation is no longer limited to Bitcoin but it is the example that can be used as a paradigm of BC (Christidis & Devetsikiotis, 2016). The data which is stored inside a block determines its type, for example, a Bitcoin block contains information about the sender, receiver, and the number of bitcoins to be transferred. Although the technology of BC is famously linked to Bitcoin, in the past two years, its utilization is exceedingly gone beyond financial jurisdiction. The financial world has been the first to approach but its use has been evident in providing a non-counterfeited digital identity to immigrants or refugees who have lost their documentation to smart contracts implemented in automatic mode when the specified conditions are met and agreed to the contract (Bartolome at al., 2017).

This disruptive technology has been shown to have a substantial impact on national governance, institutional functions, trades, and daily lives in the 21st century (Chung & Kim, 2016). It has the potential to transform the current Internet from "the Internet of information sharing" to "the Internet of value exchange" (Chen et al., 2018). Owing to its trustworthiness, immutability, and transparency it is expected that it will revolutionize the operating modes of commerce and industry promoting the rapid development of a knowledge-based economy on a global scale (Underwood, 2016).

It is evolving and disseminating itself each day and now the development of BC applications could be divided into three stages- BC 1.0, 2.0, and 3.0 (Swan, 2015). But the majority of the masses do not know even the basics of BC, not to mention the potential applications of using this emerging technology. Although researchers discussed the usage of BC in the commercial area (Swan, 2015) but several studies focused on how this cutting-edge technology can be applied in education (Chen et al., 2018; Sharples & Domingue, 2016; Devine, 2015; Bartolome et al., 2017).

The current study is a further extension in this trail considering nuances of a developing country like India ensuing two basic research questions:

1. How could BC technology usher in a bright future for Indian schooling?
2. What could be the possible implementation framework for its adoption?

LITERATURE REVIEW

Basics on Blockchain

As already told, concept of BC is relatively new that is believed to have appeared with the advent and development of Bitcoin (Alahmari et al., 2019). It is a shared, immutable ledger that facilitates the process of recording transactions and tracking assets in a business network (Tschorsch & Scheuermann, 2016). Zheng et al. (2018) defined it as an immutable, decentralized database – a chain of 'blocks' that store information such as transactions' dates, times, amounts, and/or participants (participants on the BC are usually not personally identifiable). Virtually anything of value can be tracked and traded on a BC network, reducing risk and cutting costs for all involved, any *asset* tangible (a house, car, cash, land) or intangible (intellectual property, patents, copyrights, branding).

BC is a collection of generated blocks using cryptographic hash functions and connected with the previous block in the form of a secure chain using algorithms (Aamir et al., 2020). The technology is an intricate web of several tech-innovations working together. One such vital BC puzzle is 'hashing' which is a cryptographic function that converts a string of characters of any length into a unique output, or hash, of fixed length (Seebacher & Schuritz, 2017). It is a mathematical process that generates a value or values from a string of text. It is produced using an algorithm, which helps to protect the transmission from tampering. Hashing is also a tool for efficiently sorting key values in a database table (Fraga-Lamas & Fernandez-Carames, 2019).

Business runs on information and the faster it is received accurately produces better output. BC is ideal for delivering that information because it provides immediate, shared, and completely transparent information stored on an immutable ledger that can be accessed only by permission network members (Oh & Shong, 2017). Zet and Dumitriu (2021) in the latest study pointed out three types of BC structures:

- Public BC, like Bitcoin, are large distributed networks and anyone at any level is welcome to engage in this forum.
- Private BC is a smaller network that does not employ tokens. The number of people who can join private BC is limited.

- A permissioned BC are a hybrid of public and private blockchains where anyone can access them if they have permission from the administrators to do so.

Dobrovnik et al. (2018) have found the main advantage of a public BC is its autonomy where all operators have similar liberties, and no party can control stored data, which means that operators do not have to trust and rely on a third party. Swan (2015) found public BC are extremely bulky and consume a big amount of energy as no user has access restrictions. Besides, as Steiu (2020) has pointed out private BC tend to be smaller and flexible, as only a limited number of users are allowed access, and they have different permissions and access privileges.

A BC network can track orders, payments, accounts, production and much more (Meunier, 2018). And because members share a single view of the truth, they can see all details of a transaction end to end, building greater confidence among them, as well as new efficiencies and opportunities (Zet & Dumitriu, 2021).

This technology ensures full data integrity and availability in the BC (Jaoude & Saade, 2019). The major concepts associated with this guaranteed performance by each BC platform are- data immutability, shared ledger, consensus, permissions, and cryptography (Abeyratne & Monfared, 2016). Data immutability denotes once data is written and stored, it cannot be deleted or changed (Crosby et al., 2016). This is one of the key advantages of this technology that generates trust between the users. Here 'smart contracts' (a computer program), like a traditional contract enforceable by law, controls directly and automatically the transfer of digital assets between the parties under certain conditions.

Shared ledger means a ledger that can be operated by each member or a network where all members have an immutable copy of all transactions under capable 'smart contracts' which adds next level of accuracy, automation, Speed, and Security. Besides, Lam and Dongol (2020) said transactions are recorded once to avoid any sorts of duplication. In a shared ledger, transactions are only recorded once to avoid any duplication and members can only see what they are authorized to. Consensus is a mechanism used to achieve agreement on a piece of data between/among all members of the network (Seebacher & Schuritz, 2017).

Vukolic (2015) said different algorithms have been developed to ensure this key feature, such as proof-of-work, where specific nodes, referred to as miners, compete against each other to complete transactions on a network for a reward, and the transaction is then independently verified by all nodes on the BC before it is added to the BC. Permissions denote members of a BC either have restricted access on a blockchain or not (Cheriguene et al., 2022).

Due to its uniqueness BC is deemed one of the most promising financial breakthroughs, with the potential to decrease fraud, provide speedy and safe transactions and exchanges, and assist in risk management within the global financial systems.

Benefits of Blockchain

Alahmari et al. (2019) have made a beautiful introduction to the notion of why we should go for BC or how we may be benefitted from this technology? They told operations often get slow or even waste efforts on duplicate record keeping and third-party validations. Besides the record-keeping systems can be vulnerable to fraud and cyberattacks. The arrival of IoT or *Internet of Things* (it involves the interconnectivity of physical objects and data input and output, while the *Internet of Everything* is a comprehensive term that refers to the interconnectivity of various technologies, processes and people), transaction volumes have exploded but limited transparency can slow data verification. All of this puts negative impacts on transactions, slows business, drains the bottom line that necessitates a better way out like BC (Salah & Khan, 2017).

The potential of BC technology has been noticed and being used in different industries and services (Figure 1) including financial due to the advantages it brings to the working process by shortening the transaction process, protecting copyright by 'tokenization', and the efficient and secure tracking of transactions (Lohmer & Lasch, 2020). From the perspective of financial institutions, the emergence of BC not only has technical significance but also creates the possibility that if the business model of existing financial institutions or financial intermediaries disappears, the financial services that are based on them may disappear entirely or be partially replaced, and consumer financial transaction patterns may change (Reynolds & Irwin, 2017).

Figure 1. BC applications in different fields
Source: Casino et al. (2019)

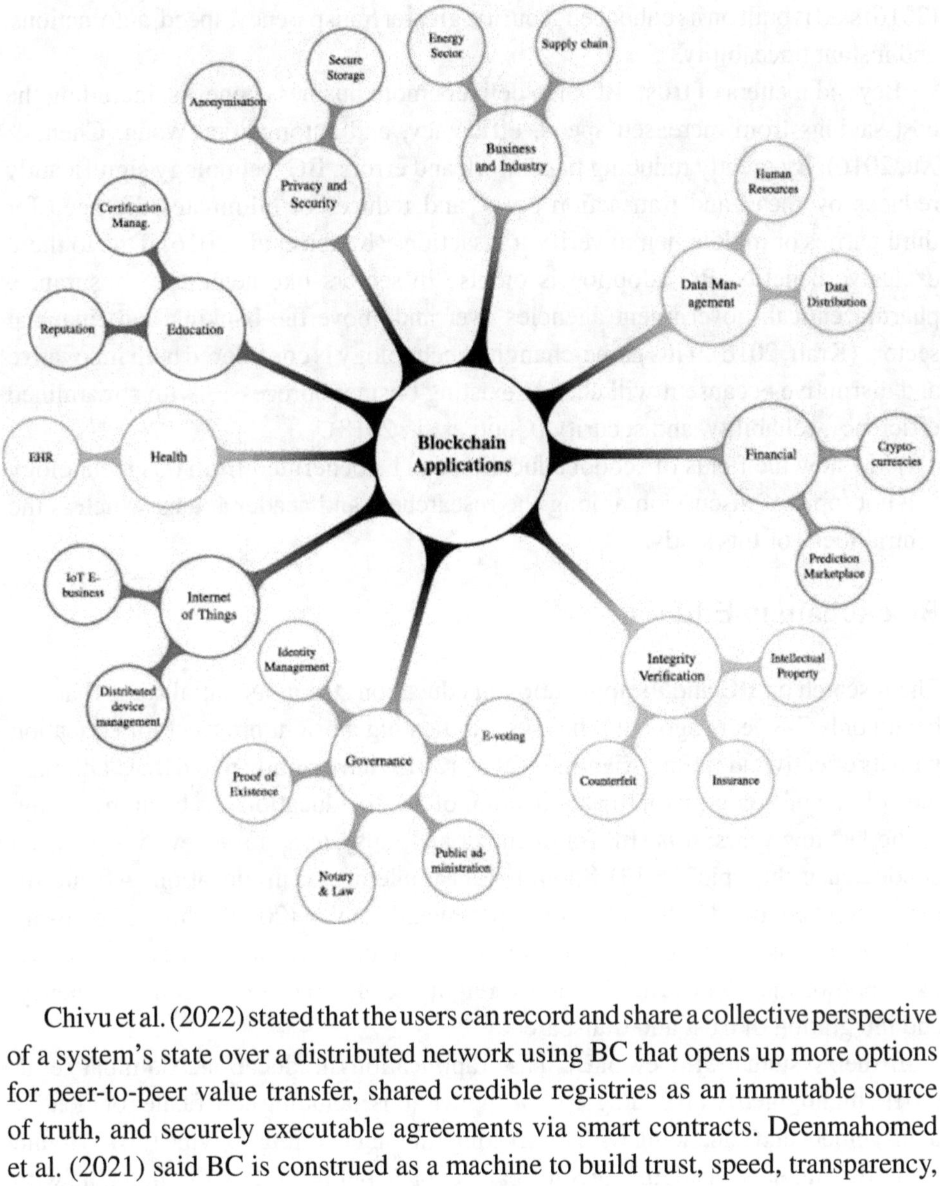

Chivu et al. (2022) stated that the users can record and share a collective perspective of a system's state over a distributed network using BC that opens up more options for peer-to-peer value transfer, shared credible registries as an immutable source of truth, and securely executable agreements via smart contracts. Deenmahomed et al. (2021) said BC is construed as a machine to build trust, speed, transparency, reliability, and effectiveness in automated and peer-to-peer transactions.

BC increases trust, security, transparency, and the traceability of the shared data shared across a business network and delivers cost savings with new efficiencies (Hawlitschek, Notheisen, & Teubner, 2018). It uses a shared and immutable ledger strictly among the members on what information each organization or member may see, and what actions each can take (Tama, et al. 2017). Blockchain is sometimes called a 'trustless network' – not because business partners do not trust each other, but because they don't have to (Cai & Zhu, 2016). This trust as Zhao, Fan, and Yan (2016) said is built on its enhanced security, greater transparency, speed, automations, and instant traceability.

Beyond matters of trust, BC also delivers more business benefits, including the cost savings from increased speed, efficiency, and automation (Wang, Chen, & Xu, 2016). By greatly reducing paperwork and errors, BC technology significantly reduces overhead and transaction costs, and reduces or eliminates the need for third parties or middlemen to verify transactions (Kosba et al., 2016). Due to these exclusive benefits, BC adoption is on rise in sectors like healthcare, insurance, pharmaceutical, government agencies over and above the banking and financial sectors (Kraft, 2016). This game-changing technology is considered both innovative and disruptive because it will change existing business processes with streamlined efficiency, reliability, and security (Conti et al., 2018).

Now how the fields of school education can be benefitted from this technology is a hot topic of discussion among the researchers and academicians which is the central focus of this study.

Blockchain in Education

The research on BC and its application in education still in its initial stage that had begun only 7-8 years ago but it has been indicating a bright prospect for education with its effective adoption. Alammary et al. (2019) have rightly noted that "although the volume of literature on the application of BC to education has been increasing in the last few years, it is still fragmented, and no systematic review has yet been conducted on the topic" (p. 13). Should BC be implemented in education? A beautiful reply might be found in the study of Tapscott and Kaplan (2019) who said by using BC, the processes of teaching and learning can be improved across key dimensions like empowerment for learners' self-sovereignty, security and institutional efficiency, and integration of trust and transparency.

In their systematic review of BC-based applications in education, Alammary et al. (2019) highlighted twelve categories of applications including certificates or identity management and enhancing with motivating lifelong learning. Devine (2015) found using BC students' academic records can be made public and easily shareable with employers and universities for further personal development opportunities. This way, he said the accredited educational timeline could be used to make projections of future potential based upon individual student learning histories.

Thayer & Yanckello (2019) claimed that most administrative systems within educational institutions perform poorly on the key metric of engagement not only within the learning process, but also across multiple levels like recruitment, enrollment, retention, and alumni advancement. Thus, they recommended BC technologies to offer personalized services to students throughout the entire educational cycle.

Few other studies have unfolded hints for BC adoption in education with assured advantages although most of these are concentrated in higher education scenario. Education has adopted the process of digitization to its fullest and is on the verge of getting into it by the next decade (Kundu, et al. 2019). BC technologies have emerged as a valued support for the digitization of this learning activities (Zhong, et al., 2018). Li et al. (2021) found due to the storage capacity of BC technology, the interests of the educational environment in this direction were focused on storing learning materials (course materials and interactive materials in electronic format) in a secure form and with the possibility of sharing them without the need for sophisticated cryptographic calculations.

Its contributions can be seen in both aspects- in the formal learning context, this includes learning contents and outcomes as well as students' achievements and academic certificates, and in the informal learning context like information about research experience, skills, online learning experience as well as individual interests (Zhu & Xu, 2021). These data can be safely stored and accessed on a BC network in appropriate ways (Chen et al., 2018). Nowadays, some universities and institutes have applied BC technology into education, and most of them use it to support academic degree management and summative evaluation for learning outcomes (Sharples & Domingue, 2016; Skiba, 2017).

Skinner (1938) in his famous learning reinforcement theory stated that behaviors are shaped by their consequences, and that individual behaviors can be changed through reinforcement, punishment and extinction. From this perspective, presently digital badges and micro-credentials are used to identify and honor a learner's accomplishments and specialized abilities (Chukowry et al., 2021). Due to the short lifespan of micro-credentials, it is often difficult to issue and validate the micro-credential promptly (Atzei et al., 2017).

Theses digital badges could also be fabricated by users, making them untrustworthy (a user can create their own digital badge) where BC technology could help by providing greater security in the way of making an inviolable accreditation (Fedorova & Skobleva, 2020).

In addition, this cutting-edge technology demonstrates a high level of security and transparency, ensuring that data cannot be tampered with and that its records remain intact (Wan et al., 2021). Thus, BC is ideal for developing a micro-accreditation system that may be used as an alternative to standard professional development (Jang & Han, 2021).

There are typical databases that can replicate the apparent advantages of BC for storing educational materials but it has several limitations that could be overcome using BC technology. In database, the administrator is in charge of the data that reduces confidence in the system while in BC, there is confidence among unknown users since every node has the same priority, and any node may check newly added transactions (Bucea-Manea-Tonis et al., 2021).

In database if students provide feedback, the course owner can see who commented which is absent in BC technology where students can provide feedback anonymously without worrying about the reaction of the course holder. Others with access to the database can edit data, or people can hack the database to change data, and the modifications will be undetectable since there is no proof but these disadvantages are absent in BC (Bucea-Manea-Tonis et al., 2021).

If the database crashes, all data will be lost but here in BC since, each user has a copy of the BC, the BC will continue to exist even if some nodes are gone. Li et al. (2021) stressed over another issue that the intellectual property of the uploaded materials is not ensured through the databases while teachers may use BC to secure their intellectual property (IP), such as courses, ideas, innovations, and patents.

Grech and Camilleri (2017) have pointed out few more potentials of BC technology, few specific aspects of education where it might put considerable contributions like:

- Accusing the demise of the paper-based certification system,
- Automatically check the validity of certificates by eliminating the need to contact the issuing organization,
- To issue and validate certificates automatically all of which are publicly available and verifiable by any user against a BC,
- For intellectual property management, such as recording first publications and citations, without requiring a central authority to keep track of these databases,
- Saving organizations' data management expenses and liability risk due to data management difficulties, and
- Institutions may adopt BC-based cryptocurrencies to ease payments.

Sharples and Domingue (2016) have decoded another prospects with regard to students and said there are still some negative subjective or objective factors causing poor learning outcomes, such as the lack of motivation and financial pressure. Due to the trait of currency property, BC can be used to motivate students by implementing "learning is earning". They hinted at 'the smart contract' between teachers and students in educational scenario where real-time awards (a certain number of digital currencies according to smart contract) can be given to students through some simple clicks by the instructors. Such money as Park (2021) also upholds can be stored in the education wallet, used as tuition, even exchanged with real currencies.

Evaluation has been a focal point in education where BC can help as well. Formative assessment has been advocated for a long time, and received lot of research yet it is still not mature enough because it is not easy to track every detail of teaching and learning (Kundu & Bej, 2020). BC can help here with its promise of being immutable, traceable, and reliable meaning the data recorded on BC are more specific, authentic, and anti-theft. The observation of Gibson et al. (2015) in this context is worth mentioning that each student submits his/her work to the learning platform through his/her unique account, the smart contract running on it will review student's performance, and the results will be recorded into blocks.

This technology could be helpful for "collaborative learning" as well which is regarded as an excellent way to carry out constructivism and cultivate students' ability to work with others (Yamada et al., 2016). But is often criticized for being accompanied by the problem of free-riding, hindering fair evaluation that could be curbed using BC technology (McArthur, 2018). All behaviors during collaboration will be saved into blocks as evidence for evaluation as well. Thus, as nodes in BC network, students' opinions would be taken into consideration when assessing them that ensures the fairness of the evaluation (Winstone et al., 2021).

From the perspective of teachers, a new assessment system can be constructed based on BC network and smart contract where teachers need to submit preplanned instructional activities as a smart contract to the schools, all teaching activities will be recorded in the BC network, the smart contract will verify the consistency of the teaching design and practice, a smart contract between teachers and schools, as well as the one between teachers and students can be verified and supplemented with each other, and finally, teachers who meet the standards will get digital currency as a reward (Chen et al., 2018).

It serves as both an appreciation and encouragement for teachers' teaching skills. Winstone et al. (2021) wrote smart contract and BC technology is useful for supervisor or academic advisor who are responsible for the supervision of the student's program. Owing to unique potential of this technology in traceability and immutability, every minute details can be traced here such as how many times the supervisor has discussed with students, how many times has the supervisor reviewed the thesis both in draft and final form, or whether they provide appropriate guidance to the students in course selection and research design, etc.

BC could be especially helpful for online education, also known as distance learning or e-learning, using data and internet technology to deliver information, provide feedback, and facilitate learning (Caldarelli & Ellul, 2021). With BC intervention, this web-based learning technique could be more effective since it offers an ideal solution to online learning issues, such as legitimacy and protection (Mahankali & Chaudhary, 2020). In this context Nabil et al. (2020) remarked BC will also create non-modifiable learning documents for online teaching without the need for third-party oversight, ensuring that course credits are adequately recognized.

Over the past few years, there have been several attempts to make use of BC for educational applications and most of the existing work concern the use of BC to store educational credentials and to verify transcripts and certificates (Alfandi, 2021). Ta and Yang (2021) wrote in an institute of higher learning, most of the time course material development and delivery follow a centralized model which is fully lecturer-controlled. They found in this model, engaging students as partners in learning is a challenging problem as students are usually hesitant to contribute due to the fear of getting it wrong, not much incentive for them to put in the extra effort, and current online learning systems lack adequate facilities to support seamless and anonymous interactions between students. These limitations could be overcome by leveraging the rising BC technology to decentralize and incentivize learning resource development and formative assessments.

Figure 2. Few applications of BC in education
Source: Chen et al. (2018)

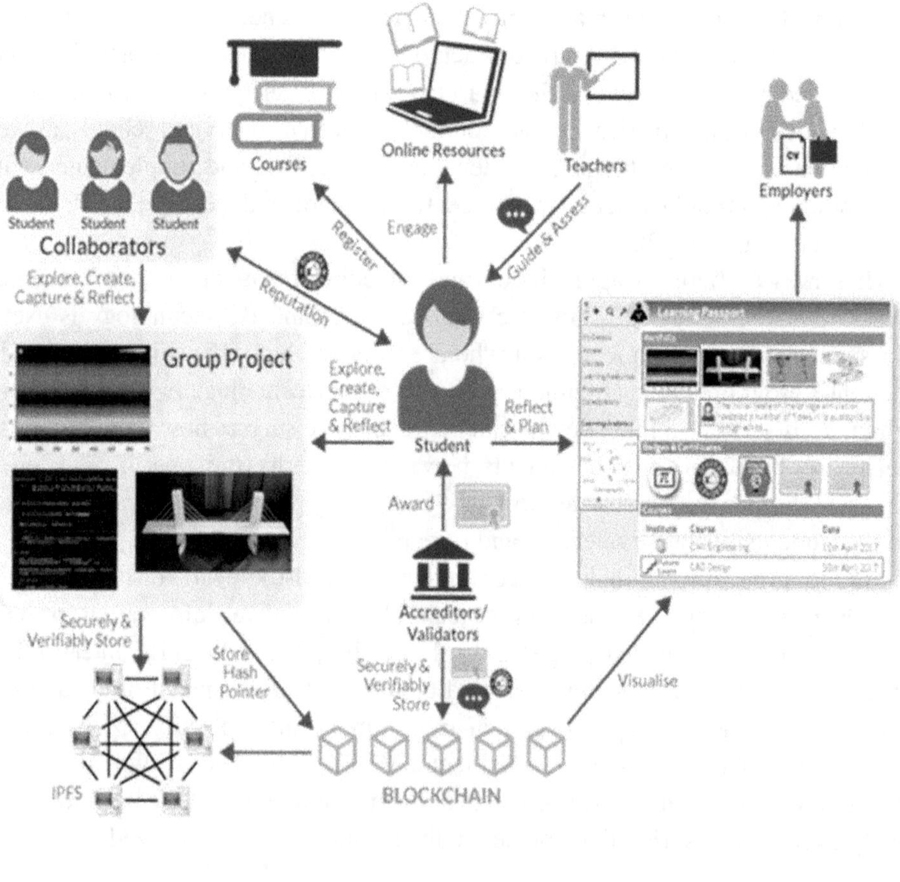

Surveying Indian Context

Indian education system is all set to enter into a new era with thoughtful use of cutting-edge technologies as evident in the ambitious agenda in National Education Policy (NEP) 2020. It took the goal to achieve 100 percent Gross Enrolment Ratio (GER) in all school levels by 2030 and to double the GER in higher education (50 percent) by 2035. The country is hopeful of providing universal access to quality education to its students at all levels of education which is the key to its continued ascent, and leadership on the global stage in terms of economic growth, social justice and equality, scientific advancement, national integration, and cultural preservation (Kundu & Bej, 2021). It is envisioned that students will have increased flexibility and subject choices, with no rigid separations across disciplines at the same time to achieve SDG4- (Agenda for Sustainable Development Goals) adopted in 2015 to ensure inclusive and equitable quality education with lifelong learning opportunities for all by 2030 (Kundu & Bej, 2022).

A renewed focus on digital education and related technologies has been made to achieve these ambitious targets towards making education holistic and multidisciplinary (Dandona et al., 2017). Such a lofty goal will require the entire education system to be reconfigured to support and foster learning so that all of the critical targets and goals in SDGs can be achieved. In this move, the country is all set to adopt educational technology including cutting-edge technologies like AI, VR, or AR that could play an important role in the improvement of educational processes and outcomes. In this line, BC could be a good addition.

Sharma et al. (2019) found there are several aspects to building a robust Digital Education Ecosystem (DEE) - content development, teaching, evaluations, grading, attendance recording, achievements, certificates, degrees, and diplomas. Stakeholders such as educational institutions, prospective employers, mentors, and certification agencies can be integrated into a DEE. With greater digitization, there is an inherent need for more secure and fool-proof systems that can track students' academic activities as well as provide the required information to all stakeholders (Sharma & Mishra, 2006). It has been showing a positive gesture toward the disruptive technologies that might assist schooling and continuing education by raising the general populace's awareness.

Data is a key fuel for AI-based technologies, and it is critical to raise awareness on issues of privacy, laws, and standards associated with data handling and data protection, etc. (Sharma, Kawachi, & Bozkurt, 2019). In this direction BC technology can emerge as a forward step toward digitization.

The country has already started taking a giant leap toward BC technology and recently, the draft strategy for National Level Blockchain Framework was formulated in 2021 by the Ministry of Electronics & Information Technology (PIB). The National

Informatics Centre (NIC) & NICSI have also established the Centre of Excellence (CoE) in BC technology intending to accelerate its adoption, offer its platform as a service to revamp design and development efforts, and execute projects focusing on different use cases.

While regulatory processes continue to evolve, India's interest in BC is expanding at an unprecedented pace. With all of the varied ways in which it is managed to express and appeal to diverse audiences and businesses, it is evident that the interest is here to stay. India's domestic BC ecosystem foresees a brighter future as it continues to consolidate its position as a leader in the tech talent landscape (Rustagi & Chakraborti, 2021). And to that end, to keep India ahead of the curve, it is critical to grasp the opportunities that BC offers, as well as steps that must be taken to realize its full potential and support the development of the necessary ecosystem.

To conclude the argument, it is clear that India has been desperately seeking cutting-edge technologies to serve quality education among students and it has fixed the thrust areas as well (Figure 3). Let's see the areas where BC technology can put a significant contribution to the Indian goal of quality education.

Figure 3. Thrust areas of technology integration in NEP 2020
Source: Kundu and Bej (2021)

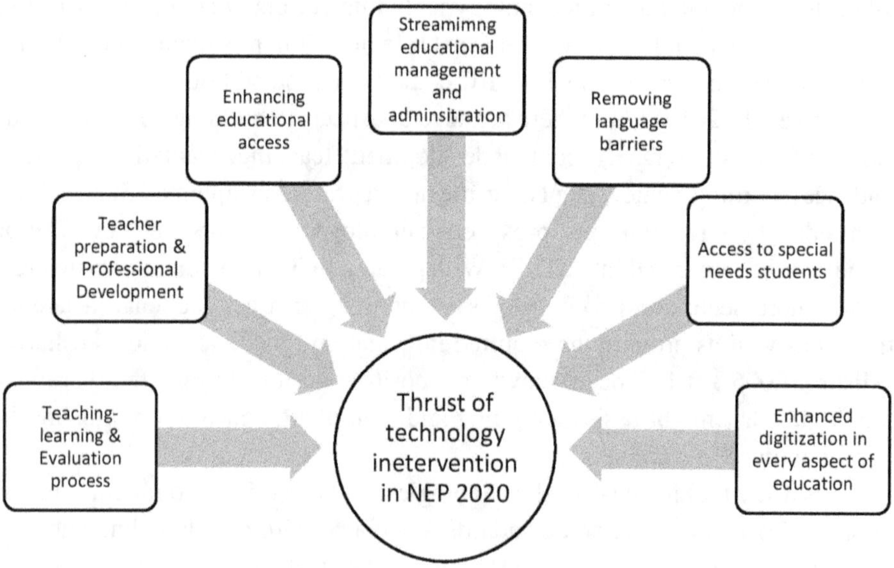

To sum up the above review of literature it appears that BC technology being an advanced database mechanism has immense potential in this data-driven world. It can be used to construct a balance to measure the learning process and academic outcomes which is reliable and proof of value for everyone. To understand what uses it can have in education or any other field, it is good to keep in mind that any transaction between parties is a target application for BC technology (Figure 2). A new domain of teaching-learning with greater freedom for the learners can be achieved through BC adoption.

But experiments on its potential for school education are very limited, especially in the context of a developing country like India. Indian aspirations in this second decade of the 21st century as reflected in NEP 2020 needs a deep and cordial adoption of BC technology in the school system without much delay. It could support the country to realize quality ideals. But how? Little research has been done on this question. And this gap has been targeted in this study to offer a solid implementation framework for BC adoption in Indian schools that might acceleration quality education through effective schooling.

DISCUSSION

Blockchain for Better Schooling in India

India is currently the home of the world's biggest school-going population (Education in India, 2023). Maintaining their records is tiresome, time-consuming, and resource-consuming. Keeping the records updated is another cumbersome task. Recently it has established the National Academic Depository (NAD), a nationwide infrastructure to provide an online storehouse of all academic awards issued by higher education institutions and school boards. It is a 24X7 online storehouse of all academic awards viz. certificates, diplomas, degrees, mark sheets, etc. duly digitized and lodged by academic institutions/boards / eligibility assessment bodies. It has been an exceptional effort not only to ensure easy access to and retrieval of an academic award but also validates and guarantees its authenticity and safe storage.

But there are possibilities of data loss from the centralized server, hacking into the servers, and tampering with the student's records. Many people forge their academic records and degrees to secure admission to higher education and employment. The verification of records is extremely difficult, time-consuming, and in some cases, impossible. Against this backdrop BC technology could provide an excellent framework to manage student records ranging from day-to-day information such as assignments, attendance, and extracurricular activities, to information about degrees and colleges they have attended.

It would be a secure system that ensures educational records remain immutable. These could be relied upon by prospective educational institutions and recruiters, who can be provided access to relevant records. Similarly, information about teachers can be safely stored and this would enable an educational institution to monitor faculty performance. The BC ledger would provide a time-stamped and tamper-proof record of faculty performance - attendance, student evaluations, number of students opting for their electives, research output and publications. These records could be linked to faculty appraisal systems, thereby ensuring greater accountability.

Storing student credentials is a major point to note. Most higher education institutions keep students' completed course records in proprietary formats. These databases are structured for exclusive access by an institution's staff and in dedicated online systems, with little or no interoperability. Further, most institutions have their own specialized system for keeping students' completed course records, which preserves the proprietary data structure of the database. In contrast, BC records are stored permanently, so documents such as degrees and course certificates can be secured and verified, regardless of whether or not a user has access to an institution's record-keeping system. Even if the institution that issued the certificates were to close, or if the entire education system collapsed, those certificates would still be verifiable against the records stored in a BC. In addition, once institutions issue a certificate, no further efforts are required to confirm the validity of that certificate to third parties, since the certificate can verify itself directly on the BC.

The NEP-2020 calls for introducing multidisciplinary education where students would be able to choose their own combination of major and minor subjects along with flexibility in course duration. Here too, BC can help implement such a multiple-entry-and-exit structure. Further, students can be assured of the quality of teachers and educators. It could enable educators to display their certified Skill Badges, allowing students to opt for courses in an informed manner. Meanwhile, students too, especially those in higher education and research, can adopt Skill Badges to indicate their proficiencies. This would enable faculty to identify the right students for projects. BC-based ecosystem could also be used to design a scholarship system incentivizing student to maintain consistency and achieve academic excellence.

The adoption of BC in school education could help improve the efficiency of the education ecosystem and optimize the use of human and physical resources. While doing so, concerns such as data privacy, cost, scalability, and integration with legacy systems will have to be addressed. Doing so is worth every penny as it would help usher in an educational system that is better equipped to handle higher enrolment while being secure, transparent, collaborative, creative and future-ready.

Micro-credentialing and ownership of learning credentials is another facet where the BC could provide a more durable and flexible system for storing student credentials as they move from course to course throughout their professional careers, as well as their secondary education. Via BC, credentials cannot be modified, providing a more reliable system of storing credentials for a lifetime of learning. It allows personal data to stay personal to the learner. Students gain control and ownership of all their education data, including accreditation and portfolios of work, in a secure place that is accessible to anyone who needs to verify it.

Public BC facilitate self-sovereignty by making individuals the final arbiter of who can access and use their data and personal information. Within an educational context, the term is on its way to becoming synonymous with the empowerment of individual learners to own, manage and share details of their credentials, without the need to call upon the educational institution as a trusted intermediary. This will be increasingly important as the industry continues its move toward skills-based education. With BC, learners could store their own evidence of formal or informal learning, share it with a desired audience, and ensure instant verification. This means students could have a self-updating curriculum vitae that can easily be shared with employers. Employers, meanwhile, could reduce their workload since they would not have to verify CVs and could simply search to see whether candidates have the skills required.

Transfer of credits has been another perennial challenge for institutions, often leaving students at a disadvantage when they find, for example, that they must repeat courses to fulfill a new institution's requirements. Students also experience difficulties transferring to another higher education institution, while still preserving and proving courses completed at a previous institution. This problem is even more vivid in cases when a student wants to transfer to an institution in another country, where language and disparate processes are likely to pose additional barriers. Moreover, standards for record storage vary, which can make inter-institutional record exchange difficult.

Intellectual property protection. Teachers often publish research and papers as part of their work. Under the traditional system, there is little way to know if a similar academic study is under way when a teacher begins his or her research. In addition, there is much piracy of the research itself. The use of BC helps address these problems. It could allow educators to publish content openly while keeping track of reuse, without putting limitations on the source material. Such a system would allow teachers to be rewarded based on the level of actual use and reuse of their teaching materials, similar to how they are rewarded based on the number of citations to their research papers. Students and institutions could then make metrics-based decisions on which teaching materials to use.

Teachers could announce the publication of their resources and link to them, as well as the other resources they used in creating the material. Crypto-coins could be awarded to educators according to the level of reuse of their respective resources. In an open scenario, coins would not be spendable but would rather be used to determine the author's prominence. In a closed scenario, coins would have monetary value and would result in monetary compensation. A more advanced implementation might automatically scan resources to identify the percentage of other resources that were reused and automatically issue awards. For example, a 'smart contract' could distribute payment to authors based on how often their material is cited or used. Authors would no longer have to go through intermediaries such as research journals, which often limit use by charging high access fees.

BC with 'smart contracts' could be especially helpful for the country's e-learning scenarios. This can aid educators in developing courses and lessons on a blockchain. After the necessary prerequisites are satisfied, the course will be automatically taught and can proceed at its own pace. Students and teachers could sign a smart contract outlining the limitations of an assignment, as well as the due date and the grading deadline.

Identity verification, another perennial problem for educational institutions requiring much manual intervention and presenting many opportunities for data tampering, could be served better with BC. With a digital process, validation of a student's identity happens once. Rather than storing the student identity document, the BC network stores information about that document. Using BC, students and job candidates can identify themselves online while maintaining control over the storage and management of their data. Within larger institutions, students need to regularly identify themselves with different parts of the organization. In such cases, either each part of the institution collects the student data for itself, or the organization uses a single sign-on, whereby one shared copy of the student data is used by all parties within the organization.

Under both models, tens if not hundreds of people might have access to a student's personal information. Keeping that data safe requires managing access rights for all those people, and ensuring their devices are also secured – a mammoth, if not impossible, undertaking. With BC, only a select few – namely the parties responsible for verifying a student's identity – have access to the data. Other than that, it is in the student's hands. This means that the organization no longer needs to manage the complex systems for access rights and only needs to secure the device or network where the initial verification takes place. This removes the need for significant investment in hardening the network against data breaches, staff training on data protection, and managing access rights (also claimed in Jain, 2019).

Students will then take the test remotely using personal computers or smartphones, with the BC performing the evaluation. Teachers would have more time to devote to other scholarly or cultural pursuits if they didn't have to grade tests. Teachers can use smart contracts and BC with defining the correct answers and scoring criteria for evaluation. Students will then appear for the examination on their PC or devices. Students' academic success and non-academic successes in education, preparation, tournaments, work, and other events outside of school can be measured using BC technologies to assess their capability, which benefits both students and businesses looking to hire them. BC-based student technical skill assessment system that can test student ability measurement methods using a clustering algorithm (Khan et al., 2020). The framework can also allow for the development of a student skill assessment ecosystem.

Figure 4. Areas for blockchain intervention for better schooling
Source: Author

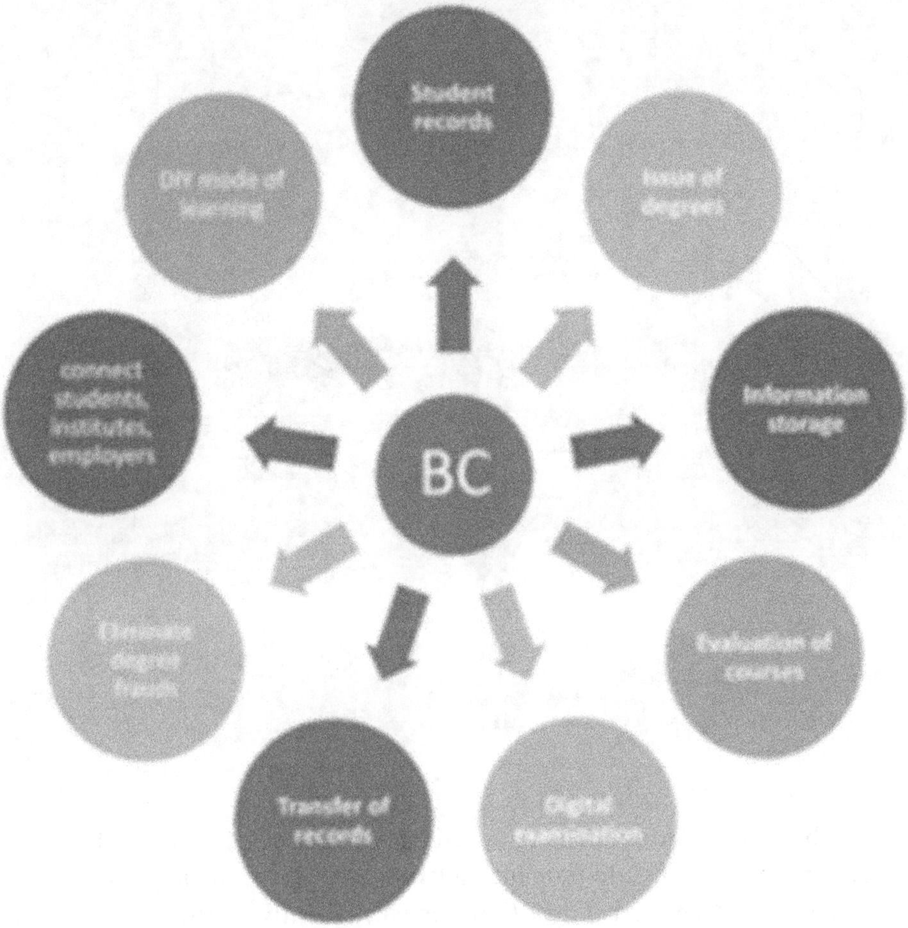

Framework for Blockchain Implementation

The advantage of this technology in education, as appeared in the above discussion, ranges from data management to data authentication without jeopardizing legitimacy. Considering the BC potential and shifting focus of Indian school policies, BC adoption is a current necessity. Responding to this need, the author has come up with a Framework for Blockchain Implementation (FBI) in schools, presented in Figure 5.

Figure 5. Framework for blockchain implementation
Source: Author

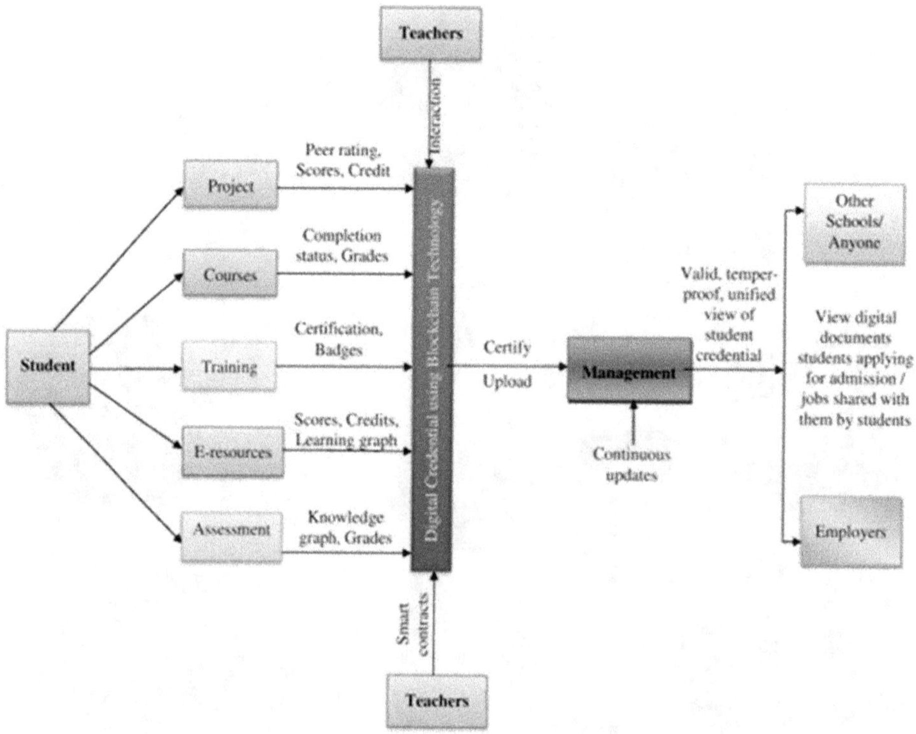

Three basic areas are covered under FBI that schools can take in the short term to begin preparing for BC: to promote understanding of BC among stakeholders, experiment with BC, be part of the BC movement, and active participation in cases.

It demands a collective effort from all three major school stakeholders – students, teachers, and management – to be benefitted from this secured, accurate, cost-saving, and constructive digital media. Students' activities, course details, certifications, assessments, etc. are put under BC domains where teachers will interact and verify the same, and the verified data will be re-verified by the management and uploaded in BC. The data needs continuous updates to make a pool of valid, temper-proof, unified views of student credentials that will be easily accessible to employers and other stakeholders.

This is a basic flow chart for working with BC. Complexities arise with enhanced utilization. It is not an isolated effort rather could be seen as an extension of educational technology integration. The process needs the basic resources as any other form of educational technology like devices, skills, and attitudes. Effective adoption needs constant improvement in all three domains available to those who act decisively, optimistically, and assertively.

CONCLUSION

Blockchain applications in education appear to be very promising and present enormous opportunities to help education providers serve students with much more digital accuracy, openness, cost-saving, and constructivist ways. The time has arrived to embrace it with sincerity even at the school level. It fits in the ambitious goal; India has set for itself in NEP 2020. The FBI model proposed in this study based on a detailed literature survey seen through the Indian lens could be helpful for BC adoption in schools and a significant addition to the 'glocal' management of BC implementation policies.

REFERENCES

Aamir, A., Qureshi, R., Khan, F. A., & Huzaifa, M. (2020). Blockchain based academic records verification in smart cities. *Wireless Personal Communications*, *113*(3), 1397–1406. doi:10.100711277-020-07226-0

Abeyratne, S. A., & Monfared, R. P. (2016). Blockchain ready manufacturing supply chain using distributed ledger. *International Journal of Research in Engineering and Technology*, *5*(9), 1–10. doi:10.15623/ijret.2016.0509001

Alahmari, M., Issa, T., Issa, T., & Nau, S. Z. (2019). Faculty awareness of the economic and environmental benefits of augmented reality for sustainability in Saudi Arabian universities. *Journal of Cleaner Production*, *226*, 259–269. doi:10.1016/j.jclepro.2019.04.090

Alammary, A., Alhazmi, S., Almasri, M., & Gilani, S. (2019). Blockchain-Based Applications in Education: A Systematic Review. *Applied Sciences (Basel, Switzerland)*, *9*(12), 2400. doi:10.3390/app9122400

Alfandi, O. (2021). Pros and cons of Blockchain-based approaches in Education. In *IEEE Global Engineering Education Conference (EDUCON)*. IEEE. https://doi.org/10.1109/EDUCON46332.2021.9454118

Atzei, N., Bartoletti, M., & Cimoli, T. (2017). A Survey of Attacks on Ethereum Smart Contracts (SoK). In Principles of Security and Trust. Springer. doi:10.1007/978-3-662-54455-6_8

Bartolomé, A., Torla, C., Castañeda, L., & Adell, J. (2017). Blockchain in Education: Introduction and Critical Review of the State of the Art. *Edutec*, *61*(61). Advance online publication. doi:10.21556/edutec.2017.61

Bucea-Manea-Tonis, R., Martins, O., Bucea-Manea-Tonis, R., Gheorghiță, C., Kuleto, V., Ilić, M. P., & Simion, V.-E. (2021). Blockchain Technology Enhances Sustainable Higher Education. *Sustainability (Basel)*, *13*(22), 12347. doi:10.3390u132212347

Cai, Y., & Zhu, D. (2016). Fraud detections for online businesses: A perspective from blockchain technology. *Financial Innovation*, *2*(20), 20. Advance online publication. doi:10.118640854-016-0039-4

Caldarelli, G., & Ellul, J. (2021). Trusted Academic Transcripts on the Blockchain: A Systematic Literature Review. *Applied Sciences (Basel, Switzerland)*, *11*(4), 1842. doi:10.3390/app11041842

Casino, F., Dasaklis, T. K., & Patsakis, C. (2019). A systematic literature review of blockchain-based applications: Current status, classification and open issues. *Telematics and Informatics*, *36*, 55–81. Advance online publication. doi:10.1016/j.tele.2018.11.006

Chen, G., Xu, B., Lu, M., & Chen, N.-S. (2018). Exploring blockchain technology and its potential applications for education. *Smart Learn. Environ*, *5*(1), 1. Advance online publication. doi:10.118640561-017-0050-x

Cheriguene, A., Kabache, T., Kerrache, C. A., Calafate, C. T., & Cano, J. C. (2022). NOTA: A novel online teaching and assessment scheme using Blockchain for emergency cases. *Education and Information Technologies*, *27*(1), 115–132. doi:10.100710639-021-10629-6 PMID:34248386

Chivu, R.-G., Popa, I.-C., Orzan, M.-C., Marinescu, C., Florescu, M. S., & Orzan, A.-O. (2022). The Role of Blockchain Technologies in the Sustainable Development of Students' Learning Process. *Sustainability*, *14*(3), 1406. doi:10.3390u14031406

Christidis, K., & Devetsikiotis, M. (2016). Blockchains and smart contracts for the internet of things. *IEEE Access : Practical Innovations, Open Solutions*, *4*, 2292–2303. doi:10.1109/ACCESS.2016.2566339

Chukowry, V., Nanuck, G., & Sungkur, R. K. (2021). The future of continuous learning–Digital badge and microcredential system using blockchain. *Global Transitions Proceedings*, *2*(2), 355–361. doi:10.1016/j.gltp.2021.08.026

Conti, M., Sandeep Kumar, E., Lal, C., & Ruj, S. (2018). A Survey on Security and Privacy Issues of Bitcoin. *IEEE Communications Surveys and Tutorials*, *20*(4), 3416–3452. doi:10.1109/COMST.2018.2842460

Crosby, M. (2016). Blockchain technology: Beyond bitcoin. *Applied Innovation, 2*, 6–10. https://j2-capital.com/blockchain-technology-beyond-bitcoin

Dandona, L., Dandona, R., Kumar, G. A., Shukla, D. K., Paul, V. K., Balakrishnan, K., ... Thakur, J. S. (2017). Nations within a nation: Variations in epidemiological transition across the states of India, 1990–2016 in the Global Burden of Disease Study. *Lancet*, *390*(10111), 2437–2460. doi:10.1016/S0140-6736(17)32804-0 PMID:29150201

Deenmahomed, H. A., Didier, M. M., & Sungkur, R. K. (2021). The future of university education: Examination, transcript, and certificate system using Blockchain. *Computer Applications in Engineering Education*, *29*(5), 1234–1256. doi:10.1002/cae.22381

Devine, P. (2015). Blockchain learning: can crypto-currency methods be appropriated to enhance online learning? *ALT Online Winter Conference*, *7*(10). https://altc.alt.ac.uk/online2015/sessions/blockch

Dobrovnik, M., Herold, D., Fürst, E., & Kummer, S. (2018). Blockchain for and in logistics: What to adopt and where to start. *Logistics*, *2*(3), 18. doi:10.3390/logistics2030018

Education in India. (2023, January 4). Wikipedia. https://en.wikipedia.org/wiki/Education_in_India

Fedorova, E. P., & Skobleva, E. I. (2020). Application of Blockchain Technology in Higher Education. *Eur. J. Contemp. Educ.*, *9*, 552–571. https://eric.ed.gov/?id=EJ1272331

Fraga-Lamas, P., & Fernandez-Carames, T. M. (2019). A Review on Blockchain Technologies for an Advanced and Cyber-Resilient Automotive Industry. *IEEE Access : Practical Innovations, Open Solutions*, *7*, 17578–17598. doi:10.1109/ACCESS.2019.2895302

George Bernard Shaw Quotes. (n.d.). *BrainyQuote.com*. Retrieved August 21, 2022, from BrainyQuote.com Web site: https://www.brainyquote.com/quotes/george_bernard_shaw_38692
3

Gibson, D., Ostashewski, N., Flintoff, K., Grant, S., & Knight, E. (2015). Digital badges in education. *Education and Information Technologies*, *20*(2), 403–410. doi:10.100710639-013-9291-7

Grech, A., & Camilleri, A. F. (2017). *Blockchain in Education*. 132 R. JRC Science for Policy Report; Publications Office of the European Union. https://publications.jrc.ec.europa.eu/repository/handle/JRC1
08255

Hawlitschek, F., Notheisen, B., & Teubner, T. (2018). The limits of trust-free systems: A literature review on blockchain technology and trust in the sharing economy. *Electronic Commerce Research and Applications*, *29*, 50–63. Advance online publication. doi:10.1016/j.elerap.2018.03.005

Jain, A. K. (2019). *Blockchain Goes to School*. Cognizant. https://www.cognizant.com/us/en/whitepapers/documents/blockc
hain-goes-to-school-codex3775.pdf

Jang, H., & Han, S. H. (2021). User experience framework for understanding user experience in blockchain services. *International Journal of Human-Computer Studies*, *158*, 102733. doi:10.1016/j.ijhcs.2021.102733

Jaoude, J., & Saade, R. (2019). Blockchain Applications – Usage in Different Domains. *IEEE Access : Practical Innovations, Open Solutions*, *7*, 45360–45381. doi:10.1109/ACCESS.2019.2902501

Khan, M. A., Quasim, M. T., Algarni, F., & Alharthi, A. (2020). *Decentralized Internet of Things: A blockchain perspective*. Springer Nature Switzerland. doi:10.1007/978-3-030-38677-1

Kosba, A., Miller, A., Shi, E., Wen, Z., & Papamanthou, C. (2016). Hawk: The Blockchain Model of Cryptography and Privacy-Preserving Smart Contracts. *2016 IEEE Symposium on Security and Privacy*, 839–858. 10.1109/SP.2016.55

Kraft, D. (2016). Difficulty control for blockchain-based consensus systems. *Peer-to-Peer Networking and Applications*, *9*(2), 397–413. doi:10.100712083-015-0347-x

Kundu, A. (2021). A Sound Framework for ICT Integration in Indian Teacher Education. *International Journal of Teacher Education and Professional Development*, *4*(1), 49–67. doi:10.4018/IJTEPD.2021010104

Kundu, A., & Bej, T. (2021). Technology Adoption in Indian National Education Policy 2020: An Analysis of Pedagogical, Institutional and Human Aspects. *Journal of Social Sciences*, *17*(1), 145–157. doi:10.3844/jssp.2021.145.157

Kundu, A., & Bej, T. (2021). Experiencing e-assessment during COVID-19: An analysis of Indian students' perception. *Higher Education Evaluation and Development*, *15*(2), 114–134. doi:10.1108/HEED-03-2021-0032

Kundu, A., & Bej, T. (2022). Reflective Teachers: A Probe into 21st Century Skills among Indian Trainee Teachers. *International Journal of Teacher Education and Professional Development*, *5*(1), 1–20. doi:10.4018/IJTEPD.2022010106

Kundu, A., Bej, T., & Rice, M. (2020). Time to Engage: Implementing Math and Literacy Blended Learning Routines in an Indian Elementary Classroom. *Education and Information Technologies*, *26*(1), 1201–1220. doi:10.100710639-020-10306-0

Lam, T. Y., & Dongol, B. (2020). A blockchain-enabled e-learning platform. Interactive learning environments, 1–23. doi:10.1080/10494820.2020.1716022

Li, D., Han, D., Zheng, Z., Weng, T.-H., Li, H., Castiglioane, A., & Li, K.-C. (2021). MOOCsChain: A Blockchain-Based Secure Storage and Sharing Scheme for MOOCs Learning. *Computer Standards & Interfaces*, *81*, 103597. doi:10.1016/j.csi.2021.103597

Lohmer, J., & Lasch, R. (2020). Blockchain in operations management and manufacturing: Potential and barriers. *Computers & Industrial Engineering*, *149*, 106789. doi:10.1016/j.cie.2020.106789

Mahankali, S., & Chaudhary, S. (2020). Blockchain in Education: A Comprehensive Approach – Utility, Use Cases, and Implementation in a University. In R. Sharma, H. Yildirim, & G. Kurubacak (Eds.), *Blockchain Technology Applications in Education* (pp. 267–293). IGI Global. doi:10.4018/978-1-5225-9478-9.ch014

McArthur, D. (2018). Will blockchains revolutionize education. *Educause Review*. https://er.educause.edu/articles/2018/5/will-blockchains-revolutionize-education

Meunier, S. (2018). Blockchain 101: What is Blockchain and how does this revolutionary technology work? In A. Mark (Ed.), *Edt.). Transforming Climate Finance and Green Investment with Blockchains* (pp. 23–34). Academic Press. doi:10.1016/B978-0-12-814447-3.00003-3

Nabil, A., Nafil, K., & Mounir, F. (2020). Blockchain Security and Privacy in Education: A Systematic Mapping Study. In Á. Rocha, H. Adeli, L. Reis, S. Costanzo, I. Orovic, & F. Moreira (Eds.), *Trends and Innovations in Information Systems and Technologies. WorldCIST 2020. Advances in Intelligent Systems and Computing* (Vol. 1161). Springer. doi:10.1007/978-3-030-45697-9_25

NAD. (n.d.). *Universities Grant Commission.* Ministry of Education. Govt. of India. https://nad.gov.in/

NEP. (2020). *National Education Policy 2020.* Ministry of Education. New Delhi. India. https://www.education.gov.in/sites/upload_files/mhrd/files/NEP_Final_English_0.pdf

Oh, J., & Shong, I. (2017). A case study on business model innovations using Blockchain: Focusing on financial institutions. *Asia Pac. J. Innov. Entrep.*, *11*(3), 335–344. doi:10.1108/APJIE-12-2017-038

Park, J. (2021). Promises and challenges of Blockchain in education. *Smart Learn. Environ.*, *8*(1), 33. doi:10.118640561-021-00179-2

PIB. (n.d.). https://pib.gov.in/PressReleseDetailm.aspx?PRID=1777711

Reynolds, P., & Irwin, A. S. (2017). Tracking digital footprints: Anonymity within the bitcoin system. *J. Money Laund. Control*, *20*(2), 172–189. doi:10.1108/JMLC-07-2016-0027

Rustagi, N., & Chakraborti, A. (2021, Dec. 3). *Blockchain in schools and colleges.* Indian Express. https://indianexpress.com/article/opinion/blockchain-technology-education-nep-7696791/

Salah, K., & Khan, M. (2017). IoT Security: Review, Blockchain Solutions, and Open Challenges. *Future Generation Computer Systems*, *82*, 395–411. doi:10.1016/j.future.2017.11.022

Seebacher, S., & Schüritz, R. (2017). Blockchain Technology as an Enabler of Service Systems: A Structured Literature Review. In S. Za, M. Drăgoicea, & M. Cavallari (Eds.), *Exploring Services Science. IESS 2017. Lecture Notes in Business Information Processing* (Vol. 279). Springer. doi:10.1007/978-3-319-56925-3_2

Sharma, R. C., Kawachi, P., & Bozkurt, A. (2019). Exploring Changing Perspectives in Distance Education. *Asian Journal of Distance Education, 14*(1), 1–6. http://www.asianjde.com/ojs/index.php/AsianJDE/article/view/285

Sharma, R. C., & Mishra, S. (2006). *Cases on Global E-Learning Practices: Successes and Pitfalls: Successes and Pitfalls*. IGI Global. https://www.researchgate.net/publication/307569541_Cases_on_Global_E-Learning_Practices_Successes_and_Pitfalls

Sharma, R. C., Yildirim, H., & Kurubacak, G. (Eds.). (2019). *Blockchain Technology Applications in Education*. IGI Global.

Sharples, M., & Domingue, J. (2016). The Blockchain and Kudos: A Distributed System for Educational Record, Reputation and Reward. Springer. doi:10.1007/978-3-319-45153-4_48

Skiba, D. J. (2017). The potential of Blockchain in education and health care. *Nursing Education Perspectives, 38*(4), 220–221. doi:10.1097/01.NEP.0000000000000190 PMID:28622267

Skinner, B. F. (1938). The Behavior of Organisms. Academic Press.

Steiu, M.-F. (2020). Blockchain in education: Opportunities, applications, and challenges. *First Monday, 25*(9). Advance online publication. doi:10.5210/fm.v25i9.10654

Swan, M. (2015). *Blockchain: Blueprint for a new economy*. O'Reilly Media, Inc.

Ta, N. B. D., & Yang, T. J. J. (2021). Ether Learn: Decentralizing learning via blockchain. *Proceedings of the IEEE TALE 2021*, 1-6. https://ink.library.smu.edu.sg/sis_research/6838

Tama, B. A., Kweka, B. J., Park, Y., & Rhee, K. H. (2017). A critical review of blockchain and its current applications. *2017 International Conference on Electrical Engineering and Computer Science (ICECOS)*, 109-113. 10.1109/ICECOS.2017.8167115

Tapscott, D., & Kaplan, A. (2019). *Blockchain Revolution in Education and Lifelong Learning*. Blockchain Research Institute and IBM Institute for Business. https://www.ibm.com/downloads/cas/93DDVAKE

Thayer, T. L., & Yanckello, R. (2019). *Education Administrative Systems Innovation Primer for 2019*. Gartner, Inc. https://www.gartner.com/en/documents/3901065/education-administrative-systemsinnovation-primer-for-2

Vukolić, M. (2016). The Quest for Scalable Blockchain Fabric: Proof-of-Work vs. BFT Replication. In J. Camenisch & D. Kesdoğan (Eds.), Lecture Notes in Computer Science: Vol. 9591. *Open Problems in Network Security. iNetSec 2015*. Springer. doi:10.1007/978-3-319-39028-4_9

Wan, H., Li, K., & Huang, Y. (2021). Blockchain: A Review from The Perspective of Operations Researchers. 2020 Winter Simulation Conference (WSC), 75-89. doi:10.1109/WSC48552.2020.9383924

Wang, H., Chen, K., & Xu, D. (2016, December). A maturity model for blockchain adoption. *Financial Innovation*, *2*(1), 12. doi:10.118640854-016-0031-z

Winstone, N. E., Ajjawi, R., Dirkx, K., & Boud, D. (2021). Measuring what matters: The positioning of students in feedback processes within national student satisfaction surveys. *Studies in Higher Education*, *47*(7), 1524–1536. doi:10.1080/03075079.2021.1916909

Yamada, Y., Nakajima, T., & Sakamoto, M. (2016). Blockchain-LI: a study on implementing activity-based micro-pricing using cryptocurrency technologies. In *Proceedings of the 14th International Conference on Advances in Mobile Computing and Multi Media* (pp. 203-207). https://doi.org/10.1145/3007120.3007151

Zet, C., & Dumitriu, G. C. (2021). Using blockchain technology for ensuring students results traceability for instrumentation classes. *Measurement: Sensors*, *18*, 100315. doi:10.1016/j.measen.2021.100315

Zhao, J.L., Fan, S., & Yan, J. (2016). Overview of business innovations and research opportunities in blockchain and introduction to the special issue. *Financ Innov*, *2*(28). doi:10.1186/s40854-016-0049-2

Zheng, Z., Xie, S., Dai, H. N., Chen, X., & Wang, H. (2018). Blockchain challenges and opportunities: A survey. *International Journal of Web and Grid Services*, *14*(4), 352. doi:10.1504/IJWGS.2018.095647

Zhong, J., Xie, H., Zou, D., & Chui, D. K. (2018). *A blockchain model for word-learning systems*. Paper presented at the 2018 5th International Conference on Behavioral, Economic, and Socio-Cultural Computing (BESC), Kaohsiung City, Taiwan. 10.1109/BESC.2018.8697299

Zhou, H., de Laat, C., & Zhao, Z. (2018). Trustworthy cloud service level agreement enforcement with blockchain based smart contract. *2018 IEEE International Conference on Cloud Computing Technology and Science (cloudcom),* 255–260. 10.1109/CloudCom2018.2018.00057

Section 4

Supply Chain and Auditing Implications

Chapter 9

Potential of Internet of Things (IoT) and Blockchain Technology in the Collaboration and Integration of the Retail Supply Chain

Xinyu Zheng
Cranfield University, UK

Syed Imran Ali
University of Huddersfield, UK

ABSTRACT

The application of Industry 4.0 technologies in the supply chain of organisations has progressively become a hot topic. These disruptive technologies have contributed to enhance efficiency, performance, and sustainability in the organizational ecosystems. However, the globalized retail industry is confronting several challenges pertaining to collaboration and integration. Hence, a thorough review is imminent to explore the potentials of Industry 4.0 technologies such as internet of things (IoT) and blockchain technology to address the challenge of achieving collaboration and integration across the retail supply chains. To address this, a systematic literature review (SLR) methodology is adopted to retrieve current academic publications from the SCOPUS database for the period of 2018-2021. Descriptive and thematic analysis is done to reveal various characteristics, themes, potentials, and relationships in achieving collaboration and integration across the retail supply chains. The findings also suggest some suggestions for small and medium enterprises and future directions.

DOI: 10.4018/978-1-6684-4153-4.ch009

Copyright © 2023, IGI Global. Copying or distributing in print or electronic forms without written permission of IGI Global is prohibited.

INTRODUCTION

The globalization of retail industry has made supply chains longer and even more complex. This has led to more technical and operational issues that has caused increased fraud and lack of trust in all aspects of the supply chains (Tan et al., 2018). With the development of Industry 4.0 technologies, retail industry has adopted various technologies such as Cyber physical systems (CPS), Internet of Things (IoT), Big data analytics (BDA), Cloud Computing technology and Blockchain, to make the supply chains more effective and integrated (Kamble et al., 2020). CPS allows to optimize supply chain decision-making by obtaining real-time data from various nodes (Zhong & Nof, 2015). BDA can facilitate supply chain performance by combining data from multiple datasets (Wamba et al., 2017). Cloud computing technology enables the transmission of information throughout the supply chains by real-time information sharing (Harris et al., 2015). IoT and blockchain have facilitated the automation of facilities by effective data monitoring and storage at various stages of the supply chain, thereby improving the efficiency and trust in the operations (Chung et al., 2015; Tan et al., 2018). Wal-Mart have implemented IBM based blockchain solution to improve traceability and food safety in its pork and mango projects in China and USA respectively (Tan et al., 2018). Brookstone have used IoT devices to recommend products to customers based on their preferences, which improves the customer shopping experience and promotes in-store sales (Caro & Sadr, 2019).

The aforementioned, literature shows that Industry 4.0 technologies have made significant contribution in various industries. The publication of articles on the application of IoT and blockchain technology have exponentially increased in recent years. Even though, many firms' pilot initiatives using IoT and blockchain technology have had some success, still large-scale adoption remains a challenge (Tan et al., 2018). Low confidence of the partners in the usage of these technologies has become a key roadblock, owing to the lack of understanding in gaining potential in the supply chains.

JUSTIFICATION OF RESEARCH

The globalization has made supply chains longer and even more complex (Khan et al., 2022). This has resulted in fragmented supply chains with less collaboration and integration (Ali et al., 2021). With the advent of new Industry 4.0 technologies such as Cyber physical systems (CPS), Internet of Things (IoT), Big data analytics (BDA), Cloud Computing technology and Blockchain, the collaboration and integration has become easier (Kamble et al., 2021). With the growth of globalized business, covid challenges, it is critical to understand as how IoT and blockchain technology in the retail industry can facilitate collaboration and integration. Therefore, an extensive literature review is required to fill the gap.

AIMS AND OBJECTIVES

The unique characteristics of IoT and blockchain technology may help in supply chain integration and collaboration, which is an opportunity for retail businesses experiencing industry-wide changes. Therefore, this chapter's aim is to explore the potentials of IoT and blockchain technology to improve the collaboration and integration in the retail supply chain. The chapter follows a Systematic Literature Review (SLR) to explore the topic. A set of objectives have been established to address, which are:

RO 1. Review the academic and practitioner literature related to the IoT and blockchain technology related to collaboration and integration of retail supply chain.
RO 2. Produce a descriptive analysis of the evolution of literature.
RO 3. Undertake a detailed review of literature to identify and summarize the potentials of IoT and blockchain technology in achieving collaboration and integration of retail supply chain.
RO 4. Provide a summary of the themes and findings on the literature examined.

METHODOLOGY

The Systematic Literature Review (SLR) method is adopted to comprehensively search and analyze the relevant literature on the potentials of IoT and blockchain technology in the collaboration and integration of retail supply chain. SLR is 'to identify, appraise and synthesize all the empirical evidence that meets pre-specified eligibility criteria to answer a specific research question' ("Cochrane Library", 2022; Tranfield et al., 2003). SLR methodology is often utilized in research across a wide range of fields, because of its good characteristics such as objectivity, transparency and reproducibility (Xiao & Watson, 2019).

To answer the major research issues, systematic assessment seeks to give a thorough review of the available literature. Many academics have proposed several mature systematic literature research stages in recent years. The number of articles screened in each stage may be reported using a Preferred Reporting Items for Systematic Reviews and Meta-Analyses (PRISMA), which shows the findings of each stage in an understandable and simple manner. This study refers to the basic framework proposed by Koutsos et al. (2019) are shown in Figure 1.

Stages of Systematic Literature Review

The systematic literature review is conducted using five stages, which are as follows:

Stage 1: Scoping

It is significant to choose a novel, contributing and crucial research direction. Reviewers need to thoroughly search on the topic and determine the research focus. During the searching it was found that more and more scholars have begun to pay attention on the application of IoT or blockchain in the supply chain i.e. from 2018, but very few scholars have studied the potentials of these two Industry 4.0 technologies in the collaboration and integration of retail supply chains.

Several academics have interviewed executives from various firms. The majority of them stated that IoT and blockchain technology have significant potential to improve supply chain performance. But whether these two technologies have the potential in achieving collaboration and integration of the retail supply chain, is yet to be explored. To explore the same, a SLR is conducted on the topic.

Figure 1. PRISMA diagram

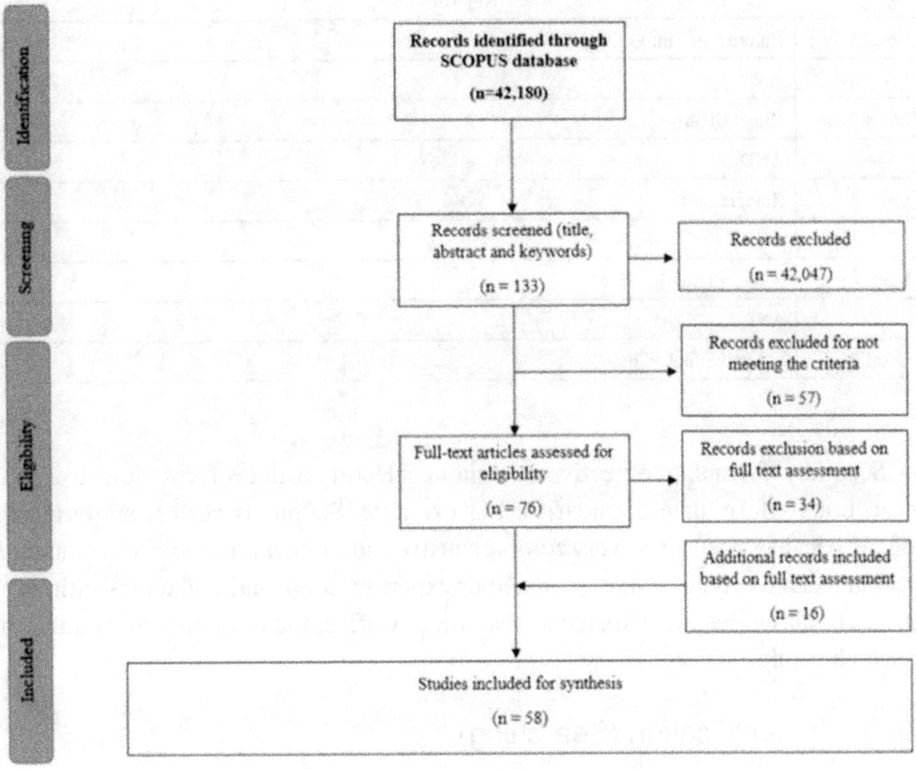

Stage 2: Planning

At this stage, there are two main steps:

Firstly, extract the keywords on the topic. The keywords for the study are as follow:

Table 1. Keywords

Keywords	
Level 1	(Internet of Things OR IoT)
	AND
Level 2	(Blockchain)
	AND
Level 3	(Integration)
	AND
Level 4	(Collaboration)
	AND
Level 5	("Retail Supply Chain")

Secondly, formulate objective criteria in selecting articles for review. It is best to accurately formulate boundaries. For example, Scopus is very friendly in this regard, because it can give a very comprehensive and authoritative objective standard. Scholars can use their own experience or expert opinion, and make pre-settings to ensure unbiased results, which can also directly affect the reliability and quality of research results.

Stage 3: Identification (Searching)

Selecting a suitable database for literature search is the key stage. The database selected for this SLR is Scopus. Utilizing suitable search keywords and boolean search operators such as "AND," "OR," and "NOT" helps to broaden or reduce the scope of the articles sought. The filters in the search are required to limit the number of articles returned.

This SLR mainly applies filters on: subject area (Business, Management and Accounting), document type (Article), source type (Journal), language (English) and publication year (since it was found in the preliminary search stage that there was no relevant document before 2018, or at most one document in 2017, the publication period limit from 2018 to 2021 in the subsequent searches.).

Stage 4: Screening

The evaluation of the selected articles is done in Stage 4. In this stage all the duplicate records are removed. Then the titles and abstracts of the selected articles are examined. Further 57 articles are removed from the sample for not meeting the criteria. The remaining 76 articles sample are thoroughly studied to determine the relevance to the study. Then further 34 articles are removed that are not aligned with the aims and objectives of the study. Finally, only 42 articles are seemed to be relevant for further analysis. At later stage further 16 articles are added by the research panel. Final selected sample after following the exclusion criteria is 58 articles.

Stage 5: Synthesis and Reporting

The final 58 selected articles are extracted into excel format with title, authors, source title, domain, volume, issue, article number, pages, cited by and DOI weblink.

DESCRIPTIVE ANALYSIS

This section delves into the details of 58 selected articles through a multi-directional investigation. The section covers year-wise distribution, articles by publishers, geographical distribution, research direction and research methodologies.

Year-Wise Distribution

Year-wise distribution of these 58 articles is shown in Figure 2, which presents the rapid growth of scholars' research on IoT and blockchain in the supply chain context from 2018-2021. The significant increase in the contributions is seen in the past two years i.e. 2021 and 2020 accounts for 80% of the total. The increase in publications represents that the potential of IoT and blockchain for coordinating and integrating retail supply chain is gradually being valued by scholars. As can be seen in the pie chart, few supply chain academics have begun to recognize the potential of the IoT and blockchain in the retail supply chain since 2018 i.e. 2%. More scholars joined the study team during the subsequent years.

Figure 2. Year-wise distribution

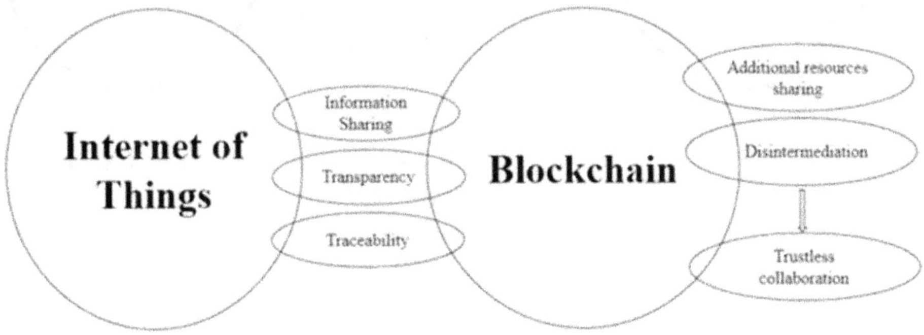

Articles by Publishers

Figure 3 describes that Emerald Group published most of the related academic literature accounting for 46%. *Elsevier* holds the second largest share of publishing at 23%. *Taylor and Francis Ltd.* and *MDPI AG* published 9% and 8% respectively. The number of publications published by Inderscience accounted for 4%, while others account for 10% of total. In terms of supply chain collaboration and integration, Emerald Group have devoted greater attention to the topic than other publishers.

Figure 3. Proportion of articles by publishers

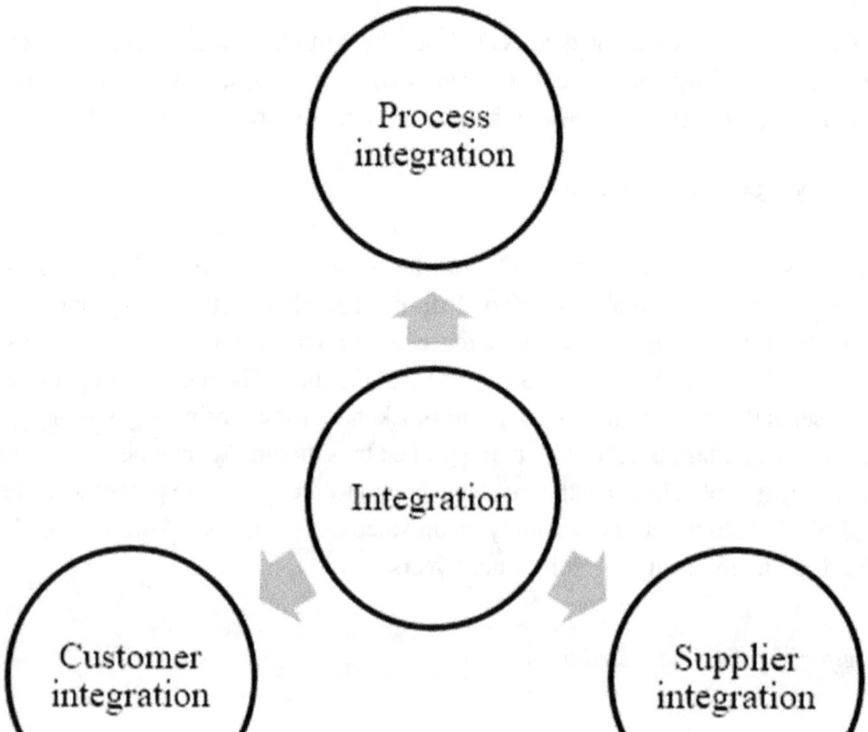

Geographical Distribution

Most of the selected articles were published in four countries i.e., India, United Kingdom, United States, and China that accounts for 72% of the total. The other countries have contributed only 28% of the total.

Figure 4. Statistics of source countries

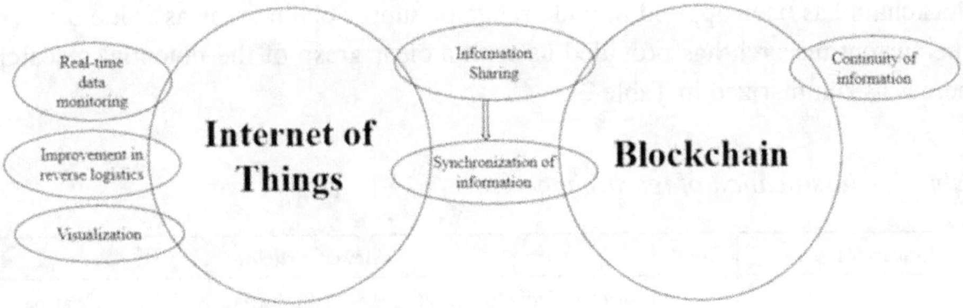

Research Direction

Table 2 shows the classification of 58 reviewed articles according to 12 research themes. The selected articles contain most of the Industry 4.0 technologies in the supply chain context, including impacts, challenges, suggestions on technology adoption and applications. The Industry 4.0 technologies such as Cyber physical systems (CPS), Internet of Things (IoT), Big data analytics (BDA), Cloud Computing technology, Artificial Intelligence (AI) and Blockchain are discussed to address various supply chains problems. The most prominent ones are sustainability, real-time data for effective decision making, efficiency of supply chains, integration and addressing challenges of covid. The articles on supply chain performance have discussed firm's efforts to integrate blockchain in supply chain systems, supply chain traceability and visibility, technology-enabled supply chain integration, digitization and its effects on enhancing supply chain performance and optimizing the overall cost. The articles on supplier selection and procurement have discussed mainly the digitization in supplier selection and procurement, supply chain quality management, procurement planning under risk and challenges in addressing the procurement across the supply chain. The articles on strategy mainly discussed on making strategies on customer and product design, smarter supply chains, digitization to bridge supply and demand. The blockchain and supply chain disruption articles have discussed applications, challenges, research opportunities, sustainability, and technology adoption challenges. The supply chain collaboration articles have discussed various approaches on enhancing operational excellence, potentials of blockchain for supply chain collaboration and achieving sustainability goals through Blockchain, IoT and RFID. The integration of the supply chain is discussed in the context of transaction cost theory and sustainability gains in order management. The sustainability articles have addressed the sustainability goals through data analytics and digitization of supply chains. Electronics is used to enhance transportation management and electronic booking in the maritime supply chain. Although the implementation of IoT and blockchain in the retail supply chain is still in its early stages, still IoT and blockchain has been applied in wide range of supply chain areas as value drivers. The current research has provided us with a clear grasp of the potential research themes as summarized in Table 2.

Table 2. Classification of research themes

Research Theme	Total	Reviewed Articles
Impacts, challenges, suggestions on technology adoption and applications	16	Sheel & Nath (2020); van Hoek (2019); Kayikci et al. (2020); Wamba & Queiroz (2020); Karamchandani et al. (2020); Wong et al. (2020); Choi & Luo (2019); Tönnissen & Teuteberg (2020); Li et al. (2020); Kamble et al. (2019); Miraz et al. (2020); Fosso Wamba et al. (2020); Tan et al. (2018); Choi (2020); Aich et al. (2019); Tian (2017)
Supply chain performance	11	Nandi et al. (2020); Roy (2021); Sheel & Nath (2020); Ganbold et al. (2020); Gupta et al. (2020); Liu et al.(2021); Saryatmo & Sukhotu (2021); Banerjee (2019); de Vass et al. (2021); Ben-Daya et al. (2019); Chen et al. (2020)
Supplier selection and procurement	5	Srai & Lorentz (2019); Sharma & Joshi (2020); Hallikas et al. (2021); Shabani-Naeeni & Ghasemy Yaghin (2021); Joseph Jerome et al. (2021)
Strategy	4	Sharma & Joshi (2020); TS & V (2020); Tripathi & Gupta (2020); Caro & Sadr (2019)
Blockchain and supply chain disruption	3	Dutta et al. (2020); Varriale et al. (2021); Aamer et al. (2021)
Supply chain collaboration	3	Cui et al. (2020); Rejeb et al. (2021); Varriale et al. (2021)
Supply chain integration	3	Schmidt & Wagner (2019); Vodenicharova (2020); Varriale et al. (2021)
Sustainability	2	Mangina et al. (2020); Sarkis et al. (2020)
Electronic transportation	2	Jović et al. (2020); Zeng et al. (2021)
Warehousing	1	Mahroof (2019)
Manufacturing	1	Raut et al. (2020)
Others	7	Esmaeilpour Ghouchani et al. (2019); Dwivedi et al. (2021); Hofmann & Rutschmann (2018); Queiroz et al. (2019); Cai & Lo (2020); Calatayud et al. (2018); Preindl et al. (2020)

Research Methodologies

The research methodologies adopted in the selected articles are mixed, qualitative and quantitative. Figure 5 shows that qualitative method is the widely used method that accounts for 72% of the articles. Quantitative method is adopted by 19% of the articles. Mixed method is only employed in 5% of the articles.

The research designs been employed in the selected articles are review, survey/ interview, experimental/model, case study and conceptual. Figure 6 shows that most of the articles are conceptual that accounts for 38% of total. Case studies are helpful in bringing new insights and observation in the study that accounts for 19%. Survey/ interview and experimental/model are equally adopted that is 17% each. Review articles are only 9% of the total.

Figure 5. Research methodologies

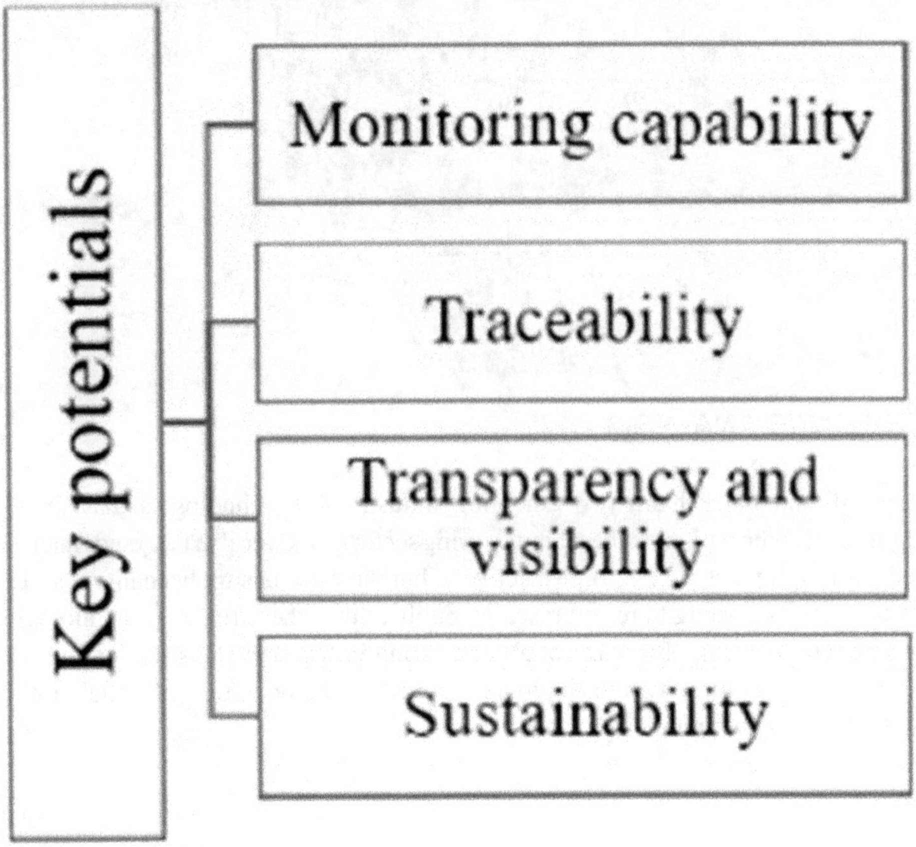

Figure 6. Research design

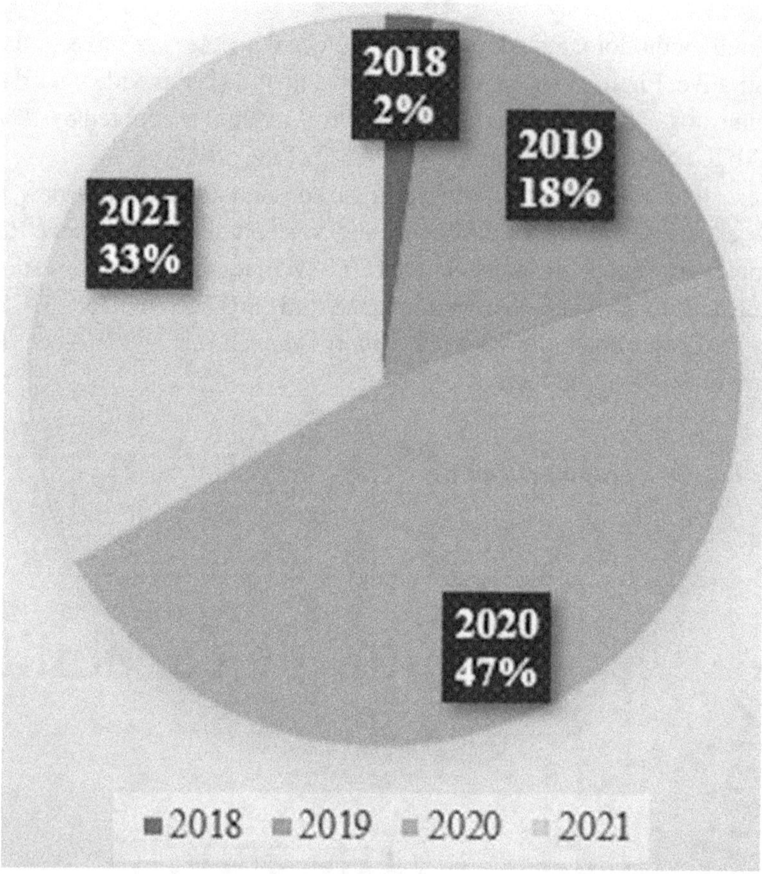

THEMATIC ANALYSIS

The IoT and blockchain, two emerging industry 4.0 technologies, have made significant contributions to the manufacturing sector (Sarkis et al., 2020) and financial industries (Dutta et al., 2020), respectively, but their use in supply chains is still in its infancy. According to recent research publications, these two new technologies can help in retail supply chain collaboration and integration. This section looks at ways to support these two technologies as well as the potentials of collaborating and integrating the retail supply chains.

Characteristics of Internet of Things (IoT) and Blockchain Technology in the Retail Supply Chain

Internet of Things

The Internet of Things (IoT) are sensor devices that uses wireless communication to link objects to the Internet (Gupta et al., 2020). IoT has ushered in a new IT revolution and elevated retail supply chain connectivity to a new level. IoT enhances retail supply chain's flexibility, visibility, and agility (Gupta et al., 2020).

Gupta et al. (2020) also went into great depth on how IoT manifests itself in supply chains. IoT can send data without human involvement, making the information transmission process much easier and supply chain more flexible. It also connects objects to the supply chain network directly through a variety of sensors, allowing staff and customers to view relevant product information and status on real time basis. This visibility helps in improving product visibility across the supply chain, increases effective decision-making, less dependability on each supply chain node, eliminates the uncertainty brought on by prediction, and increases the retail supply chain's reliability. The real-time data allows for more accurate judgments and precise control over the timing and quantity of resource distribution. The cost of the retail supply chain has also been substantially lowered because of efficient decision-making and appropriate resource allocation. Furthermore, the ability to respond to the changes has been improved because of lower operational expenses. These main five potentials of IoT are summarized in Figure 7.

Figure 7. Potentials of IoT in supply chains

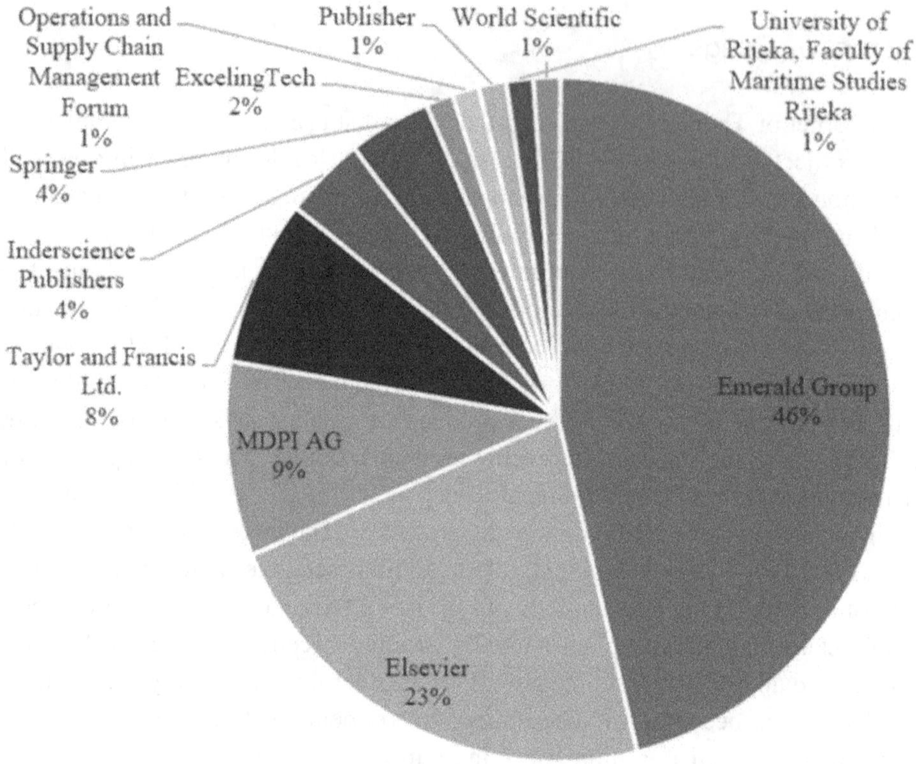

Blockchain

Blockchain is a novel technology that has recently been adopted in the field of retail supply chain. In blockchain the supply chain partner data is saved in the form of blocks and is shared in real- time across the supply chain to improve transparency (Sheel & Nath, 2020; Pilkington, 2016). The data stored in the block can form a chain, allowing the enterprise to track any data recorded on the block. The data cannot be tampered with or deleted, improving the security of information shared between partners, solving the trust problem, reducing the need for an intermediary, simplifying the operation process, and increasing efficiency (Sheel & Nath, 2020). Figure 8 summaries some of the blockchain's key achievements (Gupta et al., 2020).

To establish consensus, the entries on the blockchain will be collectively checked before adding the block containing new transaction information to the end of the chain (Efanov & Roschin, 2018). The transaction data and information on the blockchain cannot be tampered with, ensuring that the information viewed by partners is consistent and invariant. It necessitates the public ledger to record encrypted transactions. The distributed consensus algorithm makes it impossible to alter or corrupt. It is becoming more difficult for attackers to assault the blockchain as hash processing power improves (Aggarwal & Kumar, 2021). It is thought to be resistant to fraud and hacker assaults. The use of blockchain eliminates the requirement for an intermediary to guarantee transactions. Instead, each transaction is confirmed simultaneously by miners (Efanov & Roschin, 2018). Trading partners no longer need to trust each other, reducing the requirement for a third party, simplifying the transaction process, and saving time and money. Disintermediation is also aided by the smart contract, a blockchain function. The contract will be automatically executed by the blockchain when the criteria is met. Aside from the intermediary, all parties may make decisions at the same time, which saves time and reduce the error rate. For example, when inventory in the supply chain falls below a certain level, the system may automatically contact the supplier and make orders using the smart contract, eliminating the need for manual inventory monitoring and order placement. With a large range of items, this is extremely advantageous to the retail supply chain. The use of blockchain in the supply chain is revolutionary, since it saves a good deal of time and resources, streamlines numerous operational procedures, and encourages supply chain reorganization and innovation.

Figure 8. Performance of blockchain in supply chain

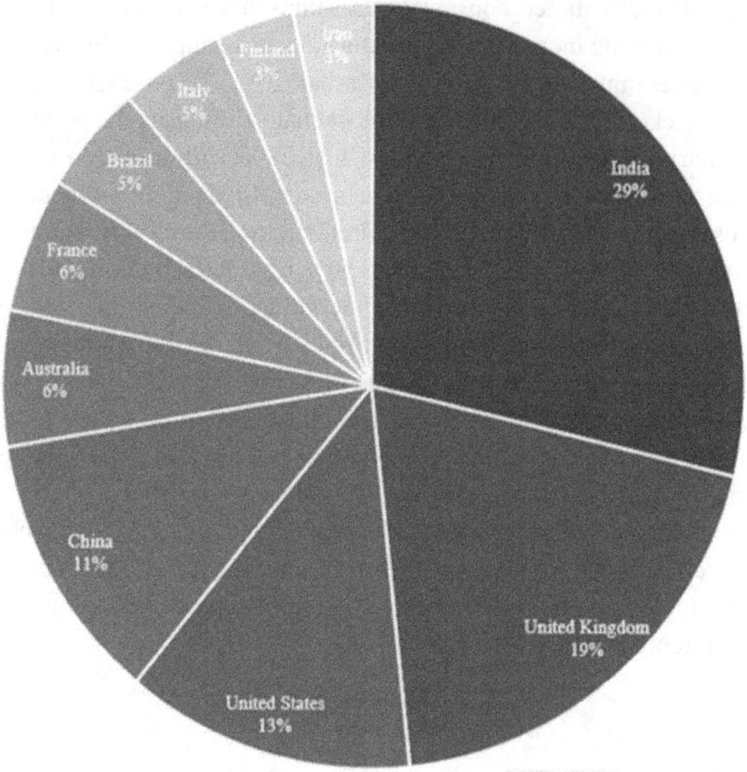

Potentials of Internet of Things (IoT) and Blockchain in Collaboration and Integration of the Retail Supply Chain

Collaboration With IoT and Blockchain

The strong competitive pressure in the retail industry requires retailers to continually squeeze earnings to stay in business, putting even more pressure on the retail supply chain to cut down the costs. Successful cooperation in the retail supply chain may substantially decrease costs, enhance supply chain efficiency, reduce product inventory, and improve service level (Francisco Luis et al., 2022).

Cooperation/collaboration is defined as the development of strategic capabilities over time in a long-term cooperative relationship between enterprises and retail supply chain partners (Nandi et al., 2020; Sanders & Premus, 2005). Collaboration may considerably increase the performance of retail supply chains, however the outcomes of cooperation among numerous retail companies are not significant.

The three main points of collaboration are process disintegration, shortage of resources and lack of trust (Figure 9): First, the enterprise management personnel's collaborative information has ambiguity in the process of transmission to the lower level, and the execution team has not received precise cooperation instructions. Second, the organization has too many tasks to fulfil and does not distribute resources appropriately, resulting in a shortage of resources throughout the cooperative process. Third, supply chain partners have little faith in the cooperative process and are hesitant to provide information, particularly private information. For example, the company's detailed product design drawings are unwillingly transferred to manufacturing partners, resulting in a lack of trust.

Figure 9. Collaboration issues

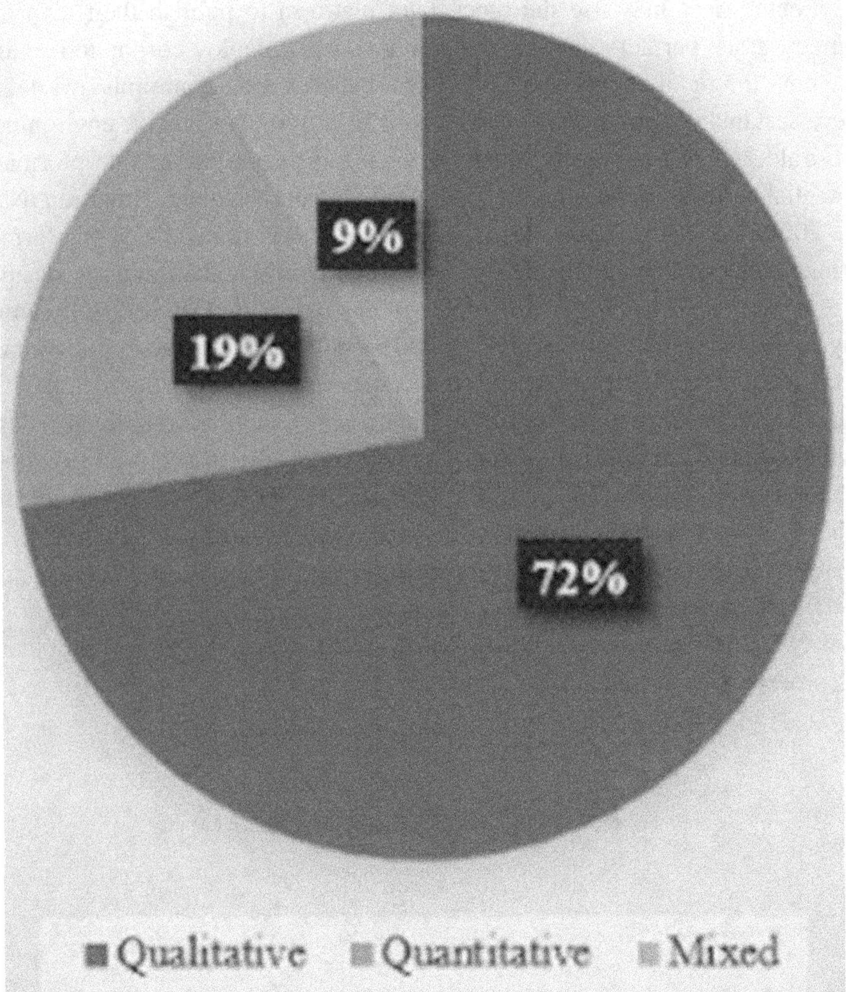

The retail supply chain's vertical and horizontal information exchange is enabled by IoT and blockchain. Each partner of supply chain have access to real-time data from other nodes, allowing them to enhance the accuracy of their decisions and the consistency of their behaviour. He et al. (2020) pointed out that IoT helps in increasing supply chain transparency and makes middle-level managers' and front-line workers' decisions more accurate. IoT accomplishes traceability by storing product data and information for retrieval by supply chain partners or customers in a short or long period of time. While the blockchain achieves traceability more effectively. The data saved on the blockchain cannot be changed or erased, allowing the whole process of product sourcing, design, manufacture, shipping, and sales in the retail supply chain to be traced. Problems can be tracked rapidly during operations, the reasons can be identified and handled, and unreasonable points can be remedied fast.

In the collaborative retail supply chain, blockchain have three distinct potentials. Firstly, enterprises may use the blockchain platform to publish their own extra resources, assist partners that are short on resources, deploy corporate resources throughout the retail supply chain, enhance resource usage, minimise waste, and achieve sustainable growth. Some businesses in extremely sensitive environments may use blockchain's information transmission security to provide private information to potential partners, increasing the likelihood of a successful collaboration (Nandi et al., 2020). Secondly, it provides a secure information transmission platform for existing partners, removing the need for a third-party middleman, and speeds up the business process. Thirdly, the trust problem may be handled immediately through the system based on blockchain for distrusted parties who have already collaborated, and cooperation activities between partners that can be expedited. Finally, for manufacturers whose products are difficult to transport, suppliers may transfer design drawings through a blockchain-enabled supply chain system and produce directly in manufacturing factory. This results in lowering transportation costs, saving time, and being environmentally friendly. Disintermediation also results in transactions that are not based on trust. This solves the challenge of supply chain management for retail companies that need to manage numerous suppliers, not only improving the safety factor of transactions and lowering transaction costs, but also automating some procedures and transactions using smart contracts.

Figure 10. Potentials in collaboration

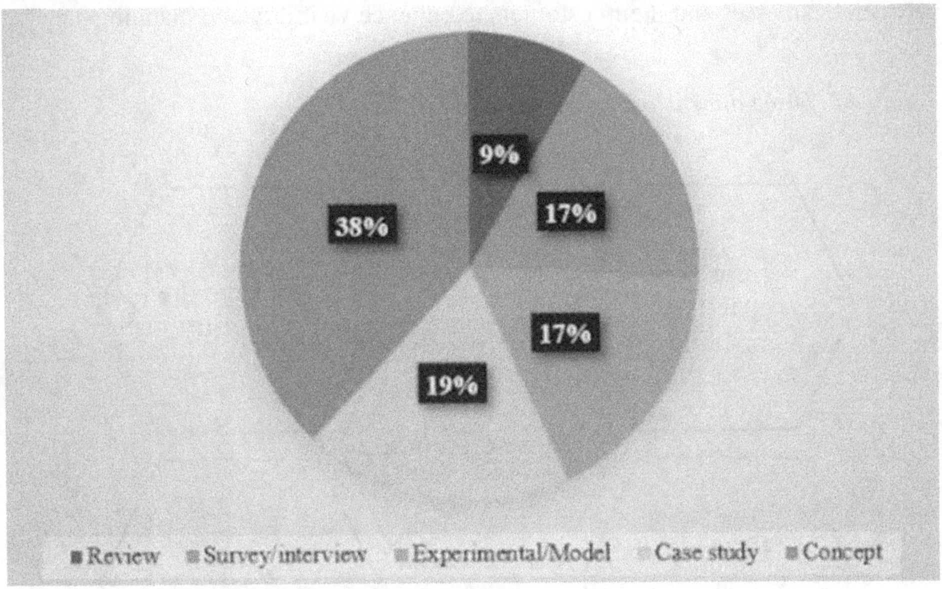

Integration With IoT and Blockchain

Low transportation costs and Internet have given retail businesses a global competitive advantage. The retail sector has evolved at a breakneck pace yet fulfilling customer requirements remains the primary objective. As a result, retail businesses must swiftly alter their operating modes to keep up with the times or risk being pushed out of the market. For retail businesses, end-to-end supply chain integration has become a must. Integration, according to BearingPoint (2018), is the key to attracting consumers.

An enterprise's integration capacity is a strategic partnership with its upstream and downstream retail supply chain partners. This capacity may be accomplished by connecting retail supply chain partners' direct information, financial, and physical flows (Nandi et al., 2020). For the retail supply chain, integration is a critical management aim. The integration is amongst, process, customer and supplier as shown in Figure 11 (Kamble et al., 2021; Ataseven & Nair, 2017; Alfalla-Luque et al., 2013; Näslund & Hulthen, 2012). The goal of process integration is to link the retail supply chain and operations to create a comprehensive system (Sutanto & Japutra, 2021). All employees have real-time access to the system's information and may follow and retrieve the supply chain data (Kamble et al., 2021; Ataseven & Nair, 2017). Supplier integration is a synchronised and controllable process between retail businesses and their numerous suppliers that provides a high level of cost efficiency and quick delivery (Flynn et al., 2010). Customer integration is the collaboration between businesses and their customers to enhance visibility and planning.

Figure 11. Three dimensions of integration

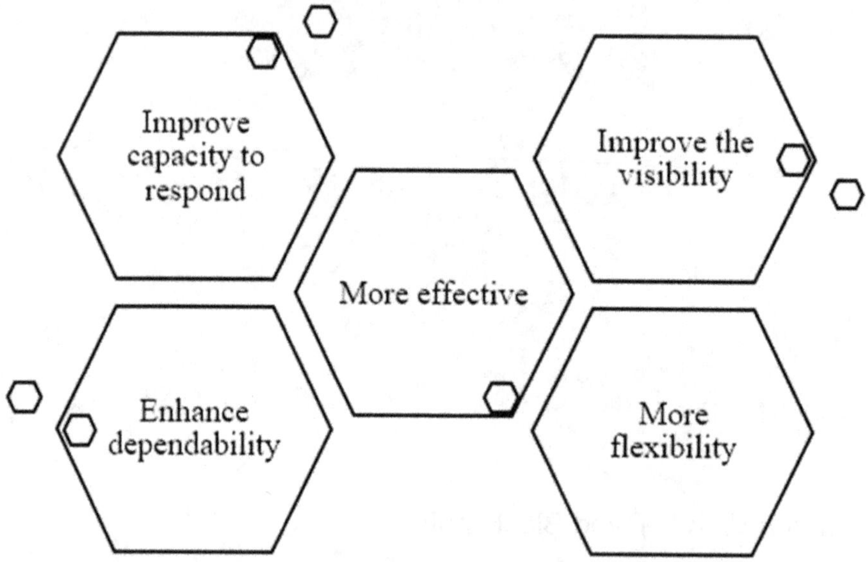

IoT and blockchain have significant potentials in terms of information transfer among retail supply chain partners as shown in Figure 12. The adoption of IoT and blockchain can fulfil the information transmission needs of each node of the supply chain, allowing end-to-end integration of the new retail supply chain. Furthermore, their capacity to synchronise information is a more stringent criterion for information sharing. When RFID-tagged products enter and exit each supply chain node, the flow information of commodities are automatically and simultaneously transmitted to the network system, allowing all participants to view the data.

Figure 12. Potentials of IoT and blockchain in integration of retail supply chain

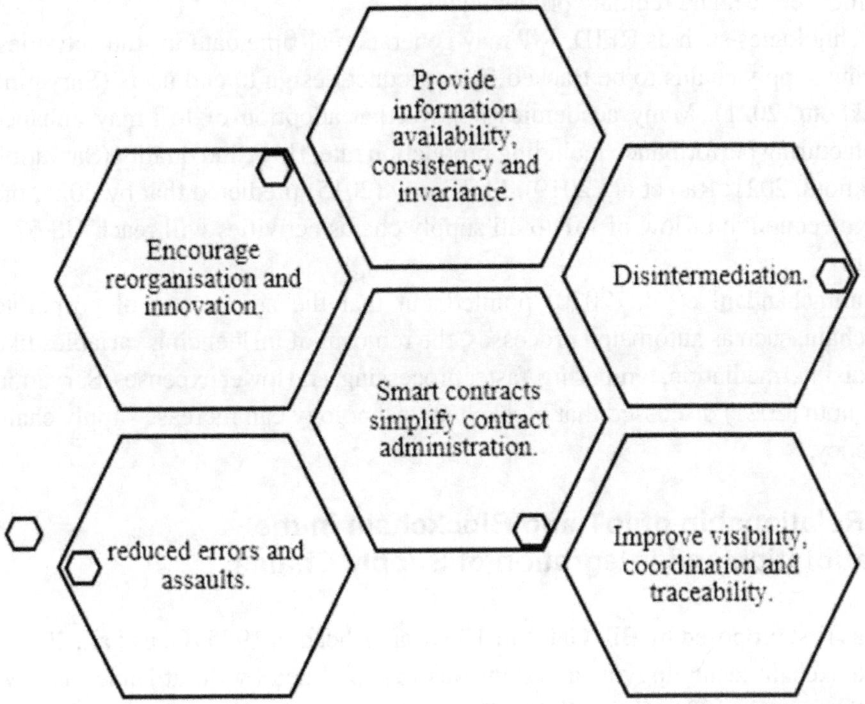

Blockchain realizes the need of supply chain information continuity because supply chain is not integrated by itself from beginning to end (Francisco Luis et al., 2022). IoT enables the retail supply chain to track product data in real time and to integrate and retrieve manufacturing, warehouse, transportation, and sales data in an integrated manner. Participants may use the system to get real-time product information and manage the whole retail supply chain. Improving reverse logistics contributes to customer integration as it helps in achieving visibility of product information during the purchase process (Deloitte, 2017).

DISCUSSION

This section determines the potential of the two new technologies i.e. IoT and blockchain. IoT and blockchain, may have played a revolutionary role in collaborating and integrating retail supply chain. Vodenicharova (2020) has discussed computer technology's significant impact on the integration of supply chains. The technology helps in increasing productivity to lower expenses, minimise time waste, improve customer service, and regulate product quality.

Technologies such as RFID, IoT may generate real-time data for the activities, allowing supply chains to be tracked from product design to end users (Saryatmo & Sukhotu, 2021). Many academics believe that adoption of IoT may enhance manufacturing performance including production rate, flow, and quality (Saryatmo & Sukhotu, 2021; Rao et al., 2019). McKinsey (2015) predicted that by 2025, the average economic inflow of IoT to all supply chains activities will reach US $7.5 trillion.

Karamchandani et al. (2020) pointed out that the advantages of corporate blockchain, such as automated processes, the removal of influencing variables like trust, disintermediation, tampering, faster processing, and lower expenses. Saryatmo & Sukhotu (2021) discussed that blockchain technology can increase supply chain efficiency.

The Relationship of IoT and Blockchain in the Collaboration and Integration of Supply Chains

IoT was first proposed by Bill Gates in The Road Ahead in 1995 (Gan et al., 2011), and blockchain technology as a system was first outlined by Surat Haber and W. Scott Stornetta in 1991 (Hyes, 2021). Both technologies have certain common key potentials to support retail supply chain collaboration and integration as shown in Figure 13.

Figure 13. The same key potentials of Internet of things and blockchain

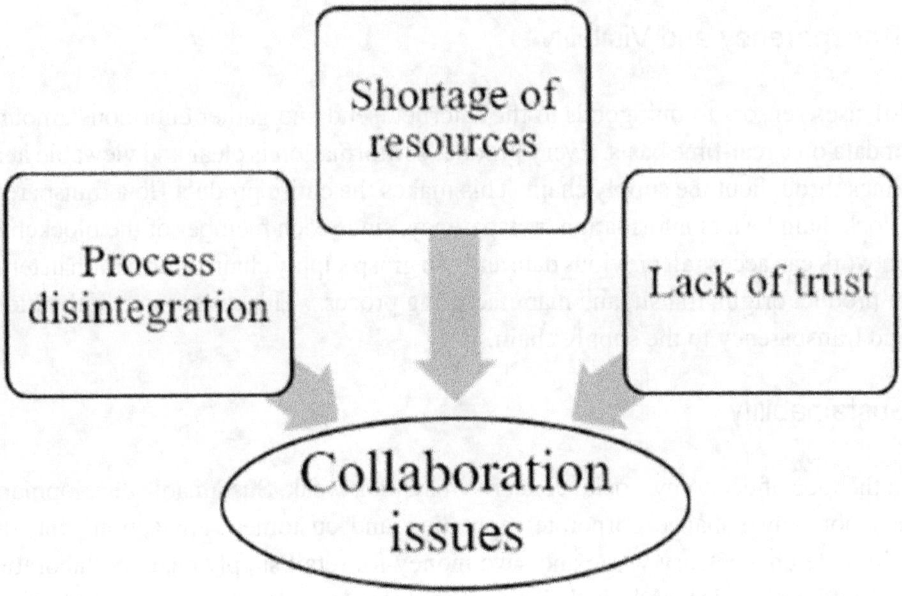

Monitoring Capability

IoT uses sensors to track the status of each object fulfilling different criteria, such as transportation, storage, and quality. The data is shared with supply chain members in real-time to monitor whether each stage of the supply chain is meeting predetermined requirements (Calatayud et al., 2018; Laskowski & Kim, 2016), such as cold chain product temperature requirements during transportation and storage, status of products on retail shelves etc. The relevant potential of the blockchain is reflected in storing of data. The regulatory authority may rapidly restore all data in the system if a node deliberately tampers with the data on the blockchain and no longer recognises the original monitoring data (Han et al., 2020). The system's security is enhanced by this regulatory capability.

Traceability

Individual goods are identified and intelligently marked using IoT technology to create traceability data. That data in IoT based systems is stored in centralized databases. The data on the blockchain is time sequenced. All of the data is stored in decentralized databases. All partners of the supply chain hold s a copy of the data that helps in achieving end-to-end traceability (Schmidt & Wagner, 2019).

Transparency and Visibility

IoT uses sensors to link goods to the Internet, send and gather enormous amounts of data on a real-time basis. Every product's information is clear and viewable at all times throughout the supply chain. This makes the entire product flow transparent. Blockchain fosters information transparency, since each member of the blockchain network can access all previous data and can grasp supply chain information in terms of product origin, transit, and manufacturing process. This gives more authenticity and transparency to the supply chain.

Sustainability

In the face of economy, social concerns may look weak. Sustainable developments can not only enhance corporate reputation and customer satisfaction, but also eliminate unnecessary waste and save money for retail supply chain collaboration and integration. IoT and blockchain technology connects the whole supply chain digitally, and information from the supply chain may well be shared throughout the network. This information is helpful in identifying the areas that needs improvement in meeting the sustainability goals of the organizations.

Hence, blockchain technology has the potential to revolutionize the organizations in making effective policies to address the challenges of collaboration and integration. Blockchain acts as a major milestone in creating an ecosystem which can not only collaborate and integrate complex supply chains but can also help in achieving the sustainability goals. The findings of this chapter could help the practitioners in enhancing business monitoring capability, traceability, transparency, visibility, and sustainability.

CONCLUSION

The globalization has extended the supply chains. This has resulted in technical and operational issues. One of the prominent one is not having effective collaboration and integration amongst the stakeholders of the supply chain. In order to address this, various Industry 4.0 technologies have been adopted by organizations such as Cyber physical systems (CPS), Internet of Things (IoT), Big data analytics (BDA), Cloud Computing technology and Blockchain (Kamble et al., 2020). These technologies have not only integrated the supply chains but have also facilitated the transmission of real-time information throughout the supply chains (Harris et al., 2015)

Internet of things (IoT) and blockchain technology are being explored in the chapter as a part of industry 4.0 technologies to enhance collaboration and integration of retail supply chains. This chapter summaries the characteristics and potentials of IoT and blockchain in achieving collaboration and integration in the retail supply chains. The findings of the study listed the key potentials such as monitoring capability, traceability, transparency, visibility and sustainability. These potentials could be useful for decision-makers and practitioners in making effective polices and strategies to address the collaboration and integration concerns of the extended supply chains.

THE LIMITATIONS, IMPLICATIONS, AND FUTURE RESEARCH

The findings of the study must be seen in the light of limitations. Firstly, as the topic is new so there are limited number of articles available on the topic. The ranking of the selected articles is not specific. Secondly, the selection of the articles is based on panel expertise which could bring some biasness in the selection of articles.

The literature discusses that IoT and blockchain model was first proposed in 2017 (Di Pietro et al., 2018), most scholars have studied the application of these two technologies separately in the supply chain context. The application of these technologies in the retail business is quite limited or insufficient. Scholars can empirically investigate the businesses and their relevant supply chains that employ these technologies in the future. The advent of new technologically focussed retail has presented small and medium-sized businesses with several opportunities and problems. It is expected that adoption of latest Industry 4.0 technologies will facilitate in effective collaboration and integration of enterprises. The implications of these technologies will be on enhancing business monitoring capability, traceability, transparency, visibility and sustainability. Karamchandani et al., (2020) summarised the benefits of technologies in achieving integration across the supply chains. The issues and obstacles that retail businesses face into their industrial structures and to what extent and what type of retail enterprises are suitable for the application of these two technologies, need to be explored by future retail enterprises and scholars.

REFERENCES

Aamer, A. M., Al-Awlaqi, M. A., Affia, I., Arumsari, S., & Mandahawi, N. (2021). The internet of things in the food supply chain: Adoption challenges. *Benchmarking*, *28*(8), 2521–2541. doi:10.1108/BIJ-07-2020-0371

Aggarwal, S., & Kumar, N. (2021). Chapter Twenty—Attacks on blockchain ☆☆ Working model. In S. Aggarwal, N. Kumar, & P. Raj (Eds.), *Advances in Computers* (Vol. 121, pp. 399–410). Elsevier. doi:10.1016/bs.adcom.2020.08.020

Aich, S., Chakraborty, S., Sain, M., Lee, H., & Kim, H.-C. (2019). A Review on Benefits of IoT Integrated Blockchain based Supply Chain Management Implementations across Different Sectors with Case Study. *2019 21st International Conference on Advanced Communication Technology (ICACT)*, 138–141. 10.23919/ICACT.2019.8701910

Alfalla-Luque, R., Medina-Lopez, C., & Schrage, H. (2013). A study of supply chain integration in the aeronautics sector. *Production Planning and Control, 24*(8–9), 769–784. doi:10.1080/09537287.2012.666868

Ali, S. I., Ali, A., AlKilabi, M., & Christie, M. (2021). Optimal supply chain design with product family: A cloud-based framework with real-time data consideration. *Computers & Operations Research, 126*, 105112. doi:10.1016/j.cor.2020.105112

Ataseven, C., & Nair, A. (2017). Assessment of supply chain integration and performance relationships: A meta-analytic investigation of the literature. *International Journal of Production Economics, 185*, 252–265. doi:10.1016/j.ijpe.2017.01.007

Banerjee, A. (2019). Chapter Nine - Blockchain with IOT: Applications and use cases for a new paradigm of supply chain driving efficiency and cost. In S. Kim, G. C. Deka, & P. Zhang (Eds.), *Advances in Computers* (Vol. 115, pp. 259–292). Elsevier. doi:10.1016/bs.adcom.2019.07.007

BearingPoint. (2018). *Integrating the retail supply chain*. BearingPoint. https://www.bearingpoint.com/files/Retail_Research_Report_2017_BE.pdf?download=0&itemId=502221#:~:text=Retailers%20who%20identify%20as%20having,reduced%20stockholding%20and%20improved%20availability%20

Ben-Daya, M., As'ad, R., & Nabi, K. A. (2019). A single-vendor multi-buyer production remanufacturing inventory system under a centralized consignment arrangement. *Computers & Industrial Engineering, 135*, 10–27. doi:10.1016/j.cie.2019.05.032

Cai, Y.-J., & Lo, C. K. Y. (2020). Omni-channel management in the new retailing era: A systematic review and future research agenda. *International Journal of Production Economics, 229*, 107729. doi:10.1016/j.ijpe.2020.107729

Calatayud, A., Mangan, J., & Christopher, M. (2018). The self-thinking supply chain. *Supply Chain Management, 24*(1), 22–38. doi:10.1108/SCM-03-2018-0136

Caro, F., & Sadr, R. (2019). The Internet of Things (IoT) in retail: Bridging supply and demand. *Business Horizons*, *62*(1), 47–54. doi:10.1016/j.bushor.2018.08.002

Chen, J., Cai, T., He, W., Chen, L., Zhao, G., Zou, W., & Guo, L. (2020). A Blockchain-Driven Supply Chain Finance Application for Auto Retail Industry. *Entropy (Basel, Switzerland)*, *22*(1), 95. doi:10.3390/e22010095 PMID:33285870

Choi, T.-M. (2020). Supply chain financing using blockchain: Impacts on supply chains selling fashionable products. *Annals of Operations Research*. Advance online publication. doi:10.100710479-020-03615-7

Choi, T.-M., & Luo, S. (2019). Data quality challenges for sustainable fashion supply chain operations in emerging markets: Roles of blockchain, government sponsors and environment taxes. *Transportation Research Part E, Logistics and Transportation Review*, *131*, 139–152. doi:10.1016/j.tre.2019.09.019

Chung, M. J.-Y., Friesen, A. L., Fox, D., Meltzoff, A. N., & Rao, R. P. N. (2015). A Bayesian developmental approach to robotic goal-based imitation learning. *PLoS One*, *10*(11), e0141965. doi:10.1371/journal.pone.0141965 PMID:26536366

Cochrane Library. (2022). *Aout Cochrane Reviews*. https://www.cochranelibrary.com/about/about-cochrane-reviews

Cui, L., Gao, M., Dai, J., & Mou, J. (2020). Improving supply chain collaboration through operational excellence approaches: An IoT perspective. *Industrial Management & Data Systems*. doi:10.1108/IMDS-01-2020-0016

de Vass, T., Shee, H., & Miah, S. J. (2021). Iot in supply chain management: A narrative on retail sector sustainability. *International Journal of Logistics*, *24*(6), 605–624. doi:10.1080/13675567.2020.1787970

Deloitte. (2017). *Continuous interconnected supply chain*. https://www2.deloitte.com/content/dam/Deloitte/pt/Documents/blockchainsupplychain/lu-blockchain-internet-things-supply-chain-traceability%20(1).pdf

Di Pietro, R., Salleras, X., Signorini, M., & Waisbard, E. (2018). A blockchain-based Trust System for the Internet of Things. *Proceedings of the 23nd ACM on Symposium on Access Control Models and Technologies*, 77–83. 10.1145/3205977.3205993

Dutta, P., Choi, T.-M., Somani, S., & Butala, R. (2020). Blockchain technology in supply chain operations: Applications, challenges and research opportunities. *Transportation Research Part E, Logistics and Transportation Review*, *142*, 102067. doi:10.1016/j.tre.2020.102067 PMID:33013183

Dwivedi, Y. K., Hughes, L., Ismagilova, E., Aarts, G., Coombs, C., Crick, T., Duan, Y., Dwivedi, R., Edwards, J., Eirug, A., Galanos, V., Ilavarasan, P. V., Janssen, M., Jones, P., Kar, A. K., Kizgin, H., Kronemann, B., Lal, B., Lucini, B., ... Williams, M. D. (2021). Artificial Intelligence (AI): Multidisciplinary perspectives on emerging challenges, opportunities, and agenda for research, practice and policy. *International Journal of Information Management*, *57*, 101994. doi:10.1016/j.ijinfomgt.2019.08.002

Efanov, D., & Roschin, P. (2018). The All-Pervasiveness of the Blockchain Technology. *Procedia Computer Science*, *123*, 116–121. doi:10.1016/j.procs.2018.01.019

Esmaeilpour Ghouchani, B., Jodaki, S., Joudaki, M., Balali, A., & Rajabion, L. (2019). A model for examining the role of the Internet of Things in the development of e-business. *VINE Journal of Information and Knowledge Management Systems*, *50*(1), 20–33. doi:10.1108/VJIKMS-04-2019-0058

Flynn, B. B., Huo, B., & Zhao, X. (2010). The impact of supply chain integration on performance: A contingency and configuration approach. *Journal of Operations Management*, *28*(1), 58–71. doi:10.1016/j.jom.2009.06.001

Fosso Wamba, S., Kala Kamdjoug, J. R., Epie Bawack, R., & Keogh, J. G. (2020). Bitcoin, Blockchain and Fintech: A systematic review and case studies in the supply chain. *Production Planning and Control*, *31*(2–3), 115–142. doi:10.1080/09537287.2019.1631460

Francisco Luis, B.-M., Pedro Víctor, N.-C.-U., Valentín, M.-M., & Esteban, R.-F. (2022). Blockchain as a Service: A Holistic Approach to Traceability in the Circular Economy. In S. S. Muthu (Ed.), *Blockchain Technologies for Sustainability* (pp. 119–133). Springer. doi:10.1007/978-981-16-6301-7_6

Gan, G., Lu, Z., & Jiang, J. (2011). Internet of Things Security Analysis. *2011 International Conference on Internet Technology and Applications*, 1–4. 10.1109/ITAP.2011.6006307

Ganbold, O., Matsui, Y., & Rotaru, K. (2020). Effect of information technology-enabled supply chain integration on firm's operational performance. *Journal of Enterprise Information Management*, *34*(3), 948–989. doi:10.1108/JEIM-10-2019-0332

Gupta, H., Kumar, S., Kusi-Sarpong, S., Jabbour, C. J. C., & Agyemang, M. (2020). Enablers to supply chain performance on the basis of digitization technologies. *Industrial Management & Data Systems*, *121*(9), 1915–1938. doi:10.1108/IMDS-07-2020-0421

Hallikas, J., Immonen, M., & Brax, S. (2021). Digitalizing procurement: The impact of data analytics on supply chain performance. *Supply Chain Management*, *26*(5), 629–646. doi:10.1108/SCM-05-2020-0201

Han, Y., Li, C., Li, P., Wu, M., Zhou, D., & Long, F. (2020). Shrec: Bandwidth-efficient transaction relay in high-throughput blockchain systems. *Proceedings of the 11th ACM Symposium on Cloud Computing*, 238–252. 10.1145/3419111.3421283

Harris, I., Wang, Y., & Wang, H. (2015). ICT in multimodal transport and technological trends: Unleashing potential for the future. *International Journal of Production Economics*, *159*, 88–103. doi:10.1016/j.ijpe.2014.09.005

He, L., Xue, M., & Gu, B. (2020). Internet-of-things enabled supply chain planning and coordination with big data services: Certain theoretic implications. *Journal of Management Science and Engineering*, *5*(1), 1–22. doi:10.1016/j.jmse.2020.03.002

Hofmann, E., & Rutschmann, E. (2018). Big data analytics and demand forecasting in supply chains: A conceptual analysis. *International Journal of Logistics Management*, *29*(2), 739–766. doi:10.1108/IJLM-04-2017-0088

Hyes, A. (2021). *Blockchain Explained*. Investopedia. https://www.investopedia.com/terms/b/blockchain.asp

Joseph Jerome, J. J., Saxena, D., Sonwaney, V., & Foropon, C. (2021). Procurement 4.0 to the rescue: Catalysing its adoption by modelling the challenges. *Benchmarking: An International Journal*. doi:10.1108/BIJ-01-2021-0030

Jović, M., Tijan, E., Perić Hadžić, A., & Karanikić, P. (2020). Economic aspects of automation innovations in electronic transportation management systems. *Pomorstvo*, *34*(2), 417–427. doi:10.31217/p.34.2.22

Kamble, S. S., Gunasekaran, A., & Gawankar, S. A. (2020). Achieving sustainable performance in a data-driven agriculture supply chain: A review for research and applications. *International Journal of Production Economics*, *219*, 179–194. doi:10.1016/j.ijpe.2019.05.022

Kamble, S. S., Gunasekaran, A., Parekh, H., & Joshi, S. (2019). Modeling the internet of things adoption barriers in food retail supply chains. *Journal of Retailing and Consumer Services*, *48*, 154–168. doi:10.1016/j.jretconser.2019.02.020

Kamble, S. S., Gunasekaran, A., Subramanian, N., Ghadge, A., Belhadi, A., & Venkatesh, M. (2021). Blockchain technology's impact on supply chain integration and sustainable supply chain performance: Evidence from the automotive industry. *Annals of Operations Research*. Advance online publication. doi:10.100710479-021-04129-6

Karamchandani, A., Srivastava, S. K., & Srivastava, R. K. (2020). Perception-based model for analyzing the impact of enterprise blockchain adoption on SCM in the Indian service industry. *International Journal of Information Management*, *52*, 102019. doi:10.1016/j.ijinfomgt.2019.10.004

Kayikci, Y., Subramanian, N., Dora, M., & Bhatia, M. S. (2020). Food supply chain in the era of Industry 4.0: Blockchain technology implementation opportunities and impediments from the perspective of people, process, performance, and technology. *Production Planning and Control*, *0*(0), 1–21. doi:10.1080/09537287.2020.1810757

Khan, S. A., Mubarik, M. S., Kusi-Sarpong, S., Gupta, H., Zaman, S. I., & Mubarik, M. (2022). Blockchain technologies as enablers of supply chain mapping for sustainable supply chains. *Business Strategy and the Environment*, *31*(8), 1–15. doi:10.1002/bse.3029

Koutsos, T. M., Menexes, G. C., & Dordas, C. A. (2019). An efficient framework for conducting systematic literature reviews in agricultural sciences. *The Science of the Total Environment*, *682*, 106–117. doi:10.1016/j.scitotenv.2019.04.354 PMID:31108265

Laskowski, M., & Kim, H. M. (2016). Rapid Prototyping of a Text Mining Application for Cryptocurrency Market Intelligence. *2016 IEEE 17th International Conference on Information Reuse and Integration (IRI)*, 448–453. 10.1109/IRI.2016.66

Li, M., Shao, S., Ye, Q., Xu, G., & Huang, G. Q. (2020). Blockchain-enabled logistics finance execution platform for capital-constrained E-commerce retail. *Robotics and Computer-integrated Manufacturing*, *65*, 101962. doi:10.1016/j.rcim.2020.101962

Liu, W., Zhang, J., & Wang, S. (2021). Factors influencing the smart supply chain innovation performance of commodity distribution enterprises: An investigation from China. *Industrial Management & Data Systems*, *121*(10), 2073–2099. doi:10.1108/IMDS-12-2020-0753

Mahroof, K. (2019). A human-centric perspective exploring the readiness towards smart warehousing: The case of a large retail distribution warehouse. *International Journal of Information Management*, *45*, 176–190. doi:10.1016/j.ijinfomgt.2018.11.008

Mangina, E., Narasimhan, P. K., Saffari, M., & Vlachos, I. (2020). Data analytics for sustainable global supply chains. *Journal of Cleaner Production*, *255*, 120300. doi:10.1016/j.jclepro.2020.120300

Miraz, M. H., Hassan, M. G., & Sharif, K. I. M. (2020). Factors Affecting Implementation of Blockchain in Retail Market in Malaysia. *Inetranational Journal of Supply Chain Management*, *9*(1), 385–391.

Nandi, M. L., Nandi, S., Moya, H., & Kaynak, H. (2020). Blockchain technology-enabled supply chain systems and supply chain performance: A resource-based view. *Supply Chain Management*, *25*(6), 841–862. doi:10.1108/SCM-12-2019-0444

Näslund, D., & Hulthen, H. (2012). Supply chain management integration: A critical analysis. *Benchmarking*, *19*(4/5), 481–501. doi:10.1108/14635771211257963

Pilkington, M. (2016). Blockchain technology: Principles and applications. *Research Handbook on Digital Transformations*. https://www.elgaronline.com/view/edcoll/9781784717759/9781784717759.00019.xml

Preindl, R., Nikolopoulos, K., & Litsiou, K. (2020). Transformation strategies for the supply chain: The impact of industry 4.0 and digital transformation. *Supply Chain Forum: An International Journal, 21*(1), 26–34. 10.1080/16258312.2020.1716633

Queiroz, M. M., Pereira, S. C. F., Telles, R., & Machado, M. C. (2019). Industry 4.0 and digital supply chain capabilities: A framework for understanding digitalisation challenges and opportunities. *Benchmarking*, *28*(5), 1761–1782. doi:10.1108/BIJ-12-2018-0435

Rao, S., Ellis, S. C., Goldsby, T. J., & Raju, D. (2019). On the "Invisible Inventory Conundrum" in RFID-Equipped Supply Chains: A Data Science Approach to Assessing Tag Performance. *Journal of Business Logistics*, *40*(4), 339–358. doi:10.1111/jbl.12232

Raut, R. D., Gotmare, A., Narkhede, B. E., Govindarajan, U. H., & Bokade, S. U. (2020). Enabling Technologies for Industry 4.0 Manufacturing and Supply Chain: Concepts, Current Status, and Adoption Challenges. *IEEE Engineering Management Review*, *48*(2), 83–102. doi:10.1109/EMR.2020.2987884

Rejeb, A., Keogh, J. G., Simske, S. J., Stafford, T., & Treiblmaier, H. (2021). Potentials of blockchain technologies for supply chain collaboration: A conceptual framework. *International Journal of Logistics Management*, *32*(3), 973–994. doi:10.1108/IJLM-02-2020-0098

Roy, V. (2021). Contrasting supply chain traceability and supply chain visibility: Are they interchangeable? *International Journal of Logistics Management*, *32*(3), 942–972. doi:10.1108/IJLM-05-2020-0214

Sanders, N. R., & Premus, R. (2005). Modeling the Relationship Between Firm It Capability, Collaboration, and Performance. *Journal of Business Logistics*, *26*(1), 1–23. doi:10.1002/j.2158-1592.2005.tb00192.x

Sarkis, J., Kouhizadeh, M., & Zhu, Q. S. (2020). Digitalization and the greening of supply chains. *Industrial Management & Data Systems*, *121*(1), 65–85. doi:10.1108/IMDS-08-2020-0450

Saryatmo, M. A., & Sukhotu, V. (2021). The Influence of the Digital Supply Chain on Operational Performance: A Study of the Food and Beverage Industry in Indonesia. *Sustainability (Basel)*, *13*(9), 5109. doi:10.3390u13095109

Schmidt, C. G., & Wagner, S. M. (2019). Blockchain and supply chain relations: A transaction cost theory perspective. *Journal of Purchasing and Supply Management*, *25*(4), 100552. doi:10.1016/j.pursup.2019.100552

Shabani-Naeeni, F., & Ghasemy Yaghin, R. (2021). Incorporating data quality into a multi-product procurement planning under risk. *Journal of Business and Industrial Marketing*, *36*(7), 1176–1190. doi:10.1108/JBIM-02-2020-0108

Sharma, M., & Joshi, S. (2020). Digital supplier selection reinforcing supply chain quality management systems to enhance firm's performance. *The TQM Journal*. doi:10.1108/TQM-07-2020-0160

Sheel, A., & Nath, V. (2020). Blockchain technology adoption in the supply chain (UTAUT2 with risk) – evidence from Indian supply chains. *International Journal of Applied Management Science*, *12*(4), 324–346. doi:10.1504/IJAMS.2020.110344

Srai, J. S., & Lorentz, H. (2019). Developing design principles for the digitalisation of purchasing and supply management. *Journal of Purchasing and Supply Management*, *25*(1), 78–98. doi:10.1016/j.pursup.2018.07.001

Sutanto, J. E., & Japutra, A. (2021). The Impact of Supply Chain Integration and Trust on Supply Chain Performance: Evidence from Indonesia Retail Sector. *International Journal of Economics & Business Administration*, *IX*(1), 211–224. doi:10.35808/ijeba/668

Tan, B., Yan, J., Chen, S., & Liu, X. (2018). The Impact of Blockchain on Food Supply Chain: The Case of Walmart. In M. Qiu (Ed.), *Smart Blockchain* (pp. 167–177). Springer International Publishing. doi:10.1007/978-3-030-05764-0_18

Tian, F. (2017). A supply chain traceability system for food safety based on HACCP, blockchain amp; Internet of things. *2017 International Conference on Service Systems and Service Management*, 1–6. 10.1109/ICSSSM.2017.7996119

Tönnissen, S., & Teuteberg, F. (2020). Analysing the impact of blockchain-technology for operations and supply chain management: An explanatory model drawn from multiple case studies. *International Journal of Information Management*, *52*, 101953. doi:10.1016/j.ijinfomgt.2019.05.009

Tranfield, D., Denyer, D., & Smart, P. (2003). Towards a Methodology for Developing Evidence-Informed Management Knowledge by Means of Systematic Review. *British Journal of Management*, *14*(3), 207–222. doi:10.1111/1467-8551.00375

Tripathi, S., & Gupta, M. (2020). Transforming towards a smarter supply chain. *International Journal of Logistics Systems and Management*, *36*(3), 319–342. doi:10.1504/IJLSM.2020.108694

Ts, D., & v, R.TS. (2020). An integrated ANP–QFD approach for prioritization of customer and design requirements for digitalization in an electronic supply chain. *Benchmarking*, *28*(4), 1213–1246. doi:10.1108/BIJ-06-2020-0298

van Hoek, R. (2019). Exploring blockchain implementation in the supply chain: Learning from pioneers and RFID research. *International Journal of Operations & Production Management, 39*(6/7/8), 829–859. doi:10.1108/IJOPM-01-2019-0022

Varriale, V., Cammarano, A., Michelino, F., & Caputo, M. (2021). Sustainable Supply Chains with Blockchain, IoT and RFID: A Simulation on Order Management. *Sustainability (Basel)*, *13*(11), 6372. doi:10.3390u13116372

Vodenicharova, M. S. (2020). Supply chain study in food industry in Bulgaria. *International Journal of Retail & Distribution Management*, *48*(9), 921–938. doi:10.1108/IJRDM-03-2019-0080

Wamba, S. F., Gunasekaran, A., Akter, S., Ren, S. J., Dubey, R., & Childe, S. J. (2017). Big data analytics and firm performance: Effects of dynamic capabilities. *Journal of Business Research*, *70*, 356–365. doi:10.1016/j.jbusres.2016.08.009

Wamba, S. F., & Queiroz, M. M. (2020). Industry 4.0 and the supply chain digitalisation: A blockchain diffusion perspective. *Production Planning and Control*, *0*(0), 1–18. doi:10.1080/09537287.2020.1810756

Wong, L.-W., Tan, G. W.-H., Lee, V.-H., Ooi, K.-B., & Sohal, A. (2020). Unearthing the determinants of Blockchain adoption in supply chain management. *International Journal of Production Research*, *58*(7), 2100–2123. doi:10.1080/00207543.2020 .1730463

Xiao, Y., & Watson, M. (2019). Guidance on Conducting a Systematic Literature Review. *Journal of Planning Education and Research*, *39*(1), 93–112. doi:10.1177/0739456X17723971

Zeng, F., Chan, H. K., & Pawar, K. (2021). The effects of inter- and intraorganizational factors on the adoption of electronic booking systems in the maritime supply chain. *International Journal of Production Economics*, *236*, 108119. doi:10.1016/j.ijpe.2021.108119

Zhong, H., & Nof, S. Y. (2015). The dynamic lines of collaboration model: Collaborative disruption response in cyber–physical systems. *Computers & Industrial Engineering*, *87*, 370–382. doi:10.1016/j.cie.2015.05.019

Chapter 10
Internal Audit Functions in Cyber Security Governance:
Turkey's Bank Sector Case

Seval Kardeş Selimoğlu
Anadolu University, Turkey

Mustafa Hakan Saldı
Anadolu University, Turkey

ABSTRACT

Cyber-attacks have been brought into the critical risk category with rising technology usage and emerging developments in digital transformation that have been dominating service sectors like banking. The assurance of the validity of the activities, which are being coordinated in cyber security governance mechanisms of the banks, is needed to ensure the financial transactions are being driven between acceptable risk levels. The control of risk management operations for cyber security processes should be performed with the support of audit units of the banks. Because of the forementioned reasons, the responsibility fields of internal audit teams in cyber security management processes are deeply examined through the Delphi technique in this chapter, which is derived from the doctoral thesis research study. On this basis, blockchain technology is going to be explored from the point of technical concepts and policy framework to propose a set of solutions for continuous audit methods in cyber security governance.

DOI: 10.4018/978-1-6684-4153-4.ch010

Copyright © 2023, IGI Global. Copying or distributing in print or electronic forms without written permission of IGI Global is prohibited.

INTRODUCTION

The banking operations have been involving high risks due to the common nature of the finance industry which heavily covers funding transactions to provide asset protection, relocation, and allocation operations of the institutional and retail investors. In this context, the account activities of the persons have been fundamentally being processed through technological tools, applications, and digitally connected networks for nearly the last four decades. Cyber attacks, which target to critical infrastructures of the nations, have been rising because of the tightening economic conditions in the globe and adversary goals of hackers for stealing critical information and knowledge assets, which are kept in databases and warehouses of the governmental authorities, public and private enterprises, are named as data theft operations also known as a class of cyber risks. Likewise, the threats in cyberspace, which have been advancing with the innovations in software and programming industries, have significant effects on banking processes that have been being functioned in this uncertain environment that is tried to be controlled through digital tools, applications, and connections. Especially, the banking industry, which is considered as a part of critical infrastructure, has been mostly being affected by cyber risks because of the monetary operations which are being performed in cyberspace through information systems that cover databases, computers, communication, and network systems. In this context, the conventional role of internal audit has been evolving with technology trends in terms of supporting risk management and control processes in information systems to assure effective cyber security governance in corporations specifically in banks where almost all of the business processes are being sustained through electronic and communication systems with information technology types of equipment and tools. According to these reasons, this article is designed to present the research methodology which was utilized in the doctoral thesis study to show the activities of internal auditors in cyber security governance processes of the banks in Turkey through the usage of the delphi technique. In connection with the research problem and goal, at first, the scholar intends to clarify the question marks about the fundamental terms that are related to the internal audit, cyber security, and blockchain in terms of setting the applicative part which is edited on the delphi technique and exploratory sequential mixed methods research design.

Fundamentally, an audit can be defined as a comprehensive and autonomous examination of the enterprise activities such as financing, marketing, producing, transporting, hiring, firing, and information technology processing which are performed through methodical ways with the contributions of the licensed experts who are recognized as auditors (Kumar & Sharma, 2013, p. 2). In general, audit activities can be categorized according to the way it's made as internal and external. In this context, internal audit is defined as counseling activities that include systematic and scientific approaches for the evaluations and enhancement of validity and efficiency of control processes in risk management and governance-related operations through the own employees of the organizations with objective style (Gantz, 2014, p. 45). In practice, the interaction between risk assessment, internal control, and governance can be optimized in terms of determining the risk limitations, corporate management principles, and legislative framework with the supportive role of internal auditors (Pickett, 2011). The risk term is described as unanticipated throughputs of happenings that can either tolerably or intolerably affect the organizational objectives. As a result, the unsymmetrical behaviors of the expected outputs can be shown as representations for measuring the risk (ISACA, 2012). In accordance with the subject of the research project, risks in electronic and communication systems can be brought to tolerable levels through planning, acting, controlling, and taking precautions consistently with organizational culture, international standards, and regulatory environment (Olson & Wu, 2015).

The assurance of confidentiality, integrity, and availability requirements for the organizational assets in digital space can be defined as cyber security (Ryder & Madhavan, 2019). Therefore, applicable defensive systems are needed to control the validity of sensitive assets' protection by structuring appropriate risk management strategies in terms of acting promptly on time manner for threats and vulnerabilities which cause cyber risks. In essence, the main goal of cyber security infrastructures is to provide maximum safety for nations' and corporations' strategic data sources and critical properties mostly with preventive approaches (Westby, 2004). Cyber threats can be explained as events or conditions which arise in the cyber ecosphere and cause losses and harmful effects to the business goals of the corporations. Also, the uncertain nature of technology management puts extra pressure on the complexities in risk management methods for the information systems protection in terms of setting the cyber threats and organizational vulnerabilities for ensuring maximum cyber hygiene inside the organizations. Accordingly, the concerns in cyberspace can be sorted into two classes firstly as ordinary technical malfunctions such as system issues which usually emerge inadvertently and have harmed operational parts of the organizations, and secondly as malicious attempts which are planned and organized strategically to trigger negative impacts for all the parts of business functions (Progrebna & Skilton, 2019).

Cyber security awareness can be maximized with traditional applications as threat matrices, heat maps, and severity schedulings to identify the occurrence probability of cyber risks and their potential impacts on critical assets (Schreider, 2017). From the point of theoretical framework, the weak points of an infrastructure that are related to cyber security can be considered through stochastic approaches with simulations and scenarios-based events in terms of understanding the risk perception of an organization for cyber attacks. Nevertheless, these methods typically may not be useful for factual conditions in the matter of estimating the coincidence moment of organizational vulnerabilities and cyber threats (Progrebna & Skilton, 2019).

The major root causes of cyber risks can be considered from the standpoints of organizational awareness and culture. The control environment can be improved with educational programs in connection with the vulnerabilities caused by the internal parts of the enterprises. However, criminal justice comes into force if there is an attack that expresses itself deliberately with a planned fashion to hurt the brand image of a firm. Based on this mindset, the legislations and international standards have been being developed and framed for the highest protection of corporations to inhibit hacking acts (Jain, 2005).

In addition to these aspects, the workforce who controls the cyber security processes should have both strong qualifications in technical subjects and robust managerial relations in terms of proposing future solutions for information security. In previous times, technicians in enterprises were accepted as alternatives to information technology experts, who have been performing in data security part to control cyber hygiene. However, this logical stance in businesses has been being modified to a much extensive mindset because of the rising strategic importance of protection for information-related assets. Therefore, the development of cyber security culture has been beginning to come ahead for spreading organizational awareness into all levels of business functions (Tipton & Krause, 2007).

Educational requirements can be defined according to the sectors of the organizations in terms of providing much feasible information security functioning. Thus, technical training programs can be used to enrich both the personal and functional skills of the employees of the corporations. For instance, firewalls can be employed to monitor network interchanges either permitting or halting the actions which are being executed through electronic and communication systems due to the set of security principles. Furthermore, software patches can be carried out to organizations' information technology infrastructure to overhaul bugs, integrate new computer program packages, review new security breaches, and upgrade operating systems. As a result, these kinds of precautions can be supported with the tutoring programs as mentioned before for the continuous improvement of cyberculture in the organizations (Penuel & Statler, 2013).

Cyber security has been evolving as a significant subject for not only nations but also enterprises. Unfortunately, they have been facing data theft issues, which cause both reputation loss and financial damage because of the cyber attacks, many of them are still carrying out outdated approaches. Due to the absence of security updates and the lack of emerging technology adaptations, hackers can much comfortably penetrate the information systems. As a matter of digitalization, most of the electronics and communication infrastructures are being controlled through developing technologies like industry 4.0, smart city, artificial intelligence, and big data applications that trigger the higher occurrence probability for cyber risks. Therefore, the audit operations, which are being performed both internally and externally, should be reconsidered from the point of cyber security, to measure the validity of risk management level of organizations (Selimoğlu & Altunel, 2019).

Despite that many companies are aware of the cyber attacks and security issues they will face, they are inadequate in precautions which are being taken against cyber risks because the actions that are taken can not be fully standardized and the effects of the destruction that are caused by these attacks on their information systems can not be repaired agilely in a short period of time (Maleh, Baddi, Alazab, Tawalbeh, & Romdhani, 2021). In addition, while most businesses keep their security systems ready for external cyberattacks, only a small number of them have a defense mechanism that is safe from malicious attacks that may be caused by their own employees (Groenfeldt, 2014). Correspondingly, the deliberate and unintentional threats can be minimized by the applications of emerging technology tools such as blockchain. In this context, the unique infrastructure of blockchain technology can be considered to eliminate the non-value added activities and loss of hours in control operations that are being traditionally performed in cyber security and information technology audits. However, there are some question marks in the minds like the subjects from secrecy in finance to controlling personal data transmitting as what kind of fresh instruments are being presented through the distributed ledger technology and how the reinforcements should be implemented into the policy framework in terms of adapting the enterprises to this innovation (OECD, 2019). As a result, the proposals are going to be presented in the conclusion part of the project by doing the swot analysis of blockchain technology in respect of internal audit and cyber security governance.

RESEARCH OUTLINE

Firstly, this research article is framed to project the application of the conceptual structure of thesis study in terms of summarizing the practice phase which was performed through the delphi technique with three rounds to gather qualitative data from panel experts by semi-structured interviews with open-ended questions for the first round and a questionnaire for second and third rounds.

Figure 1. The road map of research plan

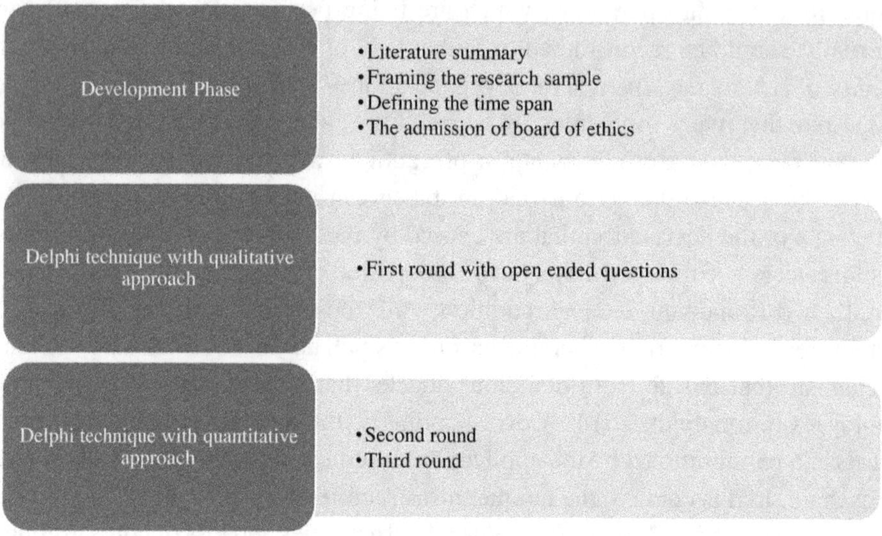

First of all, a critical literature review is made to deeply investigate the past practices of the delphi method in parallel with the thesis subject in terms of understanding the usage fields and application types of the method for adapting this technique to the research study. After that, sampling method and size are defined to cluster the panel members by filtering out the banking sector population. Later on, the time limits of the research study are determined for particularly restricting the data collection period in terms of managing the duration of the delphi practice. Finally, the admission is taken from the board of ethics to prove that the research study is going to be performed in compliance with research ethics.

Literature Review

In this stage, the previous academic studies are investigated by focusing on the research methodology of the thesis which was predefined as the delphi technique. Therefore, the literature review is made critically by identifying purposeful and meaningful research studies which are contextualizing and most related to this project. Particularly, two types of researches, which had been named as the delphi and analysis of the future: the delphi method, are analyzed much comprehensively to discover the fundamental principles and application processes of this method. In this direction, the delphi technique had been stated as a process for aiming to estimate the forthcoming events through gathering the data from persons who have expert knowledge and experience about a research subject.

Table 1. Background articles about Delphi technique

Author and Year	Article Name	Subject and Scope	Findings
Dalkey (1967)	Delphi	An outline of the delphi methodology had been explained.	Delphi method had been introduced for long-term forecasting through filtering the opinions of a category of advisers or experts with the statistical calculations of the group responses.
Helmer (1967)	Analysis of the Future: The Delphi Method	Delphi methodology had been described with basic principles.	Delphi technique could be used to attempt the processing of informed intuitional judgments for long-term predictions.
Kurubacak (2007)	Identify Research Priorities and Needs for Mobile Learning Technologies in Open and Distance Education: A Delphi Study	Delphi technique was applied to demonstrate the identification, categorization, and classification of the research opportunities for mobile learning technologies in open and distance education.	Briefly, public responsibility, management of online society, and specialists, who are engaging in digital transformation, are determined as the major research fields in the paper subject.
Kurubacak (2011)	E-learning for Pluralism: The Culture of E-learning in Building a Knowledge Society	The main objective of this research was to determine and classify the upcoming forces and needs for e-learning.	Delphi method was used with three rounds for gathering data from twenty-eight e-learning experts who evaluated major problematics and criticisms in this subject.
Omari, Barnes, and Pitman (2012)	An Exploratory Study into Audit Challenges in IT Governance: A Delphi Approach	Examining the IT governance in the Australian public sector with an empirical approach by using the delphi survey.	Ten major IT governance audit concerns were presented according to the delphi study.
Vasarheyli, Lombardi, ansd Bloch (2014)	The Future of Audit: A Modified Delphi Approach	Discussing the future of the audit profession through modified delphi technique with gathering the consensus of expert opinions which provide foresight in audit methods, standards, and mindsets.	Forecastings were proposed by the author for future functions of audit as database management, sampling, extensible markup language processing, and clustering.
Davidson and Hasledalen (2014)	Cyber Threats to Online Education: A Delphi Study	To investigate the cyber threats and vulnerabilities which could damage online learning systems by applying an e-Delphi survey with the classical design of the delphi methodology.	The leadership was noticed in terms of understanding the cyber security risks and their costs as a general solution.
Daniel Smits, Jos Van Hillegersberg(2015)	IT Governance Maturity: Developing a Maturity Model Using the Delphi Method	A description of developing an information technology (IT) governance maturity model was proposed with both the hard part which covers structural processes and soft elements that include organizational culture and behavior.	One maturity model was developed for the hard side of information technology governance, on the flip side, the soft part requires more specific competencies for each focus field.
Karabacak, Yıldırım, and Baykal (2016)	Regulatory Approaches for Cyber Security of Critical Infrastructures: The Case of Turkey	First, the critical infrastructures data of Turkey were analyzed qualitatively with the grounded theory method, second, the delphi survey was applied with six experts to derive regulations, and finally, a focus group interview was carried out.	Particularly, the critical infrastructure employees of Turkey which includes privately held contractors are mainly in favor of compliance.
Haqaf and Koyuncu (2018)	Understanding Key Skills for Information Security Managers	To investigate the major capabilities required for the information security management role with the adoption of the delphi technique.	Project and risk management were determined as two major categories from the sixteen skills that were accepted as the key classes for information security management by the implementation of the delphi methodology.

Continued on following page

Table 1. Continued

Author and Year	Article Name	Subject and Scope	Findings
Altınpulluk, Kesim and Kurubacak (2020)	The Usability of Augmented Reality in Open and Distance Learning Systems: A Qualitative Delphi Study	The primary goal of this research is to define the functionality of augmented reality in open and distance learning ecosystems in compliance with international design laws for predicting the outlook through taking the opinions of experts with Delphi methodology.	Delphi technique was carried out to fourteen experts with three rounds by using a qualitative questionnaire that covers open-ended questions as a primary data collection tool and a six-point Likert scale questionnaire as a secondary data gathering method.

The Definition of Variables and Sample

Relative to the theoretical part of the thesis study, the mechanisms and interactions in cyber security infrastructures between governance, risk, and compliance (GRC) constraints are deeply explored through factual examples and research applications. In this context, the proofs of concepts have shown that legal and international policy frameworks, corporate management principles, ethical rules, and audit functions can be filtered out as research variables for minimizing the complexity in the discovery period of the relationship between GRC elements to focus on the objectives of internal auditors in the managerial structures of cybersecurity-related processes.

Figure 2. The representative illustration of research variables

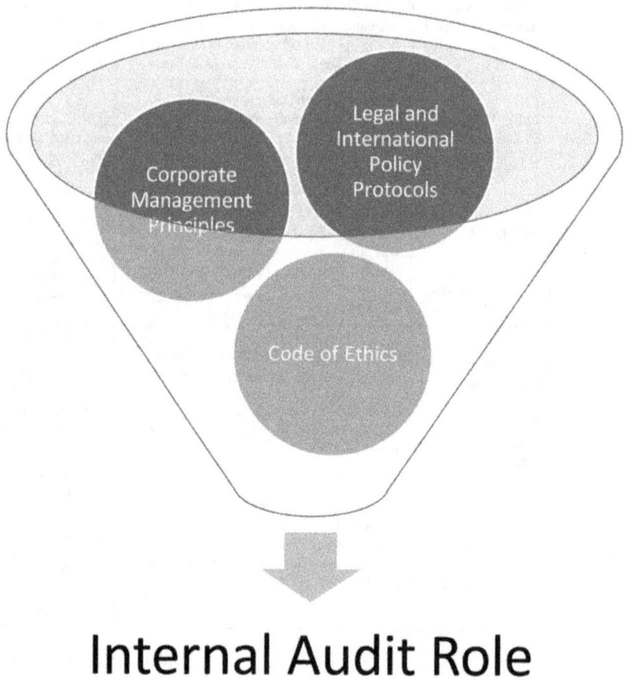

Therefore, internal auditors' functions in cyber security governance activities are explored under the spotlight of the variables which are determined such that legal and international policy protocols as the independent variable, corporate management principles and code of ethics as mediators, and internal audit role as the dependent variable. Correspondingly, the presumption is made as the standardization of cyber security governance processes is mainly linked with legal and international policy environment conditions which affect the corporate management principles and ethical conducts of the enterprises. In connection with this hypothesis, the missions of internal audit teams are broken down from the point of these elements which are thoroughly implemented to open-ended questions that are used in the first round of the delphi way to reach opinions of managers, experts, and academics.

In this study, purposeful sampling is used because it is a method that allows the researcher to select professionally experienced and knowledgeable persons intentionally by following the content of research with providing flexibility to the scholar regardless of the type of qualitative research method that will be followed. In like manner, the validity of a qualitative research study is measured by the level at which the profiles of participants have deep and extensive commands in the research subject rather than sample amount and size (Conway, 2020).

For the reasons mentioned above, it is aimed to reach the opinions of managers, team leaders, experts, relevant persons in law enforcement agencies, and academicians who have researched on this study subject, from the population which is determined as banking sector as sample space with performing the purposeful sampling method. Therefore, sampling is made from the institutions which are clustered as public, private, foreign capital, deposits, and participation from the main population that is framed as the bank sector of Turkey. In particular, it has been tried to include the sampling of banks where cyber risks are more likely to be encountered and the frequency of internet banking is high. The sample size is quantitatively limited with fifteen participants by aiming to reach their feedback relative to the research problem.

Figure 3. Framed research sample

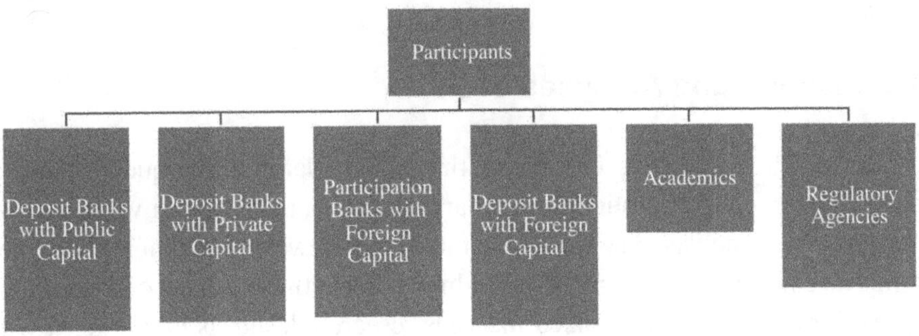

As a result, panel members are selected from the framed set of banks, academics, and regulatory institutions through the purposeful sampling method to pay much attention to the research problem.

The Assumptions and Limitations

The scholar supposed that

- The panelists, who participate in the research study are approved as recognized qualified professionals who have competency, practice, and experience in the research subject.
- The panelists, who provided feedback to open-ended questions and questionnaires, presented their ideas factually and without any partiality.
- The data, which are collected through the delphi methodology, are spanned by the control variables that were defined to get points in the direction of the research subject and goal.

The research paper is restricted with;

- The individual ideas and verdicts of participants who attended the delphi study.
- Qualitative data, which are gathered through open-ended questions and questionnaires are analyzed by both qualitative and quantitative approaches.
- The participants, who said yes to attending this scientific study from the banks, are primarily actualizing their careers in information technology audit, internal control, and cybersecurity-related disciplines.
- The academics, who accepted to participate in this study, have tutoring experiences in academia about engineering, internal audit, finance, cyber security, and management.
- The data are received through online interviews, electronic mails, and online questionnaires in the time range between 1 March 2021 and 31 December 2021.

The Collection and Analyses of Data

The data collection process is carried out through the delphi technique for the first round with open-ended questions, which are set across the research variables, are addressed to the panelists via a semi-structured interview method that is performed by online discussions and electronic mails. In parallel with the subject of the research study, exploratory sequential mixed methods research design is harmonized with the delphi technique for executing the process of the road map. Therefore, firstly, the qualitative stage of research is processed for the exploration of the data which are achieved through semi-structured interviews using open-ended questions. In this direction, the audiovisual records and textual documents are transformed into the pointed information in terms of structuring the second phase that is planned as the quantitative part of the study. As a result, the qualitative part is performed to design the tools and to sort the parameters which are going to show the way of following steps for the quantitative stage of research (Creswell, 2014, p. 16).

Fundamentally, the critical criteria can be recognized by the application of the pareto analysis, which has been used for identifying the focal causes that mostly affect the complexity, provides benefits to the researcher while sorting the qualitative data (Joiner Associates Incorporated, 1995, pp. 4, 16). Therefore, the pareto charts are carried out as visual instruments to analytically show the main standpoints of the collected data from the first round of the delphi technique, that are analyzed and classified in the Nvivo program.

The major responsibilities of internal auditors are categorized through seven activities with the densification frequencies of the evaluations of panelists in the following pareto graph. In particular, the distribution of the panelist responses demonstrates that control and compliance-related processes with reporting function are predominantly stated as primary activities of internal auditors in cyber security management and information technology control.

Figure 4. Responsibility

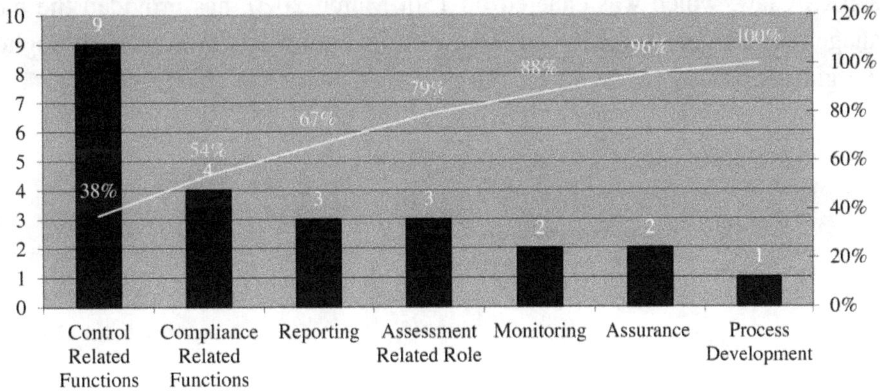

Technical knowledge, educations, and international licenses are commonly stated as needed priorities for auditors according to the discussions of panelists for the second question which is prepared to comprehend the competency elements in cyber security governance. Especially, the panelists' responses intensely showed that internal auditors have to improve their technical sufficiencies such as database processing, system and network management with basic software knowledge to monitor how the risk controls are being performed in information technology governance.

Figure 5. Competency

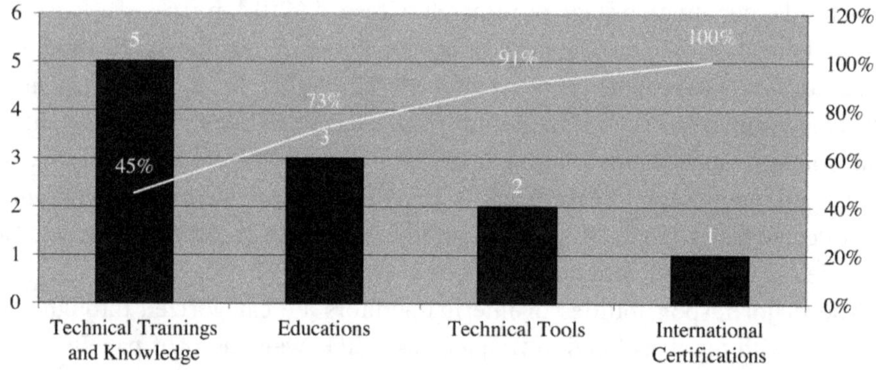

The evaluations of panel members showed that the internal auditors have been performing as review consultants who are exactly affiliated to the general manager with having the responsibility to report their observations about risk management processes in cyber security governance. Particularly, their responses proved that the BRSA law, which was enacted on 15th March 2020, has provided the bank managements to deploy the information security culture and awareness throughout the organizations.

Figure 6. Governance

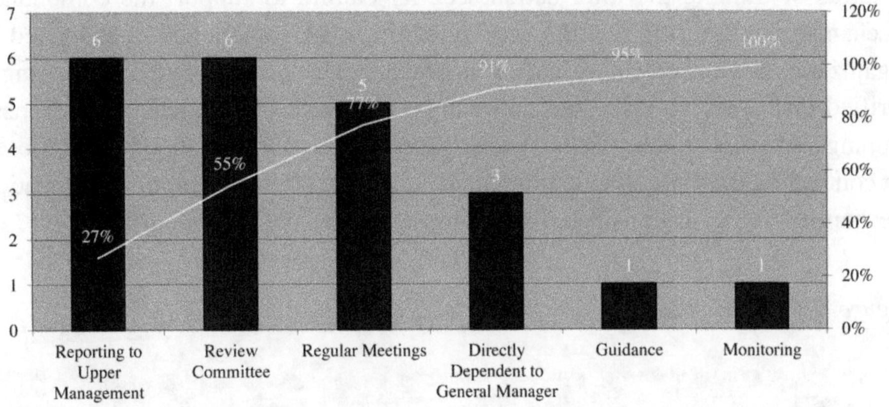

The responses of participants showed that the new BRSA law has positive effects to establish a cyber security culture in banks, from the perspective of protection and functioning mechanisms of confidentiality, integrity, and availability (CIA triad) of information systems related assets. Similarly, they emphasized that the internal auditors have a role in monitoring the compliance degree of the banks' cyber security management approaches relative to COBIT and ISO 27001 to assure the CIA triad of customers' data. Also, the authorization concept, which is commonly used as a standard while ensuring data accessibility and transfer, is foremostly highlighted for the CIA triad case by the panel members.

Figure 7. CIA triad

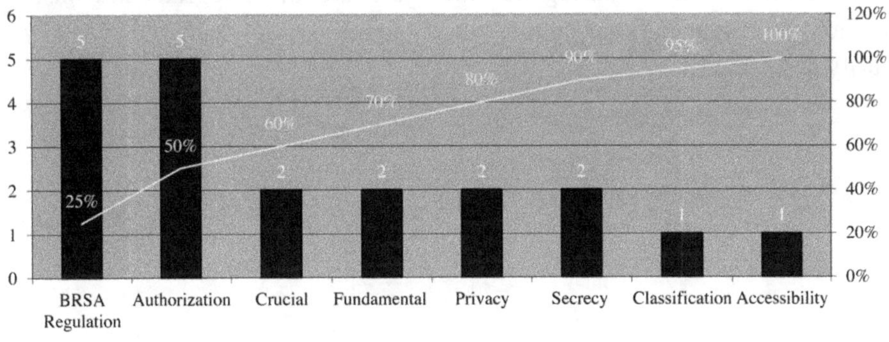

Ethical rules and moral behaviors are considered as an essential set of principles in terms of framing the information security culture to support the compliance mechanism for assuring that the validity of risk management activities related to organizational vulnerabilities and cyber threats. The panelist responses strongly verified that cyber security breaches can be minimized by organizing awareness training and event simulations. Especially, continuous audit methods are proposed for controlling the internal environment to assure the effectiveness of authorization mechanisms which is considered as the most vital part of data security.

Figure 8. Code of ethics

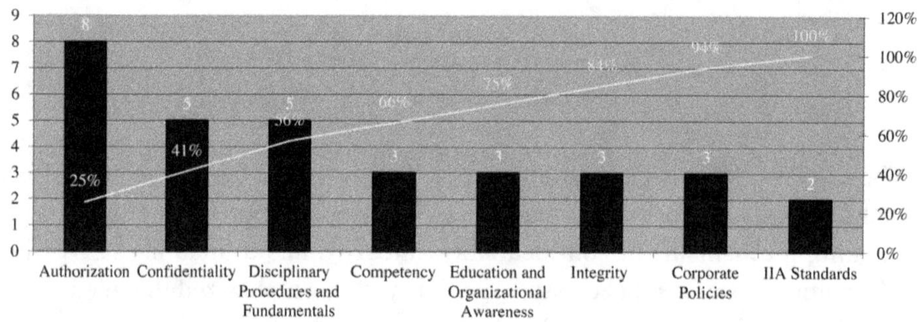

The role of internal auditors is mostly specified as reviewing and reporting the suspicious conditions which are connected with the violation of corporate management principles and code of ethics from the point of information security. The participants also stated that the internal audit committees have tasks in risk assessment and recording processes by considering BRSA law and COBIT standards into consideration while assuring the ethical rules are fully integrated into business operations and organizational culture not just to information technology and cyber security departments.

Figure 9. Corporate management principles

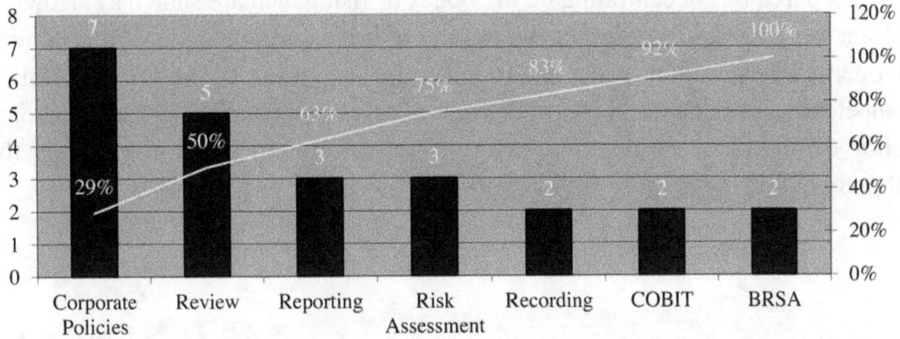

The answers of panelists showed that monitoring, compliance controls, and reporting can be evaluated as the most featured functions which internal auditors involve in cyber security governance processes for controlling the adequacy level of banks to BRSA law, COBIT, and ISO 27001.

Figure 10. Regulatory and international policy environment

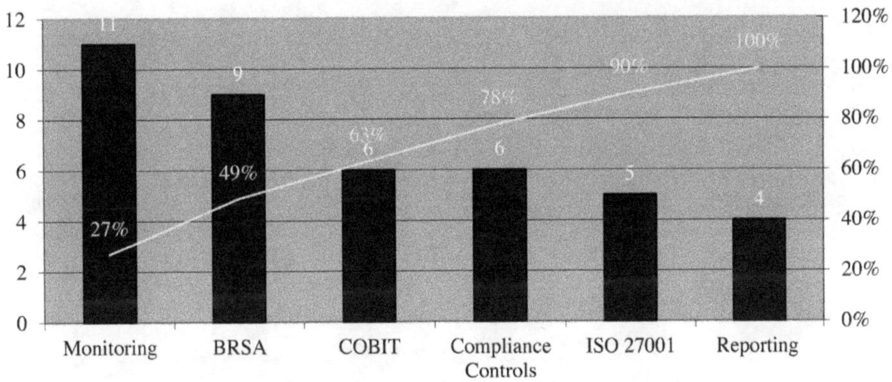

Fishbone schema is used to visualize in an organized manner for the critical pinpoints which are defined in the analyses of the first round of the Delphi through the Nvivo program and the pareto charts. Especially, the fundamental objective of applying this drawing is to pay much attention to the root causes of the research problem in terms of clustering the subfactors and their interrelations with subtitles (Joiner Associates, 1995, p. 15). Therefore, the fishbone diagram is used to outline the meaningful data which are derived from the pareto charts in the qualitative analyses part of the research. As a result, the questionnaire statements are going to be prepared by concentrating on the bones of fish which are shown as arrows to present the generic classes and main objects of brainstorming in panel discussions for open-ended questions in the first round of the Delphi study. Thus, responsibility, competency, governance, CIA triad, code of ethics, legislative and international policy environment are determined as major concepts in terms of covering the subparts within the bounds of this diagram.

Figure 11. Fishbone diagram

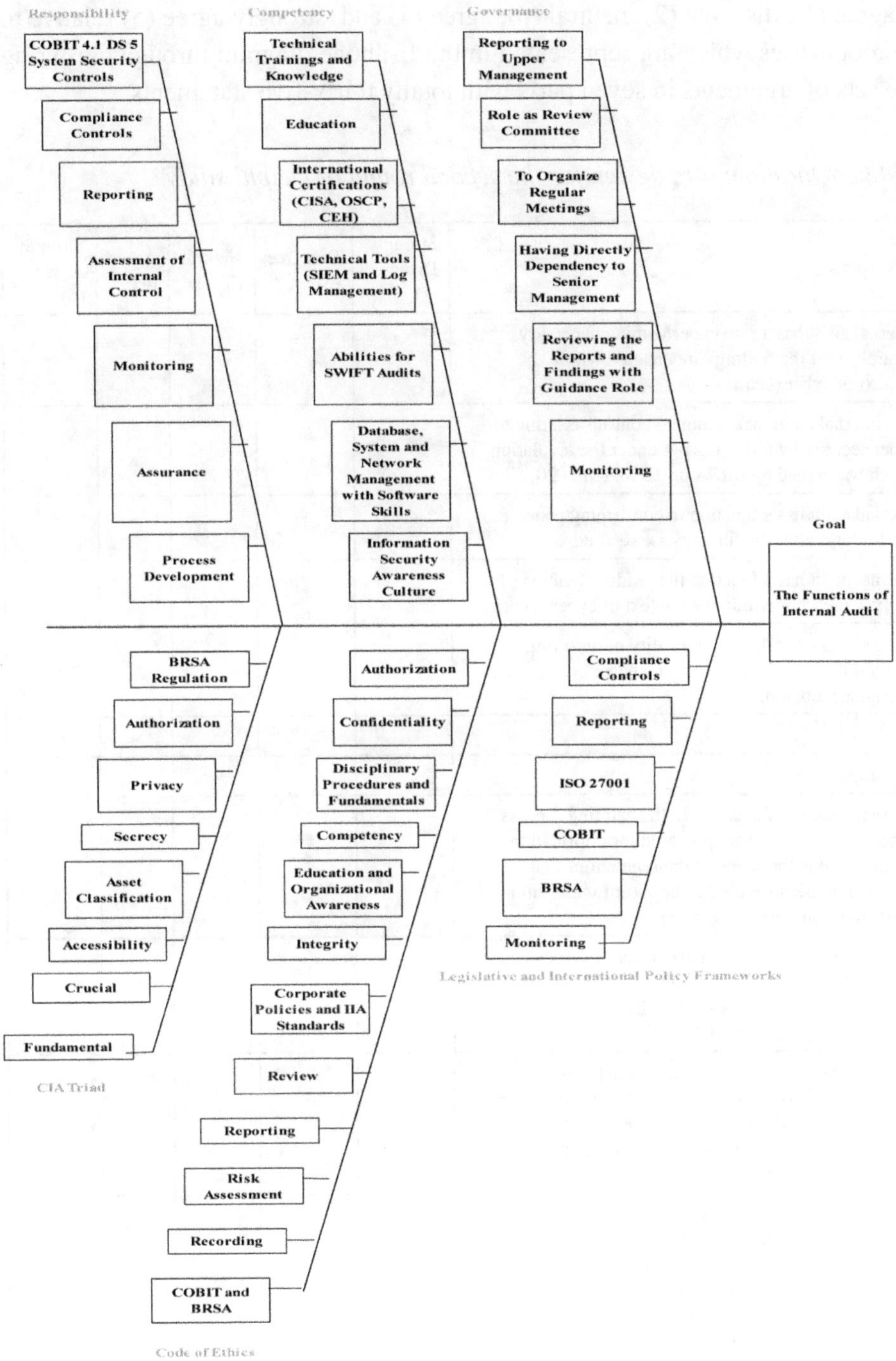

The questionnaire is formed by using the five-point Likert scale as strongly disagree (1), disagree (2), neutral (3), agree (4) and, strongly agree (5) relative to the root causes which are represented in the fishbone diagram through clustering five sets of arguments in seven parts with totally thirty-five statements.

Table 2. Questionnaire design for the second round of Delphi study

1. Responsibility	Strongly Disagree (1)	Disagree (2)	Neutral (3)	Agree (4)	Strongly Agree (5)
Internal audit has a role in performing necessary controls after the findings are ensured within the context of cyber security.					
The internal audit makes annual controls relating to cyber security by following per under the regulation which was issued by BRSA on 15 March 2020.					
Internal audit has a function in confirming penetration test findings after the findings are secured.					
Internal audit has a function in the identification of security breaches within the context of cyber security.					
Internal audit has the responsibility of reporting findings directly to the general manager or management board.					
2. Competency					
Internal audit should carry out the auditing process of control lists which are prepared for information security within the scope of the regulations that are designed by BRSA with the support of information technology auditors.					
Internal audit must have a basic level of software knowledge to maintain the audit activities which are carried out by BRSA in compliance with the information systems regulation.					
Internal audit must have system and network management knowledge to be effective in the process of controlling and conforming to the penetration test findings which are carried out by independent audit firms after they are ensured.					
Internal audit must have technical level knowledge and sufficient experience on information security architecture and infrastructure to be effective in the auditing process of control lists that are prepared for information security within the scope of the regulations that are designed by BRSA.					

Continued on following page

Table 2. Continued

Internal audit should be a master of using SIEM tools to take an active role in monitoring and surveillance activities within the scope of cyber security.					
3. Governance					
Internal audit has a consultancy role in the governance between the board of directors, general manager, and information technology department.					
Required training events should be organized for internal auditors who are performing specifically in cyber security processes.					
Internal audit must have direct authority for cyber security controls.					
Internal audit has a supportive position in cyber security controls with its advisory role.					
Internal audit has the responsibility for informing the senior management necessarily with regular interval periods in cases related to cyber security.					
4. Confidentiality, Integrity, and Availability					
Internal audit must record, maintain and report audit evidence in compliance with the principles of confidentiality, integrity, and availability when taking part in cyber security processes.					
Internal audit should carry out its activities within the scope of cyber security according to the principle of authority.					
Internal audit should ensure that the access of the findings which are obtained as a result of its activities within the scope of cyber security is realized without disorganization.					
Internal audit should provide assurance that the findings of its activities within the scope of cyber security haven't been modified or altered in any way.					
Internal audits should assure that the findings of cyber security controls are not being transmitted to unauthorized persons.					
5. Corporate Management Principles					
Internal audit should follow the legal environment and provide the necessary reportings while performing its functions within the scope of cyber security.					
Internal audit must act in accordance with the professional ethics of being not involved in any of the illegal actions while sustaining its activities within the scope of cyber security.					

Continued on following page

Table 2. Continued

Internal audit should carry out its operations in a manner that respects the ethical principles and management principles of the corporation when performing its activities within the scope of cyber security.					
Internal audit teams must engage tasks in line with their knowledge, competencies, and experiences when they are taking part in cybersecurity-related processes.					
Internal audit teams must continuously improve their capabilities, effectiveness, and quality level of service consistent with emerging technologies while they are taking part in cyber security processes.					
6. Code of Ethics					
Institutions should have written documents in which the ethical rules are defined and specified precisely and clearly.					
There should be a reporting mechanism that effectively operates in the auditing of the functioning of ethical rules.					
In case of violation of the code of ethics, reporting must be made directly to the general manager or the board of directors.					
Internal audit and cyber security teams should get findings through a coordinated approach in cases of violation of ethics.					
Internal audit teams should evaluate why the laws have been being reframed while they are controlling the compliance of institutions with the legislative framework.					
7. Legislative and International Policy Framework					
The internal audit follows the compliance of the control lists which are established within the scope of the regulations that are framed by BRSA and international standards.					
BRSA regulation, ISO 27001, and COBIT 4.1 DS5 processes are taken into consideration within the framework of compliance with legislation in the context of information security for the banking sector of Turkey.					
Internal audit reports the control deficiencies to the senior management which are defined according to the compliance framework while they are performing their monitoring activities.					

Continued on following page

Table 2. Continued

Internal audit teams carry out process controls in their activities that are related to compliance with the regulation.					
All banks, which are sustaining their businesses in Turkey, structure their information systems in accordance with the banking law no. 5411 of BRSA.					

Ordinal data, which are the kind of categorical variables, are collected through closed-ended statements that are scaled with five points likert measurement as well as tolerance conditions are decided as the interquartile range (IQR) should be less or equal to one and at a minimum seventy-five percent of the general agreement should be achieved for each of the arguments in the questionnaire.

The results, which are inferred through the set of arguments, according to the responsibility part of the questionnaire depicted that the responses of participants are consistent with each other. Therefore, control of risk management processes and compliance-related activities with reporting, evaluations, and monitoring operations are mainly accepted as the functions of internal auditors in cyber security governance.

Table 3. Responsibility

Statements for Responsibility	Mean	Standard Deviation	Median	Interquartile Range	Frequency			Consensus
					(4-5)	(3)	(1-2)	
1	4,466667	0,498887652	4	1	15 (100%)	0 (0%)	0 (0%)	Achieved
2	4,333333	0,699205899	4	1	13 (87%)	2 (13%)	0 (0%)	Achieved
3	4,466667	0,718021974	5	1	13 (87%)	2 (13%)	0 (0%)	Achieved
4	4,2	0,653197265	4	1	13 (87%)	2 (13%)	0 (0%)	Achieved
5	4,533333	0,618241233	5	1	14 (93%)	1 (7%)	0 (0%)	Achieved

The panelist responses showed that the consensus is not achieved in the seventh, eighth, ninth, and tenth arguments relative to the competency-related part of the questionnaire, so the third round for these answers is going to be performed in terms of retesting the consensus in those statements. On the other hand, their evaluations for the sixth argument are consistent with each other as the common agreement is achieved.

Table 4. Competency

Statements for Competency	Mean	Standard Deviation	Median	Interquartile Range	Frequency			Consensus
					(4-5)	(3)	(1-2)	
6	4,6	0,489897949	5	1	15 (100%)	0 (0%)	0 (0%)	Achieved
7	4,066667	0,77172246	4	2	11 (73%)	4 (27%)	0 (0%)	Not Achieved
8	4	0,894427191	4	2	11 (73%)	3 (20%)	1 (7%)	Not Achieved
9	3,933333	0,679869268	4	1	11 (73%)	4 (27%)	0 (0%)	Not Achieved
10	3,8	0,909212113	4	1	11 (73.3%)	2 (13.3%)	2 (13.3%)	Not Achieved

Panel members' answers relative to the governance part of the questionnaire showed that they are not in agreement for eleventh, thirteenth, and fifteenth arguments which are respectively related to internal audit's consultancy function, authority level of internal auditors, and their role of informing to upper management. Apart from that, their responses for the twelfth and fourteenth discussions which are respectively relevant with the organization of technical training events to internal audit part for effective governance in cyber security and advisory role of internal auditors, showed that the participants agree with these arguments. In this context, the third round is going to be utilized for the statements in which the common agreements were not reached.

Table 5. Governance

Statements for Governance	Mean	Standard Deviation	Median	Interquartile Range	Frequency			Consensus
					(4-5)	(3)	(1-2)	
11	3,666667	0,788810638	4	1	9 (60%)	5 (33.3%)	1 (6.7%)	Not Achieved
12	4,666667	0,471404521	5	1	15 (100%)	0 (0%)	0 (0%)	Achieved
13	3,733333	0,928559218	4	1	10 (66.6%)	3 (20%)	2 (13.3%)	Not Achieved
14	3,8	0,8326664	4	0	12 (80%)	1 (6.7%)	2 (13.3%)	Achieved
15	3,8	0,909212113	4	1	11 (73%)	2 (13.3%)	2 (13.3%)	Not Achieved

The answers of participants to the CIA triad part of the questionnaire showed that common agreements are achieved for all of the statements which are related to the compliance of information security relative to CIA triad principles.

Table 6. CIA triad

Statements for Confidentiality, Integrity, and Availability	Mean	Standard Deviation	Median	Interquartile Range	Frequency			Consensus
					(4-5)	(3)	(1-2)	
16	4,4	0,711805217	5	1	13 (87%)	2 (13%)	0 (0%)	Achieved
17	4,333333	0,699205899	4	1	13 (87%)	2 (13%)	0 (0%)	Achieved
18	4,066667	0,573488351	4	0	13 (87%)	2 (13%)	0 (0%)	Achieved
19	4,2	0,54160256	4	1	14 (93%)	1 (7%)	0 (0%)	Achieved
20	4,2	0,653197265	4	1	13 (87%)	2 (13%)	0 (0%)	Achieved

The inferences of panelist responses due to the corporate management principles part of the questionnaire showed that common agreements are reached for all of the statements which are connected with measuring how the internal auditors' activities are being performed in compliance with organizations' culture and how their performance can be improved.

Table 7. Corporate management principles

Statements for Corporate Management Principles	Mean	Standard Deviation	Median	Interquartile Range	Frequency			Consensus
					(4-5)	(3)	(1-2)	
21	4,2	0,54160256	4	1	14 (93%)	1 (7%)	0 (0%)	Achieved
22	4,6	0,611010093	5	1	14 (93%)	1 (7%)	0 (0%)	Achieved
23	4,533333	0,618241233	5	1	14 (93%)	1 (7%)	0 (0%)	Achieved
24	4,333333	0,596284794	4	1	14 (93%)	1 (7%)	0 (0%)	Achieved
25	4,4	0,611010093	4	1	14 (93%)	1 (7%)	0 (0%)	Achieved

The responses of panelists relative to the ethical rules part of the questionnaire showed that the consensus can not be met in the thirtieth argument which is linked with the role of investigating the regulatory framework as to explore the root causes of modifications in information security laws. Because of this reason, a third round will be organized for this statement to provide the common agreement.

Table 8. Code of ethics

Statements for Code of Ethics	Mean	Standard Deviation	Median	Interquartile Range	Frequency			Consensus
					(4-5)	(3)	(1-2)	
26	4,333333	0,596284794	4	1	14 (93%)	1 (7%)	0 (0%)	Achieved
27	4,2	0,653197265	4	1	13 (87%)	2 (13%)	0 (0%)	Achieved
28	4,133333	0,618241233	4	1	13 (87%)	2 (13%)	0 (0%)	Achieved
29	4,066667	0,442216639	4	0	14 (93%)	1 (7%)	0 (0%)	Achieved
30	3,733333	1,06249183	4	2	10 (66.6%)	2 (13.3%)	3 (20%)	Not Achieved

The answers of panelists according to the law mechanism and international policy framework part of the questionnaire showed that common agreements are provided in all of the arguments which especially prove that 5411 law of BRSA with ISO 27001 and COBIT 4.1 DS5 protocols are conformed as main guidelines while assuring that the cyber security governance activities are being executed in line with national regulations and global standards.

Table 9. Legislative and international policy framework

Statements for Legislative and International Policy Framework	Mean	Standard Deviation	Median	Interquartile Range	Frequency			Consensus
					(4-5)	(3)	(1-2)	
31	4,333333	0,471404521	4	1	15 (100%)	0 (0%)	0 (0%)	Achieved
32	4,266667	0,77172246	4	1	14 (93%)	0 (0%)	1 (7%)	Achieved
33	4,4	0,611010093	4	1	14 (93%)	1 (7%)	0 (0%)	Achieved
34	4,066667	0,679869268	4	1	12 (80%)	3 (20%)	0 (0%)	Achieved
35	4,133333	1,024152766	4	1	13 (87%)	1 (7%)	1 (7%)	Achieved

The responses of panel participants for the parts of the questionnaire which is applied in the second round of the delphi study revealed that seventh, eighth, ninth, tenth, eleventh, thirteenth, fifteenth, and thirtieth arguments should be counted into the reevaluation process as the third round. As a result, these items are sent back to the panel members whose answers are significantly differed from the median value of statistical inferences.

According to the third round, the panel members' discussions showed that the consensus for all the statements related to the competency part of the questionnaire is reached, so they mostly agreed with each other that the internal auditors' technical capacities can be improved with basic software knowledge, system and network administration, security information and event management (SIEM) applications for effective controls of risk management activities in information technology audit and cyber security governance processes.

Table 10. Competency in third round

Statements for Competency	Mean	Standard Deviation	Median	Interquartile Range	Frequency			Consensus
					(4-5)	(3)	(1-2)	
7	4,2	0,653197265	4	1	13 (87%)	2 (13%)	0 (0%)	Achieved
8	4,066667	0,853749898	4	1	12 (80%)	2 (13.3%)	1 (6.7%)	Achieved
9	4,066667	0,573488351	4	0	13 (87%)	2 (13%)	0 (0%)	Achieved
10	4	0,730296743	4	0	13 (86.6%)	1 (6.7%)	1 (6.7%)	Achieved

The discussions of panel members for the eleventh and fifteenth arguments of the questionnaire which are related to governance are showed that the common agreements are reached. However, their opinions for the thirteenth statement, which is about the authority level of internal auditors in cyber security management, showed that the consensus is not achieved because of the high critical rate of data access responsibility.

Table 11. Governance in third round

Statements for Governance	Mean	Standard Deviation	Median	Interquartile Range	Frequency			Consensus
					(4-5)	(3)	(1-2)	
11	3,866667	0,718021974	4	0	12 (80%)	2 (13.3%)	1 (6.7%)	Achieved
13	3,733333	0,928559218	4	1	10 (66.6%)	3 (20%)	2 (13.3%)	Not Achieved
15	4,066667	0,679869268	4	0	14 (93%)	0 (0%)	1 (7%)	Achieved

The answers of panelists to the thirtieth assertion in the ethical rules part of the questionnaire showed that they are mostly in agreement with the monitoring role of internal auditors as investigation of the underlying reasons of emerging regulations and standards relative to cyber security governance.

Table 12. Code of ethics in third round

Statements for Code of Ethics	Mean	Standard Deviation	Median	Interquartile Range	Frequency			Consensus
					(4-5)	(3)	(1-2)	
30	3,933333	0,928559218	4	1	12 (80%)	1 (7%)	2 (13%)	Achieved

FINDINGS AND RECOMMENDATIONS

As mentioned in the beginning part of the study, the functions of internal auditors in cyber security governance processes are deeply investigated from the viewpoints of responsibility, competency, governance, CIA triad, corporate management principles, code of ethics, and legislative and international policy frameworks which are taken into account as common titles for setting open-ended questions and questionnaires. In this context, the findings stated that internal auditors have a strategic position in ensuring the validity of cyber security governance models which the bank managements are using as well as they are playing a key role in assuring the compliance of institutions with regulatory frameworks especially as monitoring the legislative environment and international standards. The regulation, which was established by BRSA for information systems and electronic banking services of the banks at the date of 15th March 2020 in the official gazette, has been providing that internal auditors have acquired many opportunities to participate in the information technology controls for value-adding activities in terms of providing better governance with upper management, information technology, and cyber security departments. The discussions of panelists have been proving that the active support of internal auditors is needed for assuring the compliance of risk management processes' validity to regulatory environment and international standards framework in cyber security operations. In this context, both the findings which are related to the information security vulnerabilities, are formed through risk management activities and abnormal circumstances should be pursued, assessed, and reported to senior management. This has been strongly emphasized by the participants that the governance mechanisms, which are applied in the organizational structures of the banks, are harmonized with the new law which is enacted by BRSA, and by this way, the governance approaches of the banks have been transformed into much effective practices in terms of providing efficient risk controls in cyber security management.

As a result, innovations, which have been being integrated into the technology applications and tools such as continuous monitoring systems, can be practicable opportunities for the implementation of effective cyber security architectures to reduce not only the deliberate malicious acts but also the accidental mistakes. In this context, blockchain technology can be a good opportunity to change over the conventional monitoring and risk assessments mechanisms which are lagging behind to prevent the emerging cyber risks. The swot analysis is performed as follows to present the general overview of blockchain technology for cyber security management and internal audit processes;

Strengths:

- Immutability can be utilized by developing a zero-trust structure for the endpoints by deploying blockchain networks (Alevizos, Ta, & Eiza, 2021).
- Decentralization provides the availability of data to the users without a single authority (Pathak, 2021).
- Hash functions ensure the confidentiality and integrity of data through complex computations and algorithms by the facilitation of numerous forceful computers (Pathak, 2021).
- The open-source software supports transparency in terms of continuous review of system upgrades by the permitted persons (Pathak, 2021).

Weaknesses:

- Energy consumption is relatively higher than central network servers due to the need for much power generation for super-speed computers (Iredale, 2020).
- Environmental issues are caused because of the costly energy needs for operating machines that perform encryption processes (Iredale, 2020).
- The adoption of blockchain technology, which is required a full change in information systems, is hard to integrate into enterprises, from the point of completely reengineering the traditional infrastructures (Iredale, 2020).
- The workforce is not sufficient due to the lack of talented persons who both have core competencies and technical-level knowledge with deep experience (Iredale, 2020).
- New technology, which has just over ten years of history, is not enough mature that can cause major issues in standardization, legislative and policy frameworks (Iredale, 2020).

Opportunities:

- A recognized security protocol can be developed as an alternative to a throughout encryption model (Singh, 2021).
- Distributed Denial of Service (DDoS) attacks can be prevented by the decentralization of the domain name systems (DNS) accesses (Singh, 2021).
- Unauthorized access attempts can be easily monitored through encryptions with decentralized databases (Singh, 2021).

Threats:

- Regulatory concerns, which are the signs of an absent clear policy framework and international standards, can be considered as one of the critical issues in blockchain technology (University of North Dakota, 2021).
- The testing activities, which are used rather than cryptocurrency trading, are not sufficient (University of North Dakota, 2021).
- Multiple third-party dealers may be involved in a blockchain network such as payment servers, smart agreements, and deposit platforms, which may have weak information security tools and applications, can jeopardize the peer-to-peer network by opening the door to hacking (University of North Dakota, 2021).

All in all, primarily, blockchain technology is required to be improved by framing the industry standards and law enforcement in order to clear the puzzling conditions which are come into view as a result of uncertainties in the policy configuration. On the flip side, especially the sensitive sectors such as the banking industry can be adapted to this new infrastructure by using simulations and case modeling in terms of measuring how the blockchain technology can be feasible to substitute with traditional risk control tools and applications which are being utilized in cyber security and audit processes. However, technically, the weaknesses in testing methodologies, which are carried out in blockchain technology, are not satisfactory and cause serious obstacles for industry experts who particularly have a role in information technology and security departments of the banks. Therefore, banks' management, information systems, and security parts should continuously follow the new developments which can be considered as legislation, international standards, adaptation requirements, skill sets, testing tools, and operations for the blockchain technologys' infrastructure that is open to enhancement. Especially, the Banking Regulation and Supervision Agency (BRSA), which has the top responsibility in Turkey for framing regulations that are related to information security, has to establish novel standards to adapt the banks' information technology infrastructure into blockchain technology. In this context, the articles in BRSA's legislation in the seventh stage of 31069 numbered official newspaper on 15th March 2020, which are respectively shown through the numbered thirty, thirty-one, thirty-two, and thirty-three that are related to information systems' internal control, internal audit, the personal education, and asset allocation subjects, are needed to be continuously redesigned according to the developments in emerging technologies such as blockchain technology. As a result, blockchain technology can be used as an alternative or a supportive tool to conventional penetration testing, traditional risk monitoring and assessment, classical vulnerability detection by using shell-codes, horse and buggy recording, and reporting methods.

In respect of carrying the scientific quality of the study to further, emerging technologies such as robotic process automation, artificial intelligence, internet of things, machine learning, fifth-generation mobile networks, virtual reality, blockchain, and quantum computing practices can be much deeply explored from the point of information technology audit and internal controls to observe the innovative approaches and tools for defending the critical infrastructures of the nations such as banking industry or energy grids. Also, this research project can be improved with the investigation of compliance processes in critical infrastructure parts of Turkey to international standards for the feasible outline of the law framework. Additionally, the differences between the critical infrastructures of developed countries and emerging market economies can be explored from the point of how the controls in cyber security processes are being coordinated.

REFERENCES

AlOmari, L., Barnes, P., & Pitman, G. (2012). An Exploratory Study into Audit Challenges in IT Governance: A Delphi Approach. *IT Governance, Management and Audit*, 1-12. Retrieved from https://eprints.qut.edu.au/53110/22/An_Exploratory_Study_into_Audit_Challenges_in_IT_Governance_A_Delphi_Approach.pdf

Alevizos, L., Ta, V. T., & Eiza, M. H. (2021). Augmenting Zero Trust Architecture to Endpoints Using Blockchain: A State of the Art Review. *Computer Science, Cryptography and Security*, 1-17. Retrieved from https://arxiv.org/ftp/arxiv/papers/2104/2104.00460.pdf

Altınpulluk, H., Kesim, M., & Kurubacak, G. (2020, April 2). The Usability of Augmented Reality in Open and Distance Learning Systems: A Qualitative Delphi Study. *International Council for Open and Distance Education, 12*(2), 283-307. Retrieved from https://files.eric.ed.gov/fulltext/EJ1267133.pdf

Conway, C. M. (2020). *Approaches to Qualitative Research: An Oxford Handbook of Qualitative Research in American Music Education* (Vol. 1). Oxford University Press.

Creswell, J. W. (2014). *Research Design: Qualitative, Quantitative, and Mixed Methods Approaches* (4th ed.). Sage.

Dalkey, N. C. (1967). *Delphi. The Rand Corporation.* Clearinghouse for Federal Scientific & Technical Information. Retrieved from https://apps.dtic.mil/sti/pdfs/AD0660554.pdf

Davidson, P., & Hasledalen, K. (2014). Cyber Threats to Online Education: A Delphi Study. In P. Dover, S. Hariharan, & M. Cummings (Ed.), *Proceedings of the 2nd International Conference on Management, Leadership and Governance, ICMLG 2014* (pp. 68-77). Academic Conferences and Publishing International (ACPI).

Gantz, S. (2014). The Basics of IT Audit, Purposes, Practices, and Practical Information. Elsevier.

GroenfeldtT. (2014). *Forbes*. https://www.forbes.com/sites/tomgroenfeldt/2014/05/08/insiders-pose-a-serious-threat-to-corporate-information/?sh=2600bf965ca9

Haqaf, H., & Koyuncu, M. (2018). Understanding Key Skills for Information Security Managers. *International Journal of Information Management, 43*, 165–172. doi:10.1016/j.ijinfomgt.2018.07.013

Helmer, O. (1967). *Analysis of the Future: The Delphi Method.* The Rand Corporation. Retrieved from https://apps.dtic.mil/sti/pdfs/AD0649640.pdf

Iredale, G. (2020, April 17). *Disadvantages of Blockchain Technology*. Retrieved from https://101blockchains.com/disadvantages-of-blockchain/

ISACA. (2012). *COBIT 5: A Business Framework for the Governance and Management of Enterprise IT*. ISACA.

Jain, A. (2005). *Cyber Crime: Issues Threats and Management*. ISHA Books.

Joiner Associates. (1995). *Cause and Effect Diagrams: Plain and Simple*. Joiner Associates Incorporated.

Joiner Associates Incorporated. (1995). *Pareto Charts Plain & Simple Learning and Application Guide*. Joiner Associates Incorporated.

Karabacak, B., Yıldırım, S. Ö., & Baykal, N. (2016). Regulatory Approaches for Cyber Security of Critical Infrastructures: The Case of Turkey. *Computer Law & Security Report, 32*(3), 526–539. doi:10.1016/j.clsr.2016.02.005

Kumar, R., & Sharma, V. (2013). *Auditing: Principles and Practice*. PHI Learning Private Limited.

Kurubacak, G. (2007, July 1). Identify Research Priorities and Needs for Mobile Learning Technologies in Open and Distance Education: A Delphi Study. *Mobile Learning Technologies, 19*(2), 1-31. Retrieved from https://files.eric.ed.gov/fulltext/ED495997.pdf

Kurubacak, G. (2011). E-Learning for Pluralism: The Culture of E-Learning in Building a Knowledge Society. *International Journal on E-Learning, 10*(2), 145–167.

Maleh, Y., Baddi, Y., Alazab, M., Tawalbeh, L., & Romdhani, I. (2021). *Artificial Intelligence and Blockchain for Future Cybersecurity Applications*. Springer. doi:10.1007/978-3-030-74575-2

OECD. (2019). *The Policy Environment for Blockchain Innovation and Adoption*. OECD Blockchain Policy Series. Retrieved from https://www.oecd.org/finance/2019-OECD-Global-Blockchain-Policy-Forum-Summary-Report.pdf

Olson, D. L., & Wu, D. D. (2015). *Enterprise Risk Management* (Vol. 3). World Scientific. doi:10.1142/9378

Pathak, A. (2021). *Blockchain Technology: A Guide for Beginners.* https://geekflare.com/finance/blockchain-technology-for-beginners/

Penuel, K. B., & Statler, M. (2013). *Encyclopedia of Crisis Management*. Sage. doi:10.4135/9781452275956

Pickett, K. H. (2011). *The Essential Guide to Internal Auditing*. British Library.

Progrebna, G., & Skilton, M. (2019). *Navigating New Cyber Risks: How Businesses Can Plan, Build and Manage Safe Spaces in the Digital Age*. Palgrave Macmillan. doi:10.1007/978-3-030-13527-0

Ryder, R. D., & Madhavan, A. (2019). *Cyber Crisis Management Overcoming the Challenges in Cyberspace*. Bloomsbury.

Schreider, T. (2017). *Building Effective Cybersecurity Programs: A Security Manager's Handbook*. Rothstein Publishing.

Selimoğlu, S., & Altunel, M. (2019). Siber Güvenlik Risklerinden Korunmada Köprü ve Katalizör Olarak İç Denetim [Internal Audit as a Bridge and Catalyst in the Protection of Cyber Security Risks]. *Denetişim,* (19), 5-16. Retrieved from https://dergipark.org.tr/en/download/article-file/750860

Singh, S. (2021, July 16). *Potential Use Cases of Blockchain Technology for Cybersecurity*. Retrieved from https://www.itbusinessedge.com/security/potential-use-cases-of-blockchain-technology-for-cybersecurity/

Smits, D., & Hillegersberg, J. V. (2015). IT Governance Maturity: Developing a Maturity Model Using the Delphi Method. In *48th Hawaii International Conference on System Sciences* (pp. 4534-4543). Institute of Electrical and Electronics Engineers (IEEE). 10.1109/HICSS.2015.541

Tipton, H. F., & Krause, M. (2007). *Information Security Management Handbook*. Taylor&Francis Group. doi:10.1201/9781439833032

University of North Dakota. (2021). *5 Blockchain Security Issues & Tips for Solving Them*. Retrieved from https://onlinedegrees.und.edu/blog/5-blockchain-security-issues/

Vasarhelyi, M. A., Lombardi, D., & Bloch, R. (2014). The Future of Audit: A Modified Delphi Approach. In *Canadian Academic Accounting Association (CAAA) Annual Conference 2011* (pp. 1-31). Social Science Research Network (SSRN).

Westby, J. R. (2004). International Guide to Cyber Security. American Bar Association (ABA).

Chapter 11
The Impact of Blockchain Technology on Accounting, Auditing, and Assurance Practices:
Turkey Case

Seval Selimoglu
Anadolu University, Turkey

Gul Yesilcelebi
Gumushane University, Turkey

Mehtap Altunel
Anadolu University, Turkey

ABSTRACT

The purpose of this study is to examine the impact of blockchain technology on accounting, auditing, and assurance practices. For this purpose, the relevant national and international literature on the accounting, auditing, and assurance practices of blockchain technology has been reviewed. As a result of the research, the advantages and disadvantages of blockchain technology were evaluated and categorized. As a result of the study, it has been determined that the blockchain technology has a great effect on accounting, auditing, and assurance.

DOI: 10.4018/978-1-6684-4153-4.ch011

Copyright © 2023, IGI Global. Copying or distributing in print or electronic forms without written permission of IGI Global is prohibited.

INTRODUCTION

Technological developments are rapidly reflected in the business world. At this point, businesses must adapt to new technologies and change their way of doing or working. Technological developments have undoubtedly affected the fields of accounting and auditing, as in all fields. Especially in the field of accounting and auditing, it is important how the applications should be made in terms of ensuring data security and using blockchain technologies in these areas.

Although its legal regulations and standards have not yet been clearly defined, blockchain is a technology that is developing rapidly all over the world, is closely watched by both the public and private sectors and is expected to cause radical changes in current business processes (Karahan and Tufekci, 2019).

With blockchain technology and cryptocurrencies occupying the agenda of the economic world more and more each day, the rules of accounting and auditing are being rewritten at the same time. Blockchain technology is a distributed system of records with an ever-growing database called blocks, which is not dependent on a central authority in general (Kizil and Hanisoglu, 2019).

Blockchain technology facilitates accounting transactions, enables real-time reporting and real-time auditing (Alarcon and Ng, 2018). Blockchain technology has the potential to transform the accounting profession strongly and is expected to affect accounting with its functions, auditing, financial consultancy, professional ethics, and taxation (Gul, 2019).

Today, after the emergence of blockchain technology as a new research area, many researchers have conducted various studies on this technology. Researchers have based their studies mainly on conceptual analysis. Studies in the national literature are listed below.

In Guner (2021)'s study, it was aimed to determine the perceptions about the usability of blockchain technology in accounting processes and to develop a scale to determine these perceptions. The data used in the study were collected from 121 academicians working as faculty members in the field of accounting in Turkey through a questionnaire prepared. Confirmatory factor analysis (CFA) was performed to evaluate the reliability and validity of the developed scale. The CFA results revealed that the developed scale consists of a 5-dimensional structure: "transparency", "low cost", "efficiency and safety", "audit" and "compliance with legislation" and that the proposed structure as a whole is valid and safe.

In Onay (2021)'s study, blockchain technology is discussed in terms of its regulatory implications, opportunities and implementation challenges. As a result of the research, it has been found that Turkish studies dealing with the relationship between auditing and blockchain, unlike similar studies in the English literature, touch on concepts such as the definition of cryptocurrencies and taxation problems.

In Sonmez and Demir (2021)'s study, the effects of technological developments, especially in the blockchain, on the management, transfer and transformation of digital assets in the accounting and finance sector are discussed.

In Kilinc (2020)'s study, blockchain technology was introduced and the effects of this technology in the field of accounting and auditing were discussed in detail. As a result of the study, it was underlined that with the development of blockchain technology, the presence of professionals with high technical knowledge will be needed, although the members of the profession will be able to perform their jobs more easily.

In Suekinci and Catikkas (2020)'s study, it has been tried to evaluate the possible effects of blockchain technology on the accounting registration process and tax revenues, which is the most important source of income for the public.

Dogan and Ertugay (2019), in their study, conceptually discussed the functioning of the blockchain, and its potential applications related to accounting. As a result of the study, the blockchain obstacles that will take time to overcome in front of their implementation are listed.

Gul (2019), in his study, conceptually examined the working principle of blockchain technology and its effects on the accounting information system.

In Karahan and Tufekci (2019)'s study, it is aimed to present a broader perspective on the potential effects of blockchain technology by mentioning the main features of the audit profession.

In Hazar (2018)'s study, the data structures of the blockchain, how cryptography is used, smart contracts developed with the programmability feature and prominent features such as digital signature, distributed identity management, decentralization are explained in detail with its technological infrastructure. In addition, accounting applications that can be developed with this technology are presented with case studies.

Looking at the national literature in recent years, although the number of studies on blockchain technology has increased gradually, the number of academic studies that deal with accounting, auditing and assurance together is still limited.

Zhang et al. (2020) evaluated the impact of artificial intelligence and blockchain on the accounting profession in their study. The possible effects of blockchain on accounting and auditing and the blockchain applications of the four big audit firms are examined.

Liu et al. (2019) introduced two types of blockchain (i.e., permissionless and permissioned and demonstrated their technological features) in their study. They detailed the effects of blockchain on auditing and the opportunities and challenges of the two blockchains for auditors. As a result of the study, recommendations are made for auditors to adapt, adjust, and upgrade themselves to the role of strategic partners in the implementation of blockchain.

In the study of Brender et al. (2018), as a result of a face-to-face interview with 34 financial and IT auditors from 23 audit firms in Switzerland, almost all of the auditors expect a change in the audit profession in the medium term, and more than half of them expect audit functions to be much more IT. It was concluded that they expected to be oriented. Some auditors interviewed hypothesized that the auditors would be IT engineer auditors with specific IT skills such as programming, and some even mentioned the term "blockchain auditor".

Although studies in Turkey deal with blockchain-accounting and blockchain-auditing relations from various aspects, there is no study that combines the impact of blockchain technology on accounting, auditing, and assurance practices. For this reason, in this study, impact of blockchain technology on accounting, auditing and assurance practices are discussed theoretically. This study will contribute to the relevant literature with its impact on blockchain and three other concepts (accounting, auditing, assurance). In addition, both international studies were examined in detail and were revealed a framework of studies carried out within the scope of blockchain-accounting-auditing and assurance in Turkey.

Within the scope of the study, firstly the conceptual framework of blockchain technology is discussed, then legal regulations on blockchain technology in Turkey is detailed, and the impact of blockchain technology on accounting, auditing and assurance services is presented. Finally, conclusion and recommendations are given.

FRAMEWORK OF BLOCKCHAIN TECHNOLOGY

Blockchain technology is a system developed in 2008 by a person or group named Satoshi Nakamoto as a peer-to-peer network and a shared database that follows a set of rules to form the infrastructure of a digital currency Bitcoin (Bonsón and Bednárová, 2019). Blockchain is the formation of a public ledger of information and data collected on a network located on the Internet (Nakamoto, 2008). In other words, it works as a digital ledger on the network and is defined as a distributed ledger (Ozkul and Alkan, 2020).

Blockchain technology basic features are listed as near real-time settlement, distributed ledger, irreversibility, censorship resistant. These concepts are explained respectively (CPA and AICPA, 2017:):

- *Near real-time settlement,* the blockchain allows to record the transactions that take place in almost real time. However, it eliminates the risk of a payment or collection made by a user.

- *Distributed ledger,* peer-to-peer distributed network allows for historical classification of commercial transactions. A blockchain encompasses a distributed, highly accessible, and secure proof of transaction that takes place.
- *Irreversibility,* a blockchain contains a verifiable record of every single transaction made on the chain. This also prevents double/re-registration of the traceable element by the blockchain. In addition, the fact that the transactions are irreversible, unalterable, or cannot be interfered with after the record is kept, also increases the realistic presentation of the data produced by the system.
- *Censorship resistant,* on the other hand, presents the economic rules within the blockchain model as flows so that independent participants can verify new blocks. Controlling the system or censoring entry permits causes additional costs as it will require changing the technological infrastructure of the system again.

Blockchain and distributed ledger technology (DLT) is indicated in Figure 1.

Figure 1. Blockchain and distributed ledger technology (DLT)
Source: Adapted from https://www.tutorialandexample.com/blockchains-vs-distributed-ledger-technology/

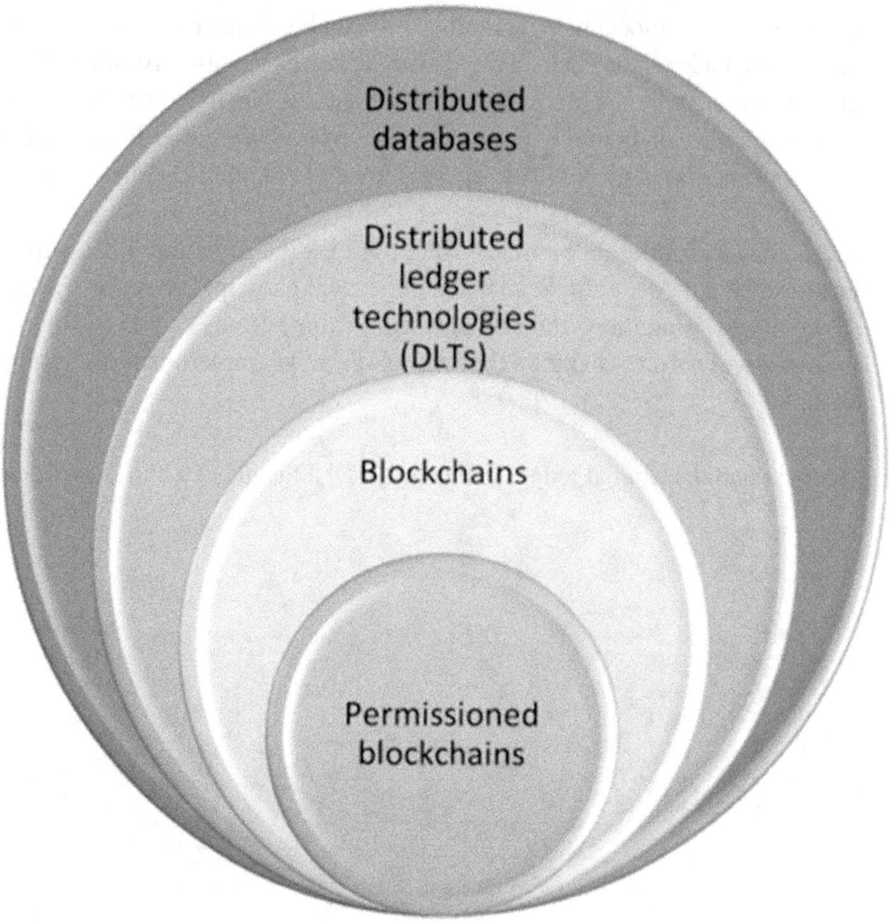

While Blockchain and DTL are sometimes used synonymously, they are not quite the same thing. DTL is an umbrella concept that describes technologies that enable the distribution of confidential or public records and information. Blockchain is the first fully functional DTL and is a subcategory of this much more comprehensive technology (Thake, 2018). Every blockchain is a distributed ledger, but not every distributed ledger is a blockchain (Karahan and Tufekci, 2019). One of the main differences between both technologies is that Blockchain is permissionless, whereas DTL is permissioned. In the Blockchain, anyone can join the network, but in DTL only selected participants can access the network (https://www.tutorialandexample.com/blockchains-vs-distributed-ledger-technology/).

There are many application areas in blockchain technology, especially in the finance sector, banking, e-government, digital identity, internet of things (IOT), health sector, education sector, retail industry, legal system, insurance, real estate sector, music industry, supply chain management, marketing, press and broadcasting, tourism, telecommunication, and automotive sectors (Kizil and Hanisoglu, 2019).

Process of blockchain technology is indicated in Figure 2.

Figure 2. Process of blockchain technology
Source: Adapted from Chartered Accountants Australia and New Zealand (2017)

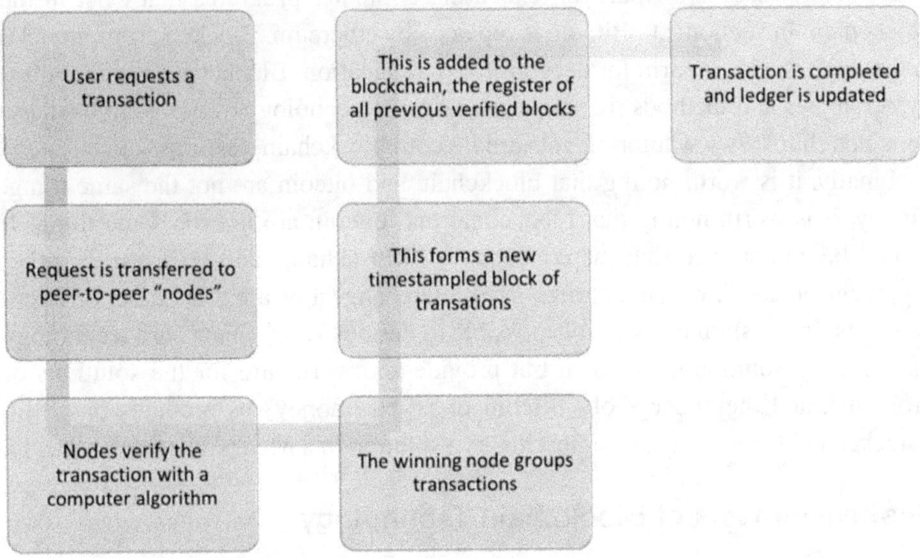

Blockchain is generally categorized in terms of the visibility of the registry and the openness of access rights, and it is commonly referred to as private and public blockchains in terms of visibility, and permissioned and permissionless blockchains in terms of access rights. Blockchains where only authorized users can see the transaction history records are called private blockchains, while the blockchains where only authorized users can take part in the mechanism of validating blocks and executing and verifying transactions are also called permission-requiring blockchains (Karahan and Tufekci, 2019).

Blockchain is not actually a single concept. After its first appearance, it developed in many ways, took different forms, and addressed unlimited usage scenarios (Lyons et al, 2018: 6). Demonstrating the evolution of blockchain technology, including the progression of blockchain technology from version 1.0 to version 4.0 are indicated in Figure 3.

Figure 3. Blockchain versions
Source: Adapted from https://www.tutorialandexample.com/blockchain-versions/

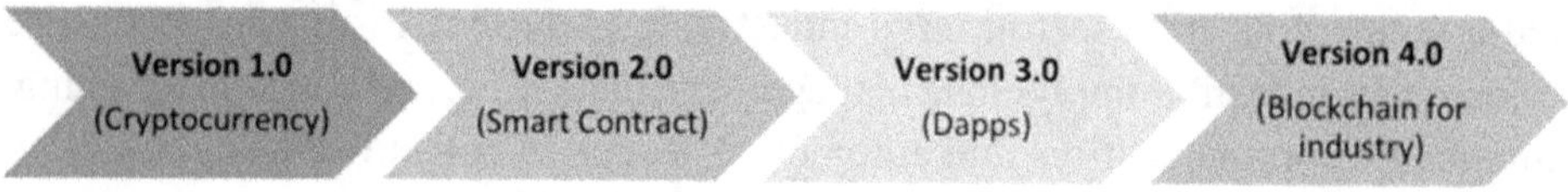

Blockchain version 1.0 is the first cryptocurrency, which records the bitcoin transactions on the public ledger in a secure method. Blockchain version 2.0 is smart contracts, it is a small self-executing computer programs that exist in the blockchain. In version 2.0, Bitcoin is replaced by ethereum. Blockchain version 3.0 is the dapp is a short form for decentralized application. Blockchain version 4.0 is the strategies and methods that make blockchain technology available to business demands (https://www.tutorialandexample.com/blockchain-versions/).

Finally, it is worth noting that blockchain and bitcoin are not the same thing. Finally, it is worth noting that blockchain and bitcoin are not the same thing. It should be emphasized that the concepts of "Blockchain" and "Bitcoin (or other cryptocurrencies)", which are often used interchangeably, are related but different concepts. In the simplest way, it is possible to define "Blockchain" as a technology that is not a solution on its own but provides infrastructure for the solutions of problems, and the concept of "Bitcoin or crypto money" as products based on 'Blockchain' technology and emerging as a solution to a need (Calisaneller, 2021).

Positive Impacts of Blockchain Technology

Being a decentralized system, blockchain technology is shown as the main benefit of not working with a third-party organization and central administrator. therefore, it is a system where participants decide without intermediaries in blockchain technology (Tenorio, 2021). Blockchain technology has many benefits in the business world. These benefits are grouped under six headings by Niranjanamurthy (2019). In addition to being reliable, blockchain also creates some benefits with its innovative approach. It is possible to list some of the benefits of blockchain technology as follows (Alarcon and Ng, 2018; Gatteschi et al., 2018):

- Peer-to-peer business transactions or information exchange without an intermediary. Blockchain technology does not need a central authority and it cannot be controlled, censored, or shut down.
- Automatic and highly secure record keeping, storing data with nodes ensures data protection against undesired events.

- Recording and tracking the ownership of assets.
- Real-time control possibilities, it provides trust between users through digital signature and verification. In addition, since users have the opportunity to examine current and past transactions, transparency is ensured between users. Everyone can access data and view transactions.
- Smart contracts: when conditions based on algorithms are met automatic contract execution and execution of transactions.
- Blockchain technology creates a repository of data that anyone can read and write.

The use of blockchain technology in the business world provides many benefits. One of the main benefits is *efficiency*. The absence of third parties when making transactions with Blockchain technology helps these transactions to be carried out quickly. Automatic execution of smart contracts and commercial actions originating from technology provides cost and time savings. On the other hand, the fact that each transaction is recorded increases the *auditability* in terms of businesses. In addition, the blockchain technology provides the ease of *traceability* in the supply chain and ensures the effective execution of goods tracking. Financial and commercial transparency is of great importance in the business world. If financial and commercial transparency is not provided, negative business relations and commercial developments are experienced. In this respect, blockchain technology has advantages in providing *transparency* for businesses of all sizes. Blockchain technology ensures that every transaction is encrypted in a complex cryptographic way, making it difficult for hackers to disrupt the traditional installation of the chain. Therefore, blockchain technology provides *security* for businesses against an attack from outside. Another benefit that blockchain technology provides to businesses is *feedback*. Provides information on feedback, installation, maintenance, shipping returns, and decommissioning. Although Blockchain technology is referred to as cryptocurrencies, it provides many benefits within businesses. For this reason, it can be seen from the above statements that the use of blockchain technologies will help businesses to stand out in the competitive environment (Niranjanamurthy, 2019).

Negative Impacts of Blockchain Technology

The diffusion and persistence features of the blockchain are essential features of this technology and are not optional. For this reason, applications created using blockchain should be compatible with these features. For example, data will go to all network users in encrypted form. If the relevant data contains sensitive information, it may not be wanted to reach everyone, even if it is encrypted. In this case, it would not be right to use blockchain technology (Hazar, 2018).

In addition to the benefits of blockchain technology, there are also some difficulties. First of all, when making a transaction, blockchain technology takes three times more time than a normal database transaction. There are basically three reasons for this. these are: digital signature, consensus mechanism, and lots of transactions. Every blockchain transaction must be signed using private or public cryptography. On the other hand, the consensus mechanisms in a distributed database every time a blockchain transaction occurs, efforts should be made to ensure that the nodes in the network reach consensus. While central databases process transactions once (or twice), in a Blockchain they must be processed independently by each node in the network. As a result, too many transactions are made. Another disadvantage of blockchain is the uncertainty about legal regulations. Since blockchain determines the regulations regarding virtual currencies, barriers to the adoption of the technology by governments' financial institutions can occur if governments do not act on the regulations. Another disadvantage of blockchain technology is that virtual currency blockchain network miners use a high level of energy during transaction verification. Although blockchain technology has a strong encryption technique, there is a cyber security risk. While the blockchain application offers solutions that require significant changes, there are integration concerns, so businesses should determine a transaction strategy for transition transactions. Blockchain offers great savings in transaction costs and time but has a high initial capital cost. The benefit of blockchain for business (Niranjanamurthy, 2019; Gatteschi et al., 2018).

LEGAL REGULATIONS ON BLOCKCHAIN TECHNOLOGY IN TURKEY

To address the issue of accounting for cryptocurrencies, there is currently no specific framework or standard in Turkey. In other words, the accounting and evaluation of cryptocurrencies is not bound to a specific legal regulation. Although there is no regulation regarding the accounting of cryptocurrencies in Turkey, procedures regarding the registration, reporting and evaluation of cryptocurrencies are required by taking into account the basic principles and concepts of accounting and reporting standards. There are various approaches to how cryptocurrencies should be accounted for. There are four approaches to how cryptocurrencies should be accounted for in Turkey (Kizil and Hanisoglu, 2019):

- Cryptocurrencies should be considered and valued as "inventory" (Other Inventories).
- Cryptocurrencies should be considered and valued as "cash and cash equivalents" (Other Cash Equivalents).

- Cryptocurrencies should be considered and valued as “stock and securities” (Other Stocks).
- Cryptocurrencies should be considered and valued as “cash/money” (Cash).

In the international arena, it is stated that cryptocurrencies should be considered and valued as “intangible assets” as a new and additional approach (Kizil and Hanisoglu, 2019).

Blockchain technology is briefly mentioned in a part of the preliminary report of the specialization commission, which was prepared within the scope of the preparations for the Eleventh Development Plan (T.C. Kalkinma Bakanligi, 2018):

Blockchain technologies, which have recently come to the fore with the use of virtual currencies such as bitcoin, offer a new opportunity that invalidates the existence of intermediary institutions with a registration system based on software technologies. It is thought that this reliable registration system can be applied to any registration system between individuals and institutions. Developments in this area are discussed in a wide range from the banking system to public records. Serious debates continue about in which areas and how these technologies can be used in public administration.

The Turkish authorities have issued guidance saying that bitcoin does not meet the standards of electronic money and that the volatility leaves users with a high level of risk; a major bitcoin exchange has ceased operations after local banks closed the main accounts of the company without prior notice. (https://www.thomsonreuters.com/en/reports/blockchain.html).

THE IMPACT OF BLOCKCHAIN TECHNOLOGY ON ACCOUNTING, AUDITING, AND ASSURANCE PRACTICES

Blockchain and Accounting Records

Blockchain is also used as an accounting technology. Accounting is the recording, reporting and analysis of financial information. In blockchain applications, it is essential that the information is accurately, and completely recorded, stored, and transmitted to the relevant parties (Hazar, 2018). Blockchain technology has the potential to affect all record keeping processes, including the way transactions are initiated, processed, authorized, recorded, and reported (CPA and AICPA, 2017).

The most important innovation brought by the blockchain technology to accounting practice is that it transforms the traditionally adopted double-entry system into a triple-entry system, which is presented as a unique feature of the system. The triple-entry system is a more developed form of the double-entry system in accounting and refers to a cryptographic system in which the third-party blockchain is included (Uysal and Kurt, 2018). As it is known, in accounting, there is a recording system in which both parties work simultaneously, consisting of assets, liabilities, and shareholders' equity accounts, in the reporting process of economic transactions. In the triple-entry system, it refers to a cryptographic system in which the block chain is included as a third party (Dai and Vasarhelyi, 2017).

Blockchain technology can be used in many ways in the field of accounting. The secure storage of accounting data, sharing information with the relevant parties as soon as possible, and considering the infrastructure of blockchain technology, increase the verifiability of business data. An important development in blockchain technology is the introduction of smart contracts (CPA and AICPA, 2017). In addition, smart contracts can be used in many areas of life, as well as in accounting (Dai and Vasarhelyi, 2017). For example, In the case of a contract for the performance of the service, the payment will be fulfilled when both parties of the transaction confirm that the service has been performed. With the widespread use of cryptocurrencies for the collection problem in financial consultancy services, the use of smart contracts in practice will also fulfill the financial advisory service and the transfer of virtual money will be possible with the approval of the taxpayer (Suekinci and Catikkas, 2020). Therefore, it is stated that smart contracts will be made in the future with blockchain technology, so that the fulfillment of obligations between individuals will be more reliable and faster (Demírhan, 2019).

Blockchain should not be considered as a ledger alternative according to the above statements. Instead, it is more accurate to express it as a tool developed for business and accounting records. The main benefit of this tool is that it is much safer. Blockchain technology will help shareholders and auditors to conduct individual transactions, collect evidence and prepare financial reports at their discretion, by keeping transaction records in an unbiased data-shared database (Wang & Kogan, 2018; Alsaqa et al., 2019). If it is desired to design blockchain technology for accounting purposes, certain factors must be taken into account. this factor is node selection, database structure, authorization and validation protocols etc. can be listed as. A privately owned blockchain environment is described by Bonsón & Bednárová (2019) in Figure 4.

Figure 4. Blockchain ecosystem in private permissioned architecture for accounting purposes
Source: Adapted from Bonsón and Bednárová (2019)

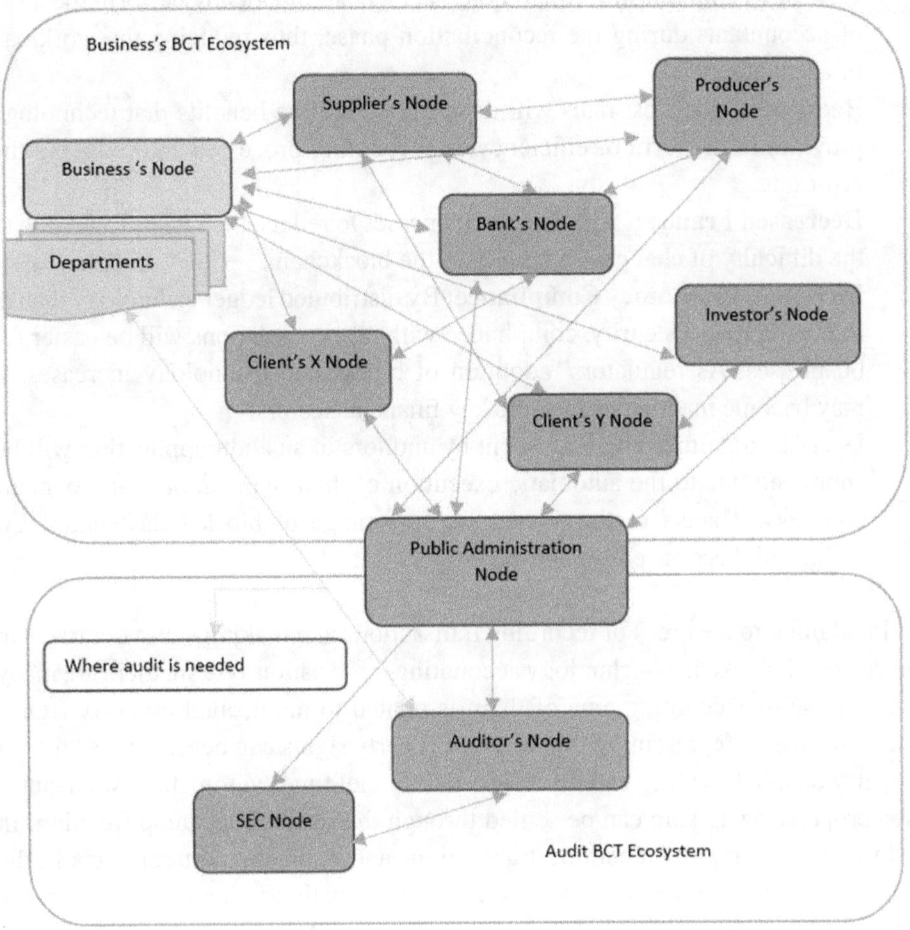

Blockchain technology has numerous benefits to accounting firms large and small. Although these benefits are similar to the issues mentioned in the advantage of blockchain section, they are explained below in terms of accounting (Martin, 2018):

- **Increased Efficiency:** A powerfully designed blockchain technology provides more efficiency compared to traditional accounting software in the process of processing data into the system and extracting data from the system, as it creates a fast and powerful database.

- **Decrease in the Rate of Errors:** It will reduce the human error because it automates the accounting function with the data recorded in the blockchain technology and smart contracts.
- **Ease of Reconciliation:** Smart contracts will automatically perform the tasks of accountants during the reconciliation phase, thus reducing the workload of accountants.
- **Reduced Cost:** Cost rates will decrease due to the benefits that technology provides in the form of efficiency in accounting processes and reducing the error rate.
- **Decreased Fraud:** It will benefit businesses to reduce the rate of fraud due to the difficulty of changing a record on the blockchain.
- **Increased Regulatory Compliance:** By distributed ledger technology thanks to the improved security, compliance with legal regulations will be easier for businesses. As regulators' adoption of blockchain technology increases, it may become mandatory in some key financial sectors.
- **Decreased Audit:** The time spent by auditors in an audit application will be shortened due to the automatic execution of their applications due to smart contracts. Thanks to the traceability advantage of blockchain technology, audits will become easier and faster.

In addition to the areas of recording transactions and making smart contracts in the design of blockchain technology accounting information system, there are many areas of use in accounting. one of them is related to intellectual property rights. Intangible assets representing intellectual property rights can be shown as "digital property contracts" which can be designed to funnel funds automatically. Disputes over property ownership can be settled through the block time stamp function. In addition to the areas of recording transactions and making smart contracts in the design of blockchain technology accounting information system, there are many areas of use in accounting. one of them is related to intellectual property rights. In addition, blockchain technology provides the opportunity to self-report financial capital. therefore, during the transfer of ownership, it provides ease of transfer between the parties, such as the exchange of cryptocurrencies today (Alsaqa et al., 2019: 69). Blockchain's impacts on accounting practices are indicated in Table 1.

Table 1. Blockchain's impacts on accounting practices

Blockchain Characteristics	Goals	Potential Impact on Accounting Practices
Non-temperable	Prove ownership	The only source of accounting data is guaranteed, and the accounting data cannot be falsified.
Distributed ledger	Enhance transaction transparency	Improvement in the transparency of accounting information.
Timestamp	Clear transaction order	Improvement in the difficulty of changing data and retain permanent accounting records.
Network consensus	Verify transaction legality	Improvement in accounting efficiency; provision of real-time transaction clearing or settlement.
Programmable	None	Intelligentization of accounting business processing through programmability, setting accounting algorithms or accounting business processing rules.

Source: Adapted from Zhang et al. (2020)

Blockchain and Auditing, Assurance Services

While traditional auditing and assurance services will continue to be essential, blockchain business practices and new accounting technology are likely to have a significant impact on the way auditors execute engagements (CPA and AICPA, 2017; Deloitte, 2021). Blockchain technology can have a significant impact on the way auditors carry out their tasks, as audit innovations in automation and data analytics and the audit and assurance profession evolve today. Also, as blockchain technology is more widely adopted, auditors may need to expand their skill sets and knowledge to meet the expected demands of the business world (CPA and AICPA, 2017).

In addition, the current assurance paradigm may fundamentally change as smart contracts are adopted. Because smart contracts are to create a verifiable and tamper-proof system. A blockchain ledger can be used as a trusted medium to store any audit-related document. As this information and documentation is continually shared with relevant parties, the role of providing assurance can be expanded primarily from auditors to a much broader range of participants, such as business partners, creditors, government agencies (Dai and Vasarhelyi, 2017).

In the blockchain-based accounting system, all transactions that are connected to the internet become a part of the chain and take place in the blocks, helping the relevant parties to fulfill the main purposes of financial reporting, and auditing, such as ensuring information security, and preventing faults, and frauds (Uysal and Kurt, 2018). In the face of increasing uncertainty in the business world, the need for auditors, both in terms of assurance and consultancy, is increasing. Therefore, it has become a necessity for them to obtain information on how to use blockchain technology in the auditing process, taking into account the various opportunities (Bonyuet, 2020). Blockchain technology could bring new challenges and opportunities to the auditing and assurance profession (CPA and AICPA, 2017).

Transparency, traceability, immutability, and integration of rules and procedures embedded in technology enrich processes and knowledge generation, significantly changing audit and control procedures, and in some cases rendering them useless. At the same time, these features create opportunities for auditors to redesign best practices, update rules and procedures, and define new standards (Brender et al., 2018). The opportunities and challenges for auditors in blockchain technology are shown in Table 2.

Table 2. Opportunities and challenges to auditors

	Opportunities	Challenges
Permissionless blockchain	Examine transaction record on blockchain.	No reversal of erroneous transactions.
	Develop novel audit process on blockchain transactions.	No centralized authority to verify the existence, ownership, and measurement of items recorded on blockchain.
	Verify the consistency between items on blockchain and in the physical world.	Data retrieval due to clients' loss of private key.
		No centralized authority to report cyberattacks.
Permissioned blockchain	Develop guidelines for blockchain implementation.	Need to be proficient in various blockchain technologies.
	Leverage industry knowledge and experience to offer advice for best practices for blockchain consensus protocols.	Difficult to reach consensus rules among all participants, when acting as an organizational agent.
	Leverage business networks to form permissioned blockchain based on market demand.	Audit transaction linked to a side agreement that is "off-chain".
	Act as planner and coordinator of potential participants of a blockchain.	Tackle the situation when central authority has power to override information on blockchain.
	Leverage their expertise on IT auditing to audit internal control of blockchain, including data integrity and security.	Cope with change of consensus protocol in a blockchain.
	Offer independent rating services to a specific blockchain.	
	Act as administrator of blockchain.	

Source: Adapted from Liu et al. (2019, p. 26)

Blockchain's impacts on auditing practices (both in terms of internal auditing and independent auditing) are indicated in Table 3.

Table 3. Blockchain's impacts on internal and external auditing practices

Auditing Practices	Blockchain's Impact	Internal Audit	External Audit
Evidence gathering	Whole-population investigation replacing the traditional sampling approach.	X	X
	Direct access to transaction history.		
Transaction validation and verification	Real-time transaction validation by a community of miners.	X	X
	Record verification and maintenance by all users.		
Compliance evaluation	Built-in compliance with most recent standards, regulations, and laws.	X	
	Instant presentation of the underlying regulation to an operator.		
	Immediate detection of violations.		
Transaction reconciliation	Automating reconciliation (if transactions take place between parties within a single blockchain network)	X	X
	Instant settlement.		
	Reduction of time spent on reconciliation and increased efficiency.		
Financial reporting	Near real-time financial reporting.	X	X
	No errors.		
	Less prone to fraud.		
Planning and advising	Providing complete, accurate records for auditors to quickly spot problems, prioritize plans, and find long-term patterns.	X	
Decision support	Offering reliable and timely information stored in blockchain to perform analytics.	X	
	Predicting consequences of actions.		
	Facilitating smart contracts by embedded analytical models (i.e., to identify trends).		

Source: Adapted from Liu et al. (2019, p. 27)

Blockchain can also reduce the labor-intensive tasks of the audit process. Adequate and appropriate audit evidence collection, which is one of the important cost elements of the audit process, can be performed at less cost and faster by recording accounting records using blockchain technology. Since there is only one record of the transactions performed and this record cannot be changed, reducing the risks of fraud, and keeping the records open to the access of independent auditors and all other relevant parties will also contribute to the increase of the efficiency of the accounting audit. In this area, if smart contracts are used and some conditional transactions are realized, the execution and control of pre-defined processes can also be provided (Dogan and Ertugay, 2019).

In terms of internal audit activities and internal auditor qualifications, it is critical for internal auditors to have sufficient knowledge of blockchain technology and to help design the necessary controls at the initial stage by taking an active role in the transformation process in order to make the most of the opportunities offered by this technology and to manage the risks as well as possible (Karahan and Tufekci, 2019).

CONCLUSION

Although public applications are still very limited, its awareness in the public is relatively low and the expectation of undertaking operational public services in the near future is low, blockchain technology has quickly taken its place in business life, considering the ever-increasing development speed of technology. It is very important to be prepared for the transformation that blockchain technology will bring, in order not to miss the opportunities of this transformation and to take the necessary precautions against the risks. In order to make this preparation effectively, the prerequisite is to have sufficient knowledge about technology and to closely follow the developments in the world (Karahan and Tufekci, 2019).

Blockchain technology will inevitably affect accounting, auditing, and assurances activities as well as all sectors. This effect can be both positive and negative.

It is suggested that the development of knowledge about blockchain technology and students as actors of the future business world should be added to the accounting and auditing issues of educators and how blockchain will affect the business world. On the other hand, researchers/academics should study how blockchain technology develops and how crypto assets will affect society (Bennet et al., 2020: 300). The study by Zhang et al. (2020) has also reached that the largest accounting firms are not late in adopting blockchain technology, while academics and students stay away from this issue. It can be said that the studies examining the relationship between accounting, auditing and blockchain in Turkey have not been sufficiently focused about blockchain, so it can be said that the same conclusion has been reached with Zhang et al. (2020).

In addition, there is currently no legal regulation in Turkey regarding the accounting of blockchain technology and cryptocurrencies.

REFERENCES

Alarcon, J. L., & Ng, C. (2018). Blockchain and the future of accounting. *Pennsylvania CPA Journal*, *90*(1), 3–7.

Alsaqa, Z. H., Hussein, A. I., & Mahmood, S. M. (2019). The impact of blockchain on accounting information systems. *Journal of Information Technology Management*, *11*(3), 62–80.

Are you ready for blockchain? (n.d.). https://www.thomsonreuters.com/en/reports/blockchain.html

Bennett, S., Charbonneau, K., Leopold, R., Mezon, L., Paradine, C., Scilipoti, A., & Villmann, R. (2020). Blockchain and cryptoassets: Insights from practice. *Accounting Perspectives*, *19*(4), 283–302. doi:10.1111/1911-3838.12238

Blockchain Versions. (n.d.). https://www.tutorialandexample.com/blockchain-versions/

Blockchains vs distributed ledger technology. (n.d.). https://www.tutorialandexample.com/blockchains-vs-distribute d-ledger-technology/

Bonsón, E., & Bednárová, M. (2019). Blockchain and its implications for accounting and auditing. *Meditari Accountancy Research*, *27*(5), 725–740. doi:10.1108/MEDAR-11-2018-0406

Bonyuet, D. (2020). Overview and impact of blockchain on auditing. *The International Journal of Digital Accounting Research*, *20*, 31–43. doi:10.4192/1577-8517-v20_2

Brender, N., Gauthier, M., Morin, J.-H., & Salihi, A. (2018). *The potential impact of blockchain technology on audit practice*. University of Applied Sciences and Arts Western Switzerland.

Calisaneller, A. (2021). *Kim korkar blokzincir'den*. https://home.kpmg/tr/tr/home/gorusler/2021/11/Kim%20korkar%20blokzincir'den.html

Chartered Accountants Australia and New Zealand. (2017). *The future of blockchain: Applications and implications of distributed ledger technology*. Chartered Accountants Australia and New Zealand.

CPA & AICPA. (2017). *Blockchain technology and its potential impact on the audit and assurance profession*. https://us.aicpa.org/content/dam/aicpa/interestareas/frc/assuranceadvisoryservices/downloadabledocuments/blockchain-technology-and-its-potential-impact-on-the-audit-and-assurance-profession.pdf

Dai, J., & Vasarhelyi, M. A. (2017). Toward blockchain-based accounting and assurance. *Journal of Information Systems, 31*(3), 5–21. doi:10.2308/isys-51804

Deloitte. (2021). *Blockchain and its potential impact on the audit profession*. https://www2.deloitte.com/za/en/pages/audit/articles/impact-of-blockchain-in-accounting.html

Demirhan, H. (2019). Vergi denetiminde yeni bir yaklasim olarak blok zinciri teknolojisi. *Bingol Universitesi Sosyal Bilimler Enstitusu Dergisi, 9*(18), 857–876. doi:10.29029/busbed.569413

Dogan, M., & Ertugay, E. (2019). Blokzinciri ve muhasebe alanindaki uygulamalari. *Third Sector Social Economic Review, 54*(4), 1654–1670. doi:10.15659/3.sektor-sosyal-ekonomi.19.11.1228

Gatteschi, V., Lamberti, F., Demartini, C., Pranteda, C., & Santamaria, V. (2018). To blockchain or not to blockchain: That is the question. *IT Professional, 20*(2), 62–74. doi:10.1109/MITP.2018.021921652

Gul, H. (2019). Blokzincir (Blockchain) teknolojisi ve muhasebe bilgi sistemine etkileri. Sosyal, Beseri ve İdari Bilimlerde Yenilikci Yaklasimlar. Ekin Yayinevi.

Guner, M. (2021). Blokzincir teknolojisinin muhasebede kullanimiyla ilgili algilarin belirlenmesine yonelik bir olcek gelistirme calismasi. *Muhasebe ve Finansman Dergisi*, 459-472.

Hazar, H. B. (2018). Blok zincir teknolojisi ile muhasebe uygulamaları. *Vergi Dunyası Dergisi, 1*(448), 45–54.

Kalkinma Bakanligi, T. C. (2018). *On birinci kalkinma plani (2019 - 2023) kamuda kurumsal yonetim ve yenilikcilik ozel ihtisas komisyonu raporu*. https://www.sbb.gov.tr/wp-content/uploads/2020/04/KamudaKurumsalYonetim_ve_YenilikcilikOzelIhtisasKomisyonuRaporu.pdf

Karahan, C., & Tufekci, A. (2019). Blokzincir teknolojisinin ic denetim faaliyetlerine etkileri: Firsatlar ve tehditler. *Denetisim, 0*(19), 55–72.

Kilinc, Y. (2020). Blockchain teknolojisi: Muhasebe ve denetim meslegi acisindan bir inceleme. *Journal of Accounting and Taxation Studies*, *13*(3), 989–1011. doi:10.29067/muvu.678681

Kizil, C., & Hanisoglu, G. S. (2019). Blok zinciri teknolojisi ve kripto paralar muhasebe ile denetimde kurallari degistiriyor [Blockchain technology and crypto currencies are changing the rules of accounting and auditing]. *Dunya Gazetesi*, p. 10. https://www.dunya.com/kose-yazisi/blokzinciri-teknolojisi-ve-kripto-paralarla-muhasebe-ile-denetimde-kurallar-degisiyor/455144

Liu, M., Wu, K., & Xu, J. J. (2019). How will blockchain technology impact auditing and accounting: Permissionless versus permissioned blockchain. *Current Issues in Auditing*, *13*(2), A19–A29. doi:10.2308/ciia-52540

Lyons, T., Courcelas, L., Grandsenne, J., Carrel, E., & Timsit, K. (2018). *Blockchain Innovation in Europe*. European Union Blockchain Observatory and Forum.

Martin, R. (2018). *How blockchain will impact accounting*. https://igniteoutsourcing.com/blockchain/blockchain-accounting-applications/

Nakamoto, S. (2008). Bitcoin: A peer-to-peer electronic cash system. *Decentralized Business Review*, 21260. https://bitcoin.org/bitcoin.pdf

Niranjanamurthy, M., Nithya, B. N., & Jagannatha, S. J. C. C. (2019). Analysis of blockchain technology: Pros, cons, and SWOT. *Cluster Computing*, *22*(6), 14743–14757. doi:10.100710586-018-2387-5

Onay, A. (2021). *Blok zincir teknolojisinin denetime etkisi ve literatüre iliskin bilimetrik bir arastirma*. In S. Karabulut (Ed.), *Muhasebe Konularinda Bilimsel Yaklasim Ve Arastirmalar* (pp. 189–220). Gazi Kitabevi.

Ozkul, F., & Alkan, B. S. (2020). Dijital çagda muhasebenin donusumu: "blockchain" teknolojisinde muhasebe ve mali kontroller. *World of Accounting Science*, *22*(2), 218–236.

Sonmez, Y., & Demir, V. (2021). muhasebe ve finansmanda dijital gelecek. *Mali Cozum*, *31*(167), 39–53.

Suekinci, C., & Catikkas, D. Ö. (2020). Blok zinciri teknolojisinin muhasebe ve vergilendirme uzerine etkileri. *Mali Cozum Dergisi*, *30*(162), 51–65.

Tenorio, E. M. (2021). *Advantages and disadvantages of Blockchain*. https://www.bbva.ch/en/news/advantages-and-disadvantages-of-blockchain/

Thake, M. (2018). *What's the difference between blockchain and DLT?* https://medium.com/nakamo-to/whats-the-difference-between-blockchain-and-dlt-e4b9312c75dd

Uysal, T. U., & Kurt, G. (2018). Muhasebede ve denetimde blok zinciri teknolojisi. *Suleyman Demirel Universitesi İktisadi ve İdari Bilimler Fakultesi Dergisi*, *23*(2), 467–481.

Wang, Y., & Kogan, A. (2018). Designing confidentiality-preserving Blockchain-based transaction processing systems. *International Journal of Accounting Information Systems*, *30*, 1–18. doi:10.1016/j.accinf.2018.06.001

Zhang, Y., Xiong, F., Xie, Y., Fan, X., & Gu, H. (2020). The impact of artificial intelligence and blockchain on the accounting profession. *IEEE Access : Practical Innovations, Open Solutions*, *8*, 110461–110477. doi:10.1109/ACCESS.2020.3000505

Compilation of References

Aamer, A. M., Al-Awlaqi, M. A., Affia, I., Arumsari, S., & Mandahawi, N. (2021). The internet of things in the food supply chain: Adoption challenges. *Benchmarking*, *28*(8), 2521–2541. doi:10.1108/BIJ-07-2020-0371

Aamir, A., Qureshi, R., Khan, F. A., & Huzaifa, M. (2020). Blockchain based academic records verification in smart cities. *Wireless Personal Communications*, *113*(3), 1397–1406. doi:10.100711277-020-07226-0

Abeyratne, S. A., & Monfared, R. P. (2016). Blockchain ready manufacturing supply chain using distributed ledger. *International Journal of Research in Engineering and Technology*, *5*(9), 1–10. doi:10.15623/ijret.2016.0509001

ABN-AMRO. (2018, January 22). *First agricultural commodity trade through blockchain completed—ABNAMRO Bank*. https://www.abnamro.com/en/news/first-agricultural-commodity -trade-through-blockchain-completed%20%20

Abooleet, S., & Kinnett, S. (2023). A Systematic Review of Blockchain-based Loyalty Programs. *Proceedings of the 56th Hawaii International Conference on System Sciences*, 5323-5332. https://hdl.handle.net/10125/103284

Agarwal, S., Basu, K., & Nath, A. (2013). Green computing and green technology based teaching learning and administration in higher education institutions. *International Journal of Advanced Computer Research*, *3*(3), 295.

Agarwal, S., & Nath, A. (2011). Green computing - A new horizon of energy efficiency and electronic waste minimization: a global perspective. *2011 International Conference on Communication Systems and Network Technologies*, 688-693, 10.1109/CSNT.2011.148

Aggarwal, S., & Kumar, N. (2021). Chapter Twenty—Attacks on blockchain☆☆Working model. In S. Aggarwal, N. Kumar, & P. Raj (Eds.), *Advances in Computers* (Vol. 121, pp. 399–410). Elsevier. doi:10.1016/bs.adcom.2020.08.020

Aghaei, H., Naderibeni, N., & Karimi, A. (2021). Designing a tourism business model on block chain platform. *Tourism Management Perspectives*, *39*, 100845. doi:10.1016/j.tmp.2021.100845

Ahmad, M. S., & Shah, S. M. (2021). Moving Beyond the Crypto-Currency Success of Blockchain: A Systematic Survey. *Scalable Computing: Practice and Experience*, *22*(3), 321–346. doi:10.12694cpe.v22i3.1853

Aich, S., Chakraborty, S., Sain, M., Lee, H., & Kim, H.-C. (2019). A Review on Benefits of IoT Integrated Blockchain based Supply Chain Management Implementations across Different Sectors with Case Study. *2019 21st International Conference on Advanced Communication Technology (ICACT)*, 138–141. 10.23919/ICACT.2019.8701910

AISHE. (2020). *All India Survey on Higher Education 2019-20*. New Delhi: Ministry of Education, Government of India. Retrieved from https://www.education.gov.in/sites/upload_files/mhrd/files/statistics-new/aishe_eng.pdf

Aksoy, M. (2013). *Satın Alma ve Diğer Departman İlişkileri*. http://www.satinalmadergisi.com/2013/03/27/satinalma-ve-diger-departman-iliskileri-2/

Al Kawasmi, E., Arnautovic, E., & Svetinovic, D. (2015). Bitcoin-based decentralized carbon emissions trading infrastructure model. *Systems Engineering*, *18*(2), 115–130. doi:10.1002ys.21291

Al Omari, L., Barnes, P., & Pitman, G. (2012). An Exploratory Study into Audit Challenges in IT Governance: A Delphi Approach. *IT Governance, Management and Audit*, 1-12. Retrieved from https://eprints.qut.edu.au/53110/22/An_Exploratory_Study_into_Audit_Challenges_in_IT_Governance_A_Delphi_Approach.pdf

Alahmari, M., Issa, T., Issa, T., & Nau, S. Z. (2019). Faculty awareness of the economic and environmental benefits of augmented reality for sustainability in Saudi Arabian universities. *Journal of Cleaner Production*, *226*, 259–269. doi:10.1016/j.jclepro.2019.04.090

Alammary, A., Alhazmi, S., Almasri, M., & Gillani, S. (2019). Blockchain-Based Applications in Education: A Systematic Review. *Applied Sciences (Basel, Switzerland)*, *9*(12), 2400. doi:10.3390/app9122400

Alarcon, J. L., & Ng, C. (2018). Blockchain and the future of accounting. *Pennsylvania CPA Journal*, *90*(1), 3–7.

Alazab, M., Alhyari, S., Awajan, A., & Abdallah, A. B. (2021). Blockchain technology in supply chain management: An empirical study of the factors affecting user adoption/acceptance. *Cluster Computing*, *24*(1), 83–101. doi:10.100710586-020-03200-4

Albizri, A., & Appelbaum, D. (2021). Trust but Verify: The Oracle Paradox of Blockchain Smart Contracts. *Journal of Information Systems*, *35*(2), 1–16. doi:10.2308/ISYS-19-024

Alevizos, L., Ta, V. T., & Eiza, M. H. (2021). Augmenting Zero Trust Architecture to Endpoints Using Blockchain: A State of the Art Review. *Computer Science, Cryptography and Security*, 1-17. Retrieved from https://arxiv.org/ftp/arxiv/papers/2104/2104.00460.pdf

Alfalla-Luque, R., Medina-Lopez, C., & Schrage, H. (2013). A study of supply chain integration in the aeronautics sector. *Production Planning and Control*, *24*(8–9), 769–784. doi:10.1080/09537287.2012.666868

Alfandi, O. (2021). Pros and cons of Blockchain-based approaches in Education. In *IEEE Global Engineering Education Conference (EDUCON)*. IEEE. https://doi.org/10.1109/EDUCON46332.2021.9454118

Ali, S. I., Ali, A., AlKilabi, M., & Christie, M. (2021). Optimal supply chain design with product family: A cloud-based framework with real-time data consideration. *Computers & Operations Research*, *126*, 105112. doi:10.1016/j.cor.2020.105112

Alladi, T., Chamola, V., Parizi, R. M., & Choo, K. K. R. (2019). Blockchain applications for industry 4.0 and industrial IoT: A review. *IEEE Access : Practical Innovations, Open Solutions*, *7*, 176935–176951. doi:10.1109/ACCESS.2019.2956748

Allam, Z. (2018). *On Smart Contracts and Organisational Performance: A Review of Smart Contracts Through the Blockchain Technology.* doi:10.1515/rebs-2018-0079

Alsaqa, Z. H., Hussein, A. I., & Mahmood, S. M. (2019). The impact of blockchain on accounting information systems. *Journal of Information Technology Management*, *11*(3), 62–80.

AlShamsi, M., Al-Emran, M., & Shaalan, K. (2022). A Systematic Review on Blockchain Adoption. *Applied Sciences (Basel, Switzerland)*, *12*(9), 4245. doi:10.3390/app12094245

Altınpulluk, H., Kesim, M., & Kurubacak, G. (2020, April 2). The Usability of Augmented Reality in Open and Distance Learning Systems: A Qualitative Delphi Study. *International Council for Open and Distance Education, 12*(2), 283-307. Retrieved from https://files.eric.ed.gov/fulltext/EJ1267133.pdf

Ammer, D. S. (1968). *Materials Management*. Richard D. Irwin.

Andoni, M., Robu, V., Flynn, D., Abram, S., Geach, D., Jenkins, D., McCallum, P., & Peacock, A. (2019). Blockchain technology in the energy sector: A systematic review of challenges and opportunities. *Renewable & Sustainable Energy Reviews*, *100*, 143–174. doi:10.1016/j.rser.2018.10.014

Androulaki, E., Barger, A., Bortnikov, V., Cachin, C., Christidis, K., De Caro, A., Enyeart, D., Ferris, C., Laventman, G., Manevich, Y., Muralidharan, S., Murthy, C., Nguyen, B., Sethi, M., Singh, G., Smith, K., Sorniotti, A., Stathakopoulou, C., Vukolić, M., ... Yellick, J. (2018). Hyperledger fabric: A distributed operating system for permissioned blockchains. *Proceedings of the Thirteenth EuroSys Conference*, 1–15. 10.1145/3190508.3190538

Antal, C., Cioara, T., Anghel, I., Antal, M., & Salomie, I. (2021). Distributed Ledger Technology Review and Decentralized Applications Development Guidelines. *Future Internet*, *13*(3), 62. doi:10.3390/fi13030062

Ante, L. (2021). Smart contracts on the blockchain – A bibliometric analysis and review. *Telematics and Informatics*, *57*, 101519. doi:10.1016/j.tele.2020.101519

Antoniadis, I., Spinthiropoulos, K., & Kontsas, S. (2020). Blockchain applications in tourism and tourism marketing: A short review. Strategic Innovative Marketing and Tourism. In A. Kavoura, E. Kefallonitis, & P. Theodoridis (Eds.), *Strategic Innovative Marketing and Tourism. Springer Proceedings in Business and Economics* (pp. 375–384). Springer. doi:10.1007/978-3-030-36126-6_41

Arbatskaya, E., & Khoreva, L. (2021). Blockchain as an innovative digitalization achievement in logistics of eco-tourism. *E3S Web of Conferences, 258*. https://doi.org/ doi:10.1051/e3sconf/202125802016

Are you ready for blockchain? (n.d.). https://www.thomsonreuters.com/en/reports/blockchain.html

Arndt, T. (2018). Empowering university students with blockchain-based transcripts. *Proceedings of the 15th International Conference on Cognition and Exploratory Learning in the Digital Age*, 399–400.

Ataseven, C., & Nair, A. (2017). Assessment of supply chain integration and performance relationships: A meta-analytic investigation of the literature. *International Journal of Production Economics*, *185*, 252–265. doi:10.1016/j.ijpe.2017.01.007

Atzei, N., Bartoletti, M., & Cimoli, T. (2017). A Survey of Attacks on Ethereum Smart Contracts (SoK). In Principles of Security and Trust. Springer. doi:10.1007/978-3-662-54455-6_8

Balasubramanian, S., Sethi, J. S., Ajayan, S., & Paris, C. M. (2022). An enabling Framework for Blockchain in Tourism. *Information Technology & Tourism*, *24*(2), 165–179. doi:10.100740558-022-00229-6

Banerjee, A. (2019). Chapter Nine - Blockchain with IOT: Applications and use cases for a new paradigm of supply chain driving efficiency and cost. In S. Kim, G. C. Deka, & P. Zhang (Eds.), *Advances in Computers* (Vol. 115, pp. 259–292). Elsevier. doi:10.1016/bs.adcom.2019.07.007

Barr, A. (2014). Artificial Intelligence in eLearning. E-learning *Guild Online Forum on New Instructional Design for New Instructional Technology*. Retrieved from https://www.elearningguild.com/showfile.cfm?id=5284

Bartoletti, M., & Pompianu, L. (2017). *An empirical analysis of smart contracts: Platforms, applications, and design patterns*. doi:10.25300/MISQ/2013/37.2.14

Bartolomé, A., Torla, C., Castañeda, L., & Adell, J. (2017). Blockchain in Education: Introduction and Critical Review of the State of the Art. *Edutec*, *61*(61). Advance online publication. doi:10.21556/edutec.2017.61

Baudier, P., Chang, V., & Arami, M. (2022). The Impacts of Blockchain on Innovation Management: Sectoral Experiments: *Journal of Innovation Economics & Management, 37*(1), 1–8. doi:10.3917/jie.037.0001

Bauer, R. A. (1960). Consumer Behavior As Risk Taking. In *Dynamic Marketing for A Changing World*. AMA.

Bdiwi, R., De Runz, C., Faiz, S., & Cherif, A. A. (2018, July). A blockchain based decentralized platform for ubiquitous learning environment. In *2018 IEEE 18th International Conference on Advanced Learning Technologies (ICALT)* (pp. 90-92). IEEE. 10.1109/ICALT.2018.00028

BearingPoint. (2018). *Integrating the retail supply chain*. BearingPoint. https://www.bearingpoint.com/files/Retail_Research_Report_2017_BE.pdf?download=0&itemId=502221#:~:text=Retailers%20who%20identify%20as%20having,reduced%20stockholding%20and%20improved%20availability%20

Bekaroo, G., Bokhoree, C., & Pattinson, C. (2016). Impacts of ICT on the natural ecosystem: A grassroot analysis for promoting socio-environmental sustainability. *Renewable & Sustainable Energy Reviews*, *57*, 1580–1595. doi:10.1016/j.rser.2015.12.147

Ben-Daya, M., As'ad, R., & Nabi, K. A. (2019). A single-vendor multi-buyer production remanufacturing inventory system under a centralized consignment arrangement. *Computers & Industrial Engineering*, *135*, 10–27. doi:10.1016/j.cie.2019.05.032

Benduch, D. (2019). *Risks And Opportunities for Tourism Using Smart Contracts. 6th Geographic Information Systems Conference and Exhibition "GIS ODYSSEY 2019"*. Retrieved from https://www.researchgate.net/publication/344340600

Bennett, S., Charbonneau, K., Leopold, R., Mezon, L., Paradine, C., Scilipoti, A., & Villmann, R. (2020). Blockchain and cryptoassets: Insights from practice. *Accounting Perspectives*, *19*(4), 283–302. doi:10.1111/1911-3838.12238

Bhadra, S., & Kundu, A. (2016). Introducing eco friendly corporate system: A green approach. *International Journal of Green Computing*, *7*(1), 1–24. doi:10.4018/IJGC.2016010101

Bhaskar, P., Tiwari, C. K., & Joshi, A. (2021). Blockchain in education management: Present and future applications. *Interactive Technology and Smart Education*, *18*(1), 1–17. doi:10.1108/ITSE-07-2020-0102

Bickerton, J. (2018, January 8). Digital Coupon Redemption Rates. *RAVEN5*. https://www.raven5.com/blog/digital-coupon-redemption-rates/

Bittencourt, D. F., Goodert, A. R., Sharma, R. C., & Bortolozzi, F. (2020). Framework blockchain education: Rupture in higher education. In R. C. Sharma, H. Yildirim, & G. Kurubacak (Eds.), *Blockchain Technology Applications in Education*. IGI Global. doi:10.4018/978-1-5225-9478-9.ch004

Blinder, M. (2018). *Making cryptocurrency more environmentally sustainable*. https://hbr.org/2018/11/making-cryptocurrency-more-environmentally-sustainable

Blockchain Versions. (n.d.). https://www.tutorialandexample.com/blockchain-versions/

Blockchains vs distributed ledger technology. (n.d.). https://www.tutorialandexample.com/blockchains-vs-distributed-ledger-technology/

Bonoma, T. V. (2008). *Satış Stratejileri*. Harvard Business Review.

Bonsón, E., & Bednárová, M. (2019). Blockchain and its implications for accounting and auditing. *Meditari Accountancy Research, 27*(5), 725–740. doi:10.1108/MEDAR-11-2018-0406

Bonyuet, D. (2020). Overview and impact of blockchain on auditing. *The International Journal of Digital Accounting Research, 20*, 31–43. doi:10.4192/1577-8517-v20_2

Borge, M., Kokoris-Kogias, E., Jovanovic, P., Gasser, L., Gailly, N., & Ford, B. (2017). Proof-of-personhood: Redemocratizing permissionless cryptocurrencies. In *2017 IEEE European Symposium on Security and Privacy Workshops (EuroS & PW)* (pp. 23-26). IEEE. 10.1109/EuroSPW.2017.46

Bouraga, S. (2021). A taxonomy of blockchain consensus protocols: A survey and classification framework. *Expert Systems with Applications, 168*, 114384. doi:10.1016/j.eswa.2020.114384

Bozkurt, A., & Ucar, H. (2020). Blockchain technology as a bridging infrastructure among informal, non-formal, and informal learning processes. In R. C. Sharma, H. Yildirim, & G. Kurubacak (Eds.), *Blockchain Technology Applications in Education*. IGI Global. doi:10.4018/978-1-5225-9478-9.ch001

Brender, N., Gauthier, M., Morin, J.-H., & Salihi, A. (2018). *The potential impact of blockchain technology on audit practice*. University of Applied Sciences and Arts Western Switzerland.

Broggi, J. D., Gallagher, M. A., Lilly, J., Duquette, J., Nimura, C., Pattenden, M., . . . Winkler, A. (2018). *Woolf - Building the first Blockchain University*. https://www.slideshare.net/eraser/woolf-building-the-first-blockchain-university-whitepaper

Broggi, J. D., Gallagher, M. A., Lilly, J., Duquette, J., Nimura, C., Pattenden, M., Richter, F., Arbide, L. S. M. N., Avin, S., Kelley, K., & Lidova, M. (2018). *Woolf - Building the first Blockchain University*. https://woolf.university/assets/doc/whitepaper.pdf

Bucea-Manea-Tonis, R., Martins, O., Bucea-Manea-Tonis, R., Gheorghiță, C., Kuleto, V., Ilić, M. P., & Simion, V.-E. (2021). Blockchain Technology Enhances Sustainable Higher Education. *Sustainability (Basel), 13*(22), 12347. doi:10.3390u132212347

Buhalis, D., & Law, R. (2008). Progress in information technology and tourism management: 20 years on and 10 years after the Internet—The state of eTourism research. *Tourism Management, 29*(4), 609–623. doi:10.1016/j.tourman.2008.01.005

Bulut, E. (2021). Blockchain-based entrepreneurial finance: Success determinants of tourism initial coin offerings. *Current Issues in Tourism*, 1–15. doi:10.1080/13683500.2021.1980505

Business Insider - India. (2018, August 21). *Here's how blockchain reduces the overall costs associated with trade*. Business Insider. https://www.businessinsider.in/heres-how-blockchain-reduces-the-overall-costs-associated-with-trade/articleshow/65484887
.cms

Business Insider. (2017, March 20). *The Blockchain in banking report: The future of blockchain solutions and technologies*. Business Insider. https://www.businessinsider.com/blockchain-in-banking-2017-3

Buterin, V. (2022). *Proof of Stake*. Seven Stories Press.

Büttgen, M., Dicenta, J., Spohrer, K., Venkatesh, V., Raman, R., Hoehle, H., De Keyser, A., Verbeeck, C., Zwienenberg, T. J., Jorgensen, K. P., ... Schar, F. (2021). Blockchain in Service Management and Service Research—Developing a Research Agenda and Managerial Implications. *Journal of Service Management Research*, *5*(2), 71–102. doi:10.15358/2511-8676-2021-2-71

Cai, Y.-J., & Lo, C. K. Y. (2020). Omni-channel management in the new retailing era: A systematic review and future research agenda. *International Journal of Production Economics*, *229*, 107729. doi:10.1016/j.ijpe.2020.107729

Cai, Y., & Zhu, D. (2016). Fraud detections for online businesses: A perspective from blockchain technology. *Financial Innovation*, *2*(20), 20. Advance online publication. doi:10.118640854-016-0039-4

Calatayud, A., Mangan, J., & Christopher, M. (2018). The self-thinking supply chain. *Supply Chain Management*, *24*(1), 22–38. doi:10.1108/SCM-03-2018-0136

Caldarelli, G., & Ellul, J. (2021). Trusted Academic Transcripts on the Blockchain: A Systematic Literature Review. *Applied Sciences (Basel, Switzerland)*, *11*(4), 1842. doi:10.3390/app11041842

Calisaneller, A. (2021). *Kim korkar blokzincir'den*. https://home.kpmg/tr/tr/home/gorusler/2021/11/Kim%20korkar%20blokzincir'den.html

Calvaresi, D., Leis, M., Dubovitskaya, A., Schegg, R., & Schumacher, M. (2019). Trust in Tourism via Blockchain Technology: Results from a Systematic Review. In J. Pesonen & J. Neidhardt (Eds.), *Information and Communication Technologies in Tourism 2019* (pp. 304–317). Springer International Publishing. doi:10.1007/978-3-030-05940-8_24

Cambridge Dictionary. (2023). *Blockchain*. Retrieved from https://dictionary.cambridge.org/dictionary/english/blockcha
in

Can, P. (2016). Endüstriyel Pazarlarda Algılanan Riskin Marka Duyarlılığına Etkisi Üzerine Bir Araştırma. *Uluslararası Alanya İşletme Fakültesi, 8,* 175-190.

Çapar, H. (2020). Using cryptocurrencies and transactions in medical tourism. *Journal of Economic and Administrative Sciences.*, *37*(4), 677–693. doi:10.1108/JEAS-07-2019-0080

Caro, F., & Sadr, R. (2019). The Internet of Things (IoT) in retail: Bridging supply and demand. *Business Horizons*, *62*(1), 47–54. doi:10.1016/j.bushor.2018.08.002

Carson, B., Romanelli, G., Walsh, P., & Zhumaev, A. (2018, June 19). *Blockchain beyond the hype: What is the strategic business value?* https://www.mckinsey.com/business-functions/mckinsey-digital/our-insights/blockchain-beyond-the-hype-what-is-the-strategic-business-value

Casey, K. (2017, April 27). *Blockchain: The Pro's and Cons of a Technology that Will Affect our Future*. Medium. https://medium.com/totvslabs/blockchain-the-pros-and-con-s-of-a-technology-that-will-affect-our-future-f67037da7d64

Casino, F., Dasaklis, T. K., & Patsakis, C. (2019). A systematic literature review of blockchain-based applications: Current status, classification and open issues. *Telematics and Informatics*, *36*, 55–81. Advance online publication. doi:10.1016/j.tele.2018.11.006

CBSE. (2021). *Academic {BlockChain} Documents (ABCD)*. New Delhi: Central Board of Secondary Education. Retrieved from https://cbse.certchain.nic.in/aboutcc

CCA. (2021). *e-Authentication guidelines for eSign- Online Electronic Signature Service: Report of the Committee on New PKI based Digital/Electronic Signature Schemes*. Controller of Certifying Authorities, MeitY, Govt. of India. Retrieved from https://cca.gov.in/sites/files/pdf/esign/CCA-EAUTH.pdf

Cedeño, J. (2020). *How blockchain could ımpact education in 2020 and beyond*. https://www.gettingsmart.com/2020/02/23/how-blockchain-could-impact-education-in-2020-and-beyond/

Cedeno, J. (2020, February 23). *How Blockchain Could Impact Education in 2020 and Beyond*. Retrieved from https://www.gettingsmart.com/2020/02/23/how-blockchain-could-impact-education-in-2020-and-beyond/

Chainalysis. (2021). *The 2021 Global Crypto Adoption Index: Worldwide Adoption Jumps Over 880% With P2P Platforms Driving Cryptocurrency Usage in Emerging Markets*. Retrieved from https://blog.chainalysis.com/reports/2021-global-crypto-adoption-index/

Chang, M., Arachchilage, C. S. M. W., Kim, M., & Lim, H. (2022). Acceptance of Tourism Blockchain Based on Utaut and Connectivism Theory. SSRN *Electronic Journal*. doi:10.2139/ssrn.4017406

Chartered Accountants Australia and New Zealand. (2017). *The future of blockchain: Applications and implications of distributed ledger technology*. Chartered Accountants Australia and New Zealand.

Chattopadhyay, S. (2014). Learner Support Services in Open Distance Learning System: Case Study on IGNOU. *Conference Proceedings 'Distance Learning and Reciprocal Library System: Is Public Library Network the better Option!'*. Retrieved from http://eprints.rclis.org/29071/

Chen, G., Xu, B., Lu, M., & Chen, N.-S. (2018). Exploring Blockchain technology and its potential applications for education. *Smart Learning Environments*, *5*(1), 1–10. doi:10.118640561-017-0050-x

Chenguel, M. (2021). Blockchain and ecological impact: Between reality and accusation? *International Conference on Global Economic Revolutions: The Era of the Digital Economy: ICGER 2021*.

Chen, J., Cai, T., He, W., Chen, L., Zhao, G., Zou, W., & Guo, L. (2020). A Blockchain-Driven Supply Chain Finance Application for Auto Retail Industry. *Entropy (Basel, Switzerland)*, *22*(1), 95. doi:10.3390/e22010095 PMID:33285870

Chen, S.-H., & Tham, A. (2023). A crypto-tourism case study of agnes water/seventeen seventy, Australia. *Tourism and Hospitality Research*, *23*(1), 108–112. doi:10.1177/14673584221085472

Chen, W., Bohloul, S. M., Ma, Y., & Li, L. (2022). A blockchain-based information management system for academic institutions: A case study of international students' workflow. *Information Discovery and Delivery*, *50*(4), 343–352. doi:10.1108/IDD-01-2021-0010

Chen, Y. (2018). Blockchain tokens and the potential democratization of entrepreneurship and innovation. *Business Horizons*, *61*(4), 567–575. doi:10.1016/j.bushor.2018.03.006

Cheriguene, A., Kabache, T., Kerrache, C. A., Calafate, C. T., & Cano, J. C. (2022). NOTA: A novel online teaching and assessment scheme using Blockchain for emergency cases. *Education and Information Technologies*, *27*(1), 115–132. doi:10.100710639-021-10629-6 PMID:34248386

Chivu, R.-G., Popa, I.-C., Orzan, M.-C., Marinescu, C., Florescu, M. S., & Orzan, A.-O. (2022). The Role of Blockchain Technologies in the Sustainable Development of Students' Learning Process. *Sustainability*, *14*(3), 1406. doi:10.3390u14031406

Choi, T.-M. (2020). Supply chain financing using blockchain: Impacts on supply chains selling fashionable products. *Annals of Operations Research.* Advance online publication. doi:10.100710479-020-03615-7

Choi, T.-M., & Luo, S. (2019). Data quality challenges for sustainable fashion supply chain operations in emerging markets: Roles of blockchain, government sponsors and environment taxes. *Transportation Research Part E, Logistics and Transportation Review*, *131*, 139–152. doi:10.1016/j.tre.2019.09.019

Christianson, S. J. (2019, April 29). *The Blockchain Game: A great new tool for your classroom*. Blockchain Pulse: IBM Blockchain Blog. https://www.ibm.com/blogs/blockchain/2019/04/the-blockchain-game-a-great-new-tool-for-your-classroom/

Christidis, K., & Devetsikiotis, M. (2016). Blockchains and Smart Contracts for the Internet of Things. *IEEE Access : Practical Innovations, Open Solutions*, *4*, 2292–2303. doi:10.1109/ACCESS.2016.2566339

Chuen, D. (2015). *Handbook of Digital Currency: Bitcoin*. Innovation, Financial Instruments, and Big Data.

Chukowry, V., Nanuck, G., & Sungkur, R. K. (2021). The future of continuous learning–Digital badge and microcredential system using blockchain. *Global Transitions Proceedings*, *2*(2), 355–361. doi:10.1016/j.gltp.2021.08.026

Chung, M. J.-Y., Friesen, A. L., Fox, D., Meltzoff, A. N., & Rao, R. P. N. (2015). A Bayesian developmental approach to robotic goal-based imitation learning. *PLoS One*, *10*(11), e0141965. doi:10.1371/journal.pone.0141965 PMID:26536366

Ciftci, G. T. (2020). Change of the learning cycle after Blockchain: Changing trust society. In R. C. Sharma, H. Yildirim, & G. Kurubacak (Eds.), *Blockchain Technology Applications in Education*. IGI Global. doi:10.4018/978-1-5225-9478-9.ch003

Clark, D. (2016). *10 ways Blockchain could be used in education*. Retrieved from https://oeb.global/oeb-insights/10-ways-blockchain-could-be-used-in-education/.

Clark, D. (2020, February 20). *Forecasted expenditure on advanced EdTech worldwide 2025*. Retrieved from https://www.statista.com/statistics/1085930/edtech-expenditure-forecast/

ClarkC. D.BaksiV. A.BraineL. (2016). Smart contract templates: Foundations design landscape and research directions. arXiv:1608.00771

Cobanoglu, C., Dogan, S., Berezina, K., & Collins, G. (2021). *Hospitality and Tourism Information Technology*. USF M3 Publishing, LLC. https://doi.org/ doi:10.5038/9781732127593

Cochrane Library. (2022). *Aout Cochrane Reviews*. https://www.cochranelibrary.com/about/about-cochrane-reviews

Cong, L. W., & He, Z. (2019). Blockchain Disruption and Smart Contracts. *Review of Financial Studies*, *32*(5), 1754–1797. doi:10.1093/rfs/hhz007

Conti, M., Sandeep Kumar, E., Lal, C., & Ruj, S. (2018). A Survey on Security and Privacy Issues of Bitcoin. *IEEE Communications Surveys and Tutorials*, *20*(4), 3416–3452. doi:10.1109/COMST.2018.2842460

Conway, L. (2021, November 4). *Blockchain Definition: What you need to know*. Retrieved from https://www.investopedia.com/terms/b/blockchain.asp

Conway, C. M. (2020). *Approaches to Qualitative Research: An Oxford Handbook of Qualitative Research in American Music Education* (Vol. 1). Oxford University Press.

Copley, T. P., & Callom, F. L. (1971). Industrial Search Behavior And Perceived Risk. *Annual Conference of the Association for Consumer Research*, 208- 231.

CPA & AICPA. (2017). *Blockchain technology and its potential impact on the audit and assurance profession.* https://us.aicpa.org/content/dam/aicpa/interestareas/frc/assuranceadvisoryservices/downloadabledocuments/blockchain-technology-and-its-potential-impact-on-the-audit-and-assurance-profession.pdf

Creswell, J. W. (2014). *Research Design: Qualitative, Quantitative, and Mixed Methods Approaches* (4th ed.). Sage.

Crosby, M. (2016). Blockchain technology: Beyond bitcoin. *Applied Innovation, 2,* 6–10. https://j2-capital.com/blockchain-technology-beyond-bitcoin

Cui, L., Gao, M., Dai, J., & Mou, J. (2020). Improving supply chain collaboration through operational excellence approaches: An IoT perspective. *Industrial Management & Data Systems.* doi:10.1108/IMDS-01-2020-0016

Curry, E. J. (1999). Uluslararası Pazarlama. BZD Yayın ve İletişim.

Cvitanović, P. L. (2018). New Technologies in Marketing as Competitive Advantage. *Proceedings of the ENTRENOVA - ENTerprise REsearch InNOVAtion Conference, 4*(1), 266–274.

Dai, J., & Vasarhelyi, M. A. (2017). Toward blockchain-based accounting and assurance. *Journal of Information Systems, 31*(3), 5–21. doi:10.2308/isys-51804

Dalkey, N. C. (1967). *Delphi. The Rand Corporation.* Clearinghouse for Federal Scientific & Technical Information. Retrieved from https://apps.dtic.mil/sti/pdfs/AD0660554.pdf

Dalton, M. (2018, November 9). *Feeless Cryptocurrency: How Do Blockchains Achieve Free Transactions?* https://www.bitrates.com/news/p/feeless-cryptocurrency-how-do-blockchains-achieve-free-transactions

Dandona, L., Dandona, R., Kumar, G. A., Shukla, D. K., Paul, V. K., Balakrishnan, K., ... Thakur, J. S. (2017). Nations within a nation: Variations in epidemiological transition across the states of India, 1990–2016 in the Global Burden of Disease Study. *Lancet, 390*(10111), 2437–2460. doi:10.1016/S0140-6736(17)32804-0 PMID:29150201

Dash, M. K., Panda, G., Kumar, A., & Luthra, S. (2022). Applications of blockchain in government education sector: A comprehensive review and future research potentials. *Journal of Global Operations and Strategic Sourcing, 15*(3), 449–472. doi:10.1108/JGOSS-09-2021-0076

Daskalakis, N., & Georgitseas, P. (2020). *An Introduction to Cryptocurrencies.* Routledge. doi:10.4324/9780429352584

DasS.RoutJ.PriyadarshiniR.MishraM. (2022). A Comparative Analysis of the Consensus Algorithms in Blockchain Technology. SSRN. Retrieved from doi:10.2139/ssrn.4091413

Davidson, P., & Hasledalen, K. (2014). Cyber Threats to Online Education: A Delphi Study. In P. Dover, S. Hariharan, & M. Cummings (Ed.), *Proceedings of the 2nd International Conference on Management, Leadership and Governance, ICMLG 2014* (pp. 68-77). Academic Conferences and Publishing International (ACPI).

Dayal, P. (2020, April 9). *Why Blockchain Occupies a Unique Seat at the Advertising Table*. https://www.martechadvisor.com/articles/blockchain/why-blockchain-occupies-a-unique-seat-at-the-advertising-table/

de Vass, T., Shee, H., & Miah, S. J. (2021). Iot in supply chain management: A narrative on retail sector sustainability. *International Journal of Logistics*, *24*(6), 605–624. doi:10.1080/13675567.2020.1787970

Deenmahomed, H. A., Didier, M. M., & Sungkur, R. K. (2021). The future of university education: Examination, transcript, and certificate system using Blockchain. *Computer Applications in Engineering Education*, *29*(5), 1234–1256. doi:10.1002/cae.22381

Deeter-Schmelz & Kennedy. (2002). An exploratory study of the internet as an industrial communication tool examining buyers perceptions. İnternational Marketing Management, 145-154.

Delisle, B. (2018, January 30). An Introduction to Basic Attention Token (BAT)—Blockchain-Based Digital Advertising. *CryptoSlate*. https://cryptoslate.com/basic-attention-token/

Deloitte & Assocham. (2017). *Blockchain technology in India: Opportunities and challenges*. Retrieved from https://www2.deloitte.com/content/dam/Deloitte/in/Documents/strategy/in-strategy-innovation-blockchain-technology-india-opportunities-challenges-noexp.pdf

Deloitte. (2016, February 24). *Blockchain: Democratized trust*. Deloitte Insights. /content/www/us/en/insights/focus/tech-trends/2016/blockchain-applications-and-trust-in-a-global-economy.html

Deloitte. (2017). *Continuous interconnected supply chain*. https://www2.deloitte.com/content/dam/Deloitte/pt/Documents/blockchainsupplychain/lu-blockchain-internet-things-supply-chain-traceability%20(1).pdf

Deloitte. (2018a). *Deloitte and Chronicled Alliance*. Deloitte United States. https://www2.deloitte.com/us/en/pages/about-deloitte/solutions/deloitte-and-chronicled-alliance.html

Deloitte. (2018b). *Global Blockchain Survey 2018*. Deloitte United Kingdom. https://www2.deloitte.com/uk/en/pages/innovation/articles/global-blockchain-survey-2018.html

Deloitte. (2021). *Blockchain and its potential impact on the audit profession*. https://www2.deloitte.com/za/en/pages/audit/articles/impact-of-blockchain-in-accounting.html

Demir, H., & Gümüşoğlu, Ş. (2009). Üretim yönetimi (İşlemler yönetimi). Beta Basım A.

Demirel, E., Karagöz Zeren, S., & Hakan, K. (2021). Smart contracts in tourism industry: A model with blockchain integration for post pandemic economy. *Current Issues in Tourism*, 1–15. doi:10.1080/13683500.2021.1960280

Demirhan, H. (2019). Vergi denetiminde yeni bir yaklasim olarak blok zinciri teknolojisi. *Bingol Universitesi Sosyal Bilimler Enstitusu Dergisi*, *9*(18), 857–876. doi:10.29029/busbed.569413

DePatie, J. (2017). *New Kid on the Blockchain – Exchange*. http://scienceandentertainmentexchange.org/article/new-kid-on-the-blockchain/

Deshmukh, S. (2020, September 27). *3 ways blockchain can accelerate sustainable development*. World Economic Forum. https://www.weforum.org/agenda/2020/09/3-ways-blockchain-can-contribute-to-sustainable-development/

Deshpande, I. (2019, July 25). *What Is Blockchain in Marketing? Strategies, Best Practices, Benefits and Examples*. https://www.martechadvisor.com/articles/blockchain/what-is-blockchain-how-to-use-it-in-marketing/

Deshpande, Y., & Unhelkar, B. (2011). Information systems for a green organisation. In B. Unhelkar (Ed.), *Handbook of research on green ICT: Technology, business and social perspectives* (pp. 116–130). IGI Global. doi:10.4018/978-1-61692-834-6.ch008

Devine, P. (2015). Blockchain learning: can crypto-currency methods be appropriated to enhance online learning? *ALT Online Winter Conference*, *7*(10). https://altc.alt.ac.uk/online2015/sessions/blockch

Dhagarra, D., Goswami, M., Sarma, P. R. S., & Choudhury, A. (2019). Big Data and blockchain supported conceptual model for enhanced healthcare coverage: The Indian context. *Business Process Management Journal*, *25*(7), 1612–1632. doi:10.1108/BPMJ-06-2018-0164

Dhaou, S., Zalan, T., & Toufaily, E. (2017). Blockchain Technology Adoption: Implications and Challenges. *ICEB 2017 Proceedings*. https://aisel.aisnet.org/iceb2017/39

Di Pietro, R., Salleras, X., Signorini, M., & Waisbard, E. (2018). A blockchain-based Trust System for the Internet of Things. *Proceedings of the 23nd ACM on Symposium on Access Control Models and Technologies*, 77–83. 10.1145/3205977.3205993

Dicker, J. (2018, May 1). *3 Steps Toward Accepting Blockchain's Place in Media*. https://www.adweek.com/performance-marketing/3-steps-toward-accepting-blockchains-place-in-media/

Dimitrov, B. (2020, April 9). *Why Fighting COVID-19 Needs More Than Blockchain To Succeed*. Forbes. https://www.forbes.com/sites/biserdimitrov/2020/04/09/why-fighting-covid-19-needs-more-than-blockchain-to-succeed/

Din, N., Haron, S., & Ahmad, H. (2013). The level of awareness on the green ICT concept and self directed learning among Malaysian Facebook users. *Procedia: Social and Behavioral Sciences*, *85*, 464–473. doi:10.1016/j.sbspro.2013.08.375

Dobrovnik, M., Herold, D. M., Fürst, E., & Kummer, S. (2018). Blockchain for and in Logistics: What to Adopt and Where to Start. *Logistics*, *2*(3), 18. doi:10.3390/logistics2030018

Dogan, M., & Ertugay, E. (2019). Blokzinciri ve muhasebe alanindaki uygulamalari. *Third Sector Social Economic Review*, *54*(4), 1654–1670. doi:10.15659/3.sektor-sosyal-ekonomi.19.11.1228

Dogru, T., Mody, M., & Leonardi, C. (2018). Blockchain Technology & its Implications for the Hospitality Industry. *Boston Hospitality Review*, 1-12. Retrieved from https://www.bu.edu/bhr/files/2018/02/Blockchain-Technology-and-its-Implications-for-the-Hospitality-Industry.pdf

Dutta, A. (2021, July 25). *Real-World applications of Blockchain Technology* [Blog]. Retrieved from https://www.analyticsinsight.net/real-world-applications-of-blockchain-technologies/

Dutta, P., Choi, T.-M., Somani, S., & Butala, R. (2020). Blockchain technology in supply chain operations: Applications, challenges and research opportunities. *Transportation Research Part E, Logistics and Transportation Review*, *142*, 102067. doi:10.1016/j.tre.2020.102067 PMID:33013183

Dwivedi, Y. K., Hughes, L., Ismagilova, E., Aarts, G., Coombs, C., Crick, T., Duan, Y., Dwivedi, R., Edwards, J., Eirug, A., Galanos, V., Ilavarasan, P. V., Janssen, M., Jones, P., Kar, A. K., Kizgin, H., Kronemann, B., Lal, B., Lucini, B., ... Williams, M. D. (2021). Artificial Intelligence (AI): Multidisciplinary perspectives on emerging challenges, opportunities, and agenda for research, practice and policy. *International Journal of Information Management*, *57*, 101994. doi:10.1016/j.ijinfomgt.2019.08.002

Education in India. (2023, January 4). Wikipedia. https://en.wikipedia.org/wiki/Education_in_India

Efanov, D., & Roschin, P. (2018). The All-Pervasiveness of the Blockchain Technology. *Procedia Computer Science*, *123*, 116–121. doi:10.1016/j.procs.2018.01.019

Eickhoff, M., Muntermann, J., & Weinrich, T. (2017). *What do FinTechs actually do? A Taxonomy of FinTech Business Models*. Academic Press.

Enescu, F. M., Bizon, N., Onu, A., Răboacă, M. S., Thounthong, P., Mazare, A. G., & Şerban, G. (2020). Implementing blockchain technology in irrigation systems that integrate photovoltaic energy generation systems. *Sustainability (Basel)*, *12*(4), 1540. doi:10.3390u12041540

English, T. (2019, January 7). *What is the Future of Blockchain and Advertising?* https://interestingengineering.com/what-is-the-future-of-blockchain-and-advertising

Ennis, D. (2019, August 6). *Walmart may tap unbanked market with cryptocurrency plan*. Marketing Dive. https://www.marketingdive.com/news/walmart-crypto-patent-banking-credit-card-fees/560318/

Erceg, A., Damoska Sekuloska, J., & Kelić, I. (2020). Blockchain in the Tourism Industry—A Review of the Situation in Croatia and Macedonia. *Informatics (MDPI)*, *7*(1), 5. doi:10.3390/informatics7010005

Erdal, M. (2018). *Tedarikçi araştırması ve örgütsel pazarlarda yeni istihbarat kaynakları: Akıllı mobil uygulamalar.* www.startupvadisi.com/...pazarlarda-yeni-istihbarat-kaynakla ri-akilli-mobil-uygulamalar

Ertemel, A. V. (2018). Implications of Blockchain Technology on Marketing. *Journal of International Trade. Logistics and Law*, *4*(2), 35–44.

Eryılmaz, M., Kaygusuz, S., & Efil, İ. (2015). *İşletme Kuram ve Pratik.* Ekin Basım yayın.

Esmaeilpour Ghouchani, B., Jodaki, S., Joudaki, M., Balali, A., & Rajabion, L. (2019). A model for examining the role of the Internet of Things in the development of e-business. *VINE Journal of Information and Knowledge Management Systems*, *50*(1), 20–33. doi:10.1108/VJIKMS-04-2019-0058

European Economic and Social Committee. (2019, July 31). *Blockchain can have multiple applications in social economy but must not create new digital economy elite, says EESC.* EESC. https://www.eesc.europa.eu/en/news-media/news/blockchain-can-have-multiple-applications-social-economy-must-not-create-new-digital-economy-elite-says-eesc

European Environment Agency. (2020). *Blockchain and the environment.* https://www.eea.europa.eu/publications/blockchain-and-the-environment

Fanning, K., & Centers, D. P. (2016). Blockchain and its coming impact on financial services. *Journal of Corporate Accounting & Finance*, *27*(5), 53–57. doi:10.1002/jcaf.22179

Farah, J. C., Vozniuk, A., Rodríguez-Triana, M. J., & Gillet, D. (2018, July). A blueprint for a blockchain-based architecture to power a distributed network of tamper-evident learning trace repositories. In *2018 IEEE 18th International Conference on Advanced Learning Technologies (ICALT)* (pp. 218-222). IEEE. 10.1109/ICALT.2018.00059

Fedorova, E. P., & Skobleva, E. I. (2020). Application of Blockchain Technology in Higher Education. *Eur. J. Contemp. Educ.*, *9*, 552–571. https://eric.ed.gov/?id=EJ1272331

Filimonau, V., & Naumova, E. (2020). The blockchain technology and the scope of its application in hospitality operations. *International Journal of Hospitality Management*, *87*, 102383. doi:10.1016/j.ijhm.2019.102383

Filipova, N. (2018). Blockchain - an opportunity for developing new business models. Business Management. *D. A. Tsenov Academy of Economics*, *20*(2), 75–92.

Flynn, B. B., Huo, B., & Zhao, X. (2010). The impact of supply chain integration on performance: A contingency and configuration approach. *Journal of Operations Management*, *28*(1), 58–71. doi:10.1016/j.jom.2009.06.001

Fosso Wamba, S., Kala Kamdjoug, J. R., Epie Bawack, R., & Keogh, J. G. (2020). Bitcoin, Blockchain and Fintech: A systematic review and case studies in the supply chain. *Production Planning and Control*, *31*(2–3), 115–142. doi:10.1080/09537287.2019.1631460

Fraga-Lamas, P., & Fernandez-Carames, T. M. (2019). A Review on Blockchain Technologies for an Advanced and Cyber-Resilient Automotive Industry. *IEEE Access : Practical Innovations, Open Solutions*, *7*, 17578–17598. doi:10.1109/ACCESS.2019.2895302

Fragnière, E., Sahut, J.-M., Hikkerova, L., Schegg, R., Schumacher, M., Grèzes, S., & Ramseyer, R. (2022). Blockchain technology in the tourism industry: New perspectives in Switzerland. *Journal of Innovation Economics Management*, *37*(1), 65–90. doi:10.3917/jie.pr1.0111

França, A. S. L., Neto, J. A., Gonçalves, R. F., & Almeida, C. M. V. B. (2020). Proposing the use of blockchain to improve the solid waste management in small municipalities. *Journal of Cleaner Production*, *244*, 118529. doi:10.1016/j.jclepro.2019.118529

Francisco Luis, B.-M., Pedro Víctor, N.-C.-U., Valentín, M.-M., & Esteban, R.-F. (2022). Blockchain as a Service: A Holistic Approach to Traceability in the Circular Economy. In S. S. Muthu (Ed.), *Blockchain Technologies for Sustainability* (pp. 119–133). Springer. doi:10.1007/978-981-16-6301-7_6

Friedlmaier, M., Tumasjan, A., & Welpe, I. M. (2018). Disrupting Industries with Blockchain: The Industry, Venture Capital Funding, and Regional Distribution of Blockchain Ventures. *Proceedings of the 51st Hawaii International Conference on System Sciences*, 10. 10.24251/HICSS.2018.445

Galvez, J. F., Mejuto, J. C., & Simal-Gandara, J. (2018). Future challenges on the use of blockchain for food traceability analysis. *Trends in Analytical Chemistry*, *107*, 222–232. doi:10.1016/j.trac.2018.08.011

Ganbold, O., Matsui, Y., & Rotaru, K. (2020). Effect of information technology-enabled supply chain integration on firm's operational performance. *Journal of Enterprise Information Management*, *34*(3), 948–989. doi:10.1108/JEIM-10-2019-0332

Gan, G., Lu, Z., & Jiang, J. (2011). Internet of Things Security Analysis. *2011 International Conference on Internet Technology and Applications*, 1–4. 10.1109/ITAP.2011.6006307

Gantz, S. (2014). The Basics of IT Audit, Purposes, Practices, and Practical Information. Elsevier.

Garg, M., Gupta, S., Goh, M., & Desouza, R. (2010). Sustaining the green information technology movement. In N. Bajgoric (Ed.), *Always-on enterprise information systems for business continuance: Technologies for reliable and scalable operations* (pp. 218–230). IGI Global. doi:10.4018/978-1-60566-723-2.ch013

Gartner. (2019, December 12). *Gartner Survey Reveals Blockchain Adoption Combined with IoT Adoption Is Booming in the U.S.* Retrieved 27 December, 2021 from https://www.gartner.com/en/newsroom/press-releases/2019-12-12-gartner-survey-reveals-blockchain-adoption-combined-with-iot-adoption-is-booming-in-the-us

Gartner. (2019, June 12). *Gartner Reveals Seven Mistakes to Avoid in Blockchain Projects*. Gartner. https://www.gartner.com/en/newsroom/press-releases/2019-06-12-gartner-reveals-seven-mistakes-to-avoid-in-blockchain

Gatteschi, V., Lamberti, F., & Demartini, C. (2020). Blockchain technology use cases. In S. Kim & G. C. Deka (Eds.), *Advanced applications of Blockchain technology* (pp. 91–114). Springer. doi:10.1007/978-981-13-8775-3_4

Gatteschi, V., Lamberti, F., Demartini, C., Pranteda, C., & Santamaria, V. (2018). To blockchain or not to blockchain: That is the question. *IT Professional, 20*(2), 62–74. doi:10.1109/MITP.2018.021921652

Gaur, N. (2020). Blockchain challenges in adoption. *Managerial Finance, 46*(6), 849–858. doi:10.1108/MF-07-2019-0328

George Bernard Shaw Quotes. (n.d.). *BrainyQuote.com*. Retrieved August 21, 2022, from BrainyQuote.com Web site: https://www.brainyquote.com/quotes/george_bernard_shaw_38692 3

Gibson, D., Ostashewski, N., Flintoff, K., Grant, S., & Knight, E. (2015). Digital badges in education. *Education and Information Technologies, 20*(2), 403–410. doi:10.100710639-013-9291-7

Gilda, S., & Mehrotra, M. (2018, January). Blockchain for student data privacy and consent. In *2018 International Conference on Computer Communication and Informatics (ICCCI)* (pp. 1-5). IEEE. 10.1109/ICCCI.2018.8441445

Giungato, P., Rana, R., Tarabella, A., & Tricase, C. (2017). Current trends in sustainability of bitcoins and related blockchain technology. *Sustainability, 9*(12), 2214. doi:10.3390/su9122214

Gökal, H., Buzun, M., & Adalier, A. (2023). Blockchain: A Content Analysis of Graduate Theses in Turkey. *The Turkish Online Journal of Educational Technology, 22*(1). http://www.tojet.net/articles/v22i1/2215.pdf

Goldin, M., Soleimani, A., & Young, J. (2017). *The AdChain Registry*. Academic Press.

Goodwin, T. (2019). Digital Darwinism/ Survival of the Fittest in the Age of Business Distruption. Dijital Darwinizm. Siyah Kitap.

Government of India. (2021). *National Strategy on Blockchain*. New Delhi: MeitY, Government of India. Retrieved from https://www.meity.gov.in/writereaddata/files/NationalStrategyBCT_%20Jan2021_final.pdf

Government of India. (n.d.). *Centre of Excellence in Blockchain Technology*. Retrieved from https://blockchain.gov.in/aboutus.html

Gräther, W., Kolvenbach, S., Ruland, R., Schütte, J., Torres, C., & Wendland, F. (2018). Blockchain for education: Lifelong learning passport. In W. Prinz & P. Hoschka (Eds.), *Proceedings of the 1st ERCIM Blockchain Workshop 2018, Reports of the European Society for Socially Embedded Technologies*. doi:10.18420/blockchain2018_07

Grech, A., & Camilleri, A. F. (2017). *Blockchain in Education*. 132 R. JRC Science for Policy Report; Publications Office of the European Union. https://publications.jrc.ec.europa.eu/repository/handle/JRC1 08255

Grech, A., & Camilleri, A. F. (2017). Blockchain in Education. EUR 28778 EN. doi:10.2760/60649

GroenfeldtT. (2014). *Forbes*. https://www.forbes.com/sites/tomgroenfeldt/2014/05/08/insiders-pose-a-serious-threat-to-corporate-information/?sh=2600bf 965ca9

Gul, H. (2019). Blokzincir (Blockchain) teknolojisi ve muhasebe bilgi sistemine etkileri. Sosyal, Beseri ve İdari Bilimlerde Yenilikci Yaklasimlar. Ekin Yayinevi.

Guner, M. (2021). Blokzincir teknolojisinin muhasebede kullanimiyla ilgili algilarin belirlenmesine yonelik bir olcek gelistirme calismasi. *Muhasebe ve Finansman Dergisi*, 459-472.

Gupta, H., Kumar, S., Kusi-Sarpong, S., Jabbour, C. J. C., & Agyemang, M. (2020). Enablers to supply chain performance on the basis of digitization technologies. *Industrial Management & Data Systems, 121*(9), 1915–1938. doi:10.1108/IMDS-07-2020-0421

Gupta, N., & Bedi, P. (2018). E-waste management using blockchain based smart contracts. In *2018 International Conference on Advances in Computing, Communications and Informatics (ICACCI)* (pp. 915-921). IEEE. 10.1109/ICACCI.2018.8554912

Guzenko, I. (2019, September 17). *Council Post: How Blockchain Is Affecting the Marketing and Advertising Industry*. Forbes. https://www.forbes.com/sites/forbestechcouncil/2019/09/27/how-blockchain-is-affecting-the-marketing-and-advertising-indu stry/

Haber, S., & Stornetta, W. S. (1991). How to timestamp a digital document. *Journal of Cryptology, 3*(2), 99–111. doi:10.1007/BF00196791

Hallikas, J., Immonen, M., & Brax, S. (2021). Digitalizing procurement: The impact of data analytics on supply chain performance. *Supply Chain Management, 26*(5), 629–646. doi:10.1108/SCM-05-2020-0201

Han, Y., Li, C., Li, P., Wu, M., Zhou, D., & Long, F. (2020). Shrec: Bandwidth-efficient transaction relay in high-throughput blockchain systems. *Proceedings of the 11th ACM Symposium on Cloud Computing*, 238–252. 10.1145/3419111.3421283

Haqaf, H., & Koyuncu, M. (2018). Understanding Key Skills for Information Security Managers. *International Journal of Information Management, 43*, 165–172. doi:10.1016/j.ijinfomgt.2018.07.013

Harris, I., Wang, Y., & Wang, H. (2015). ICT in multimodal transport and technological trends: Unleashing potential for the future. *International Journal of Production Economics, 159*, 88–103. doi:10.1016/j.ijpe.2014.09.005

Harvey, C. R., Moorman, C., & Castillo Toledo, M. (2018). How Blockchain Will Change Marketing As We Know It. SSRN *Electronic Journal*. doi:10.2139/ssrn.3257511

Harvey, C. R., Moorman, C., & Toledo, M. (2018, October 1). How Blockchain Can Help Marketers Build Better Relationships with Their Customers. *Harvard Business Review*. https://hbr.org/2018/10/how-blockchain-can-help-marketers-build-better-relationships-with-their-customers

Hasan, M. R., Deng, S., Sultana, N., & Hossain, M. Z. (2021). The applicability of blockchain technology in healthcare contexts to contain COVID-19 challenges. *Library Hi Tech*, *39*(3), 814–833. doi:10.1108/LHT-02-2021-0071

Hashimy, L., Jain, G., & Grifell-Tatjé, E. (2022). Determinants of blockchain adoption as decentralized business model by Spanish firms – an innovation theory perspective. *Industrial Management & Data Systems*. Advance online publication. doi:10.1108/IMDS-01-2022-0030

Hassija, V., Zeadally, S., Jain, I., Tahiliani, A., Chamola, V., & Gupta, S. (2021). Framework for determining the suitability of blockchain: Criteria and issues to consider. *Transactions on Emerging Telecommunications Technologies*, *32*(10). Advance online publication. doi:10.1002/ett.4334

Hawlitschek, F., Notheisen, B., & Teubner, T. (2018). The limits of trust-free systems: A literature review on blockchain technology and trust in the sharing economy. *Electronic Commerce Research and Applications*, *29*, 50–63. Advance online publication. doi:10.1016/j.elerap.2018.03.005

Hazar, H. B. (2018). Blok zincir teknolojisi ile muhasebe uygulamaları. *Vergi Dunyası Dergisi*, *1*(448), 45–54.

Hedera. (2023). Blockchain Oracles: How Smart Contracts Talk to the World. *Hedera website*. Retrieved from https://hedera.com/learning/decentralized-finance/blockchain-oracle

He, L., Xue, M., & Gu, B. (2020). Internet-of-things enabled supply chain planning and coordination with big data services: Certain theoretic implications. *Journal of Management Science and Engineering*, *5*(1), 1–22. doi:10.1016/j.jmse.2020.03.002

Helmer, O. (1967). *Analysis of the Future: The Delphi Method*. The Rand Corporation. Retrieved from https://apps.dtic.mil/sti/pdfs/AD0649640.pdf

Hernandez, A. A., & Ona, S. E. (2016). Green it adoption: Lessons from the philippines business process outsourcing industry. *International Journal of Social Ecology and Sustainable Development*, 7(1), 1–34. doi:10.4018/IJSESD.2016010101

Hewa, T., Ylianttila, M., & Liyanage, M. (2021). Survey on blockchain based smart contracts: Applications, opportunities and challenges. *Journal of Network and Computer Applications*, *177*, 102857. doi:10.1016/j.jnca.2020.102857

Himeur, Y., Sayed, A., Alsalemi, A., Bensaali, F., Amira, A., Varlamis, I., Eirinaki, M., Sardianos, C., & Dimitrakopoulos, G. (2022). Blockchain-based recommender systems: Applications, challenges and future opportunities. *Computer Science Review*, *43*, 100439. doi:10.1016/j.cosrev.2021.100439

Hiremath, G. A. K., G. M., R., & K. R., N. K. (2022). A Secure Blockchain-Powered Land Record Modernization System for Documenting Land User's Rights: Land Transparency Initiative. In P. Nijalingappa & M. Ghonge (Eds.), Blockchain Technologies and Applications for Digital Governance (pp. 37-59). IGI Global. doi:10.4018/978-1-7998-8493-4.ch002

Hofmann, E., & Rutschmann, E. (2018). Big data analytics and demand forecasting in supply chains: A conceptual analysis. *International Journal of Logistics Management*, *29*(2), 739–766. doi:10.1108/IJLM-04-2017-0088

Hussain, I., & Cakir, O. (2020). Blockchain technology in higher education: prospects, issues, and challenges. In R. Sharma, H. Yildirim, & G. Kurubacak (Eds.), *Blockchain technology applications in education* (pp. 97–112). IGI Global. doi:10.4018/978-1-5225-9478-9.ch005

Hyes, A. (2021). *Blockchain Explained.* Investopedia. https://www.investopedia.com/terms/b/blockchain.asp

Iansiti, M., & Lakhani, K. R. (2017). The Truth About Blockchain. *Harvard Business Review.* https://hbr.org/2017/01/the-truth-about-blockchain

Iansiti, M., & Lakhani, K. R. (2017, January 1). The Truth About Blockchain. *Harvard Business Review.* https://hbr.org/2017/01/the-truth-about-blockchain

IBM. (n.d.). *What are smart contracts on Blockchain?* Retrieved November, 2, 2021 from https://www.ibm.com/in-en/topics/smart-contracts#:~:text=Smart%20contracts%20are%20simply%20programs,intermediary's%20involvement%20or%20time%20loss

Ilyas, A., & Sukhrob, B. (2022). Advantages of Using and Developing Blockchain Technology in Tourism. *Journal of Advanced Research and Stability*, *2*(1), 525–526.

Ingram, D. (2018, May 9). Facebook shakes up management, launches blockchain division. *Reuters.* https://www.reuters.com/article/us-facebook-moves-idUSKBN1IA05S

Irannezhad, E., & Mahadevan, R. (2021). Is blockchain tourism's new hope? *Journal of Hospitality and Tourism Technology*, *12*(1), 85–96. doi:10.1108/JHTT-02-2019-0039

Iredale, G. (2020, April 17). *Disadvantages of Blockchain Technology.* Retrieved from https://101blockchains.com/disadvantages-of-blockchain/

ISACA. (2012). *COBIT 5: A Business Framework for the Governance and Management of Enterprise IT*. ISACA.

Isaias, P., Issa, T., Chang, V., & Issa, T. (2015). Outlining the issues of cloud computing and sustainability opportunities and risks in European organizations: A SEM study. *Journal of Electronic Commerce in Organizations*, *13*(4), 1–25. doi:10.4018/JECO.2015100101

İslamoğlu, A. H. (2009). Temel pazarlama bilgisi. Beta Basım.

İslamoğlu, A. H., Aydın, K., Candan, B., & Hacıefendioğlu, Ş. (2011). *Hizmet Pazarlaması*. Beta Basım A. Ş.

Issa, T., Tolani, G., Chang, V., & Issa, T. (2015). Awareness of sustainability, green IT, and cloud computing in Indian organisations. In Green services engineering, optimization, and modeling in the technological age (pp. 269–287). IGI Global. doi:10.4018/978-1-4666-8447-8.ch011

Jain, A. K. (2019). *Blockchain Goes to School*. Cognizant. https://www.cognizant.com/us/en/whitepapers/documents/blockchain-goes-to-school-codex3775.pdf

Jain, A. K. (2019). *Blockchain Goes to School: Cognizant 20-20 Insights*. Retrieved 27 December, 2021 from https://www.cognizant.com/whitepapers/blockchain-goes-to-school-codex3775.pdf

Jain, A. (2005). *Cyber Crime: Issues Threats and Management*. ISHA Books.

Jang, H., & Han, S. H. (2021). User experience framework for understanding user experience in blockchain services. *International Journal of Human-Computer Studies*, *158*, 102733. doi:10.1016/j.ijhcs.2021.102733

Janssen, M., Weerakkody, V., Ismagilova, E., Sivarajah, U., & Irani, Z. (2020). A framework for analysing blockchain technology adoption: Integrating institutional, market and technical factors. *International Journal of Information Management*, *50*, 302–309. doi:10.1016/j.ijinfomgt.2019.08.012

Jao, N. (2019, July 16). *Lack of understanding is blockchain's biggest hurdle in Asia: Survey*. TechNode. https://technode.com/2019/07/16/lack-of-understanding-is-blockchains-biggest-hurdle-in-asia-survey/

Jaoude, J., & Saade, R. (2019). Blockchain Applications – Usage in Different Domains. *IEEE Access : Practical Innovations, Open Solutions*, *7*, 45360–45381. doi:10.1109/ACCESS.2019.2902501

Jasmi, N., & Kamis, A. (2019). Importance of green technology, education for sustainable development (ESD) and environmental education for students and society. *Journal of Engineering Research and Application*, *9*(2), 56–59.

Joiner Associates Incorporated. (1995). *Pareto Charts Plain & Simple Learning and Application Guide*. Joiner Associates Incorporated.

Joiner Associates. (1995). *Cause and Effect Diagrams: Plain and Simple*. Joiner Associates Incorporated.

Joseph Jerome, J. J., Saxena, D., Sonwaney, V., & Foropon, C. (2021). Procurement 4.0 to the rescue: Catalysing its adoption by modelling the challenges. *Benchmarking: An International Journal*. doi:10.1108/BIJ-01-2021-0030

Jović, M., Tijan, E., Perić Hadžić, A., & Karanikić, P. (2020). Economic aspects of automation innovations in electronic transportation management systems. *Pomorstvo, 34*(2), 417–427. doi:10.31217/p.34.2.22

Juniper Research. (2017, September 26). *Ad Fraud to Cost Advertisers $19 billion in 2018, Representing 9% of Total Digital Advertising Spend*. Juniper Research. https://www.juniperresearch.com/press/press-releases/ad-fraud-to-cost-advertisers-19-billion-in-2018

Kaaru, S. (2020, March 21). *Toyota launches blockchain lab for verification, supply chain and more*. CoinGeek. https://coingeek.com/toyota-launches-blockchain-lab-for-verification-supply-chain-and-more/

Kalkinma Bakanligi, T. C. (2018). *On birinci kalkinma plani (2019 - 2023) kamuda kurumsal yonetim ve yenilikcilik ozel ihtisas komisyonu raporu*. https://www.sbb.gov.tr/wp-content/uploads/2020/04/KamudaKurumsalYonetim_ve_YenilikcilikOzelIhtisasKomisyonuRaporu.pdf

Kamath, R. (2018). Food Traceability on Blockchain: Walmart's Pork and Mango Pilots with IBM. *The Journal of the British Blockchain Association, 1*(1), 1–12. doi:10.31585/jbba-1-1-(10)2018

Kamble, S. S., Gunasekaran, A., & Gawankar, S. A. (2020). Achieving sustainable performance in a data-driven agriculture supply chain: A review for research and applications. *International Journal of Production Economics, 219*, 179–194. doi:10.1016/j.ijpe.2019.05.022

Kamble, S. S., Gunasekaran, A., Parekh, H., & Joshi, S. (2019). Modeling the internet of things adoption barriers in food retail supply chains. *Journal of Retailing and Consumer Services, 48*, 154–168. doi:10.1016/j.jretconser.2019.02.020

Kamble, S. S., Gunasekaran, A., Subramanian, N., Ghadge, A., Belhadi, A., & Venkatesh, M. (2021). Blockchain technology's impact on supply chain integration and sustainable supply chain performance: Evidence from the automotive industry. *Annals of Operations Research*. Advance online publication. doi:10.100710479-021-04129-6

Kamble, S., Gunasekaran, A., & Arha, H. (2019). Understanding the Blockchain technology adoption in supply chains-Indian context. *International Journal of Production Research, 57*(7), 2009–2033. doi:10.1080/00207543.2018.1518610

Kandaswamy, R., & Furlonger, D. (2018). *Blockchain-based transformation: A Gartner Trend Insight Report*. Gartner Research. Retrieved from https://www.gartner.com/en/doc/3869696-blockchain-based-transformation-a-gartner-trend-insight-report

Kapil, S., & Kapil, K. N. (2022). Blockchain in Hospitality and Tourism Industry Way Forward. *International Journal of Business and Economics*, *6*(2), 289–298. doi:10.5281/zenodo.5839071

Karabacak, B., Yıldırım, S. Ö., & Baykal, N. (2016). Regulatory Approaches for Cyber Security of Critical Infrastructures: The Case of Turkey. *Computer Law & Security Report*, *32*(3), 526–539. doi:10.1016/j.clsr.2016.02.005

Karabıyıkoğlu, M. (2018). *Satıcının Kariyer Yolculuğu*. Optimist Yayın.

Karaca, E. K. (2020). *Çin'in Teknoloji Liderliği, Dijital Yönetim Kırılma & Yeni Bir Dünya Düzeni*. Altınbaş Üniversitesi Yayınları.

Karahan, C., & Tufekci, A. (2019). Blokzincir teknolojisinin ic denetim faaliyetlerine etkileri: Firsatlar ve tehditler. *Denetisim*, *0*(19), 55–72.

Karakış, İ. (2020). *Dijital Satın Alma Araştırması 2020 PwC EMEA*. https://www.pwc.com.tr/tr/Hizmetlerimiz/danismanlik/tedarik-zinciri-yonetimi/dijital-sat%C4%B1nalma-ara%C5%9Ft%C4%B1rmas%C4%B1-2020/dijital-satinalma-arastirmasi-2020.pdf

Karamchandani, A., Srivastava, S. K., & Srivastava, R. K. (2020). Perception-based model for analyzing the impact of enterprise blockchain adoption on SCM in the Indian service industry. *International Journal of Information Management*, *52*, 102019. doi:10.1016/j.ijinfomgt.2019.10.004

Karisma, K. (2022). A Pragmatic Regulatory Framework for Artificial Intelligence. In P. Tehrani (Ed.), *Regulatory Aspects of Artificial Intelligence on Blockchain* (pp. 21–39). IGI Global. doi:10.4018/978-1-7998-7927-5.ch002

Kasemsap, K. (2018). Cloud computing, green computing, and green ict. In K. Munir (Ed.), *Cloud Computing Technologies for Green Enterprises* (pp. 28–50). IGI Global. doi:10.4018/978-1-5225-3038-1.ch002

Kayikci, Y., Subramanian, N., Dora, M., & Bhatia, M. S. (2020). Food supply chain in the era of Industry 4.0: Blockchain technology implementation opportunities and impediments from the perspective of people, process, performance, and technology. *Production Planning and Control*, *0*(0), 1–21. doi:10.1080/09537287.2020.1810757

Kaza, S., Yao, L., Bhada-Tata, P., & Van Woerden, F. (2018). *What a waste 2.0: a global snapshot of solid waste management to 2050*. World Bank Publications. doi:10.1596/978-1-4648-1329-0

Keagan, D. (1996). *Foundations of Distance Education* (3rd ed.). Routledge.

Kesswani, N., & Jain, S. K. (2017). Schematic classification model of green computing approaches. In Nature-inspired computing: concepts, methodologies, tools, and applications (pp. 1643-1650). IGI Global. doi:10.4018/978-1-5225-0788-8.ch063

Khan, M. A., Quasim, M. T., Algarni, F., & Alharthi, A. (2020). *Decentralized Internet of Things: A blockchain perspective*. Springer Nature Switzerland. doi:10.1007/978-3-030-38677-1

Khan, S. A., Mubarik, M. S., Kusi-Sarpong, S., Gupta, H., Zaman, S. I., & Mubarik, M. (2022). Blockchain technologies as enablers of supply chain mapping for sustainable supply chains. *Business Strategy and the Environment*, *31*(8), 1–15. doi:10.1002/bse.3029

Khan, S. N., Loukil, F., Ghedira-Guegan, C., Benkhelifa, E., & Bani-Hani, A. (2021). Blockchain smart contracts: Applications, challenges, and future trends. *Peer-to-Peer Networking and Applications*, *14*(5), 2901–2925. doi:10.100712083-021-01127-0 PMID:33897937

Khaqqi, K. N., Sikorski, J. J., Hadinoto, K., & Kraft, M. (2018). Incorporating seller/buyer reputation-based system in blockchain-enabled emission trading application. *Applied Energy*, *209*, 8–19. doi:10.1016/j.apenergy.2017.10.070

Kilinc, Y. (2020). Blockchain teknolojisi: Muhasebe ve denetim meslegi acisindan bir inceleme. *Journal of Accounting and Taxation Studies*, *13*(3), 989–1011. doi:10.29067/muvu.678681

Kılvan, P., & Barutçu, S. (2018). Endüstriyel Satın Alımlardaki Tedarikçi İlişkilerinde Güven ve Operasyonel İş Birliğinin Etkisi: Denizli Tekstil Firmalarında Bir Araştırma. *IBANESS Congress Series*.

Kizil, C., & Hanisoglu, G. S. (2019). Blok zinciri teknolojisi ve kripto paralar muhasebe ile denetimde kurallari degistiriyor [Blockchain technology and cryptocurrencies are changing the rules of accounting and auditing]. *Dunya Gazetesi*, p. 10. https://www.dunya.com/kose-yazisi/blokzinciri-teknolojisi-ve-kripto-paralarla-muhasebe-ile-denetimde-kurallar-degisiyor/455144

Koldemir, B. (2020). A Digital Transformation in International Transport and Logistics: Blockchain. In G. Ceyhun (Ed.), *Handbook of Research on the Applications of International Transportation and Logistics for World Trade* (pp. 425–453). IGI Global. doi:10.4018/978-1-7998-1397-2.ch023

Kommadi, B. (2022). Blockchain Open Issues, Research Gaps, and Road Map of Future Blockchain for 2030: Charting the Digital Future. In P. Nijalingappa & M. Ghonge (Eds.), *Blockchain Technologies and Applications for Digital Governance* (pp. 200–229). IGI Global. doi:10.4018/978-1-7998-8493-4.ch009

Kosba, A., Miller, A., Shi, E., Wen, Z., & Papamanthou, C. (2016). Hawk: The Blockchain Model of Cryptography and Privacy-Preserving Smart Contracts. *2016 IEEE Symposium on Security and Privacy*, 839–858. 10.1109/SP.2016.55

Kouhizadeh, M., Saberi, S., & Sarkis, J. (2021). Blockchain technology and the sustainable supply chain: Theoretically exploring adoption barriers. *International Journal of Production Economics*, *231*, 107831. Advance online publication. doi:10.1016/j.ijpe.2020.107831

Kouhizadeh, M., & Sarkis, J. (2018). Blockchain practices, potentials, and perspectives in greening supply chains. *Sustainability (Basel)*, *10*(10), 3652. doi:10.3390u10103652

Koutsos, T. M., Menexes, G. C., & Dordas, C. A. (2019). An efficient framework for conducting systematic literature reviews in agricultural sciences. *The Science of the Total Environment*, *682*, 106–117. doi:10.1016/j.scitotenv.2019.04.354 PMID:31108265

Kraft, D. (2016). Difficulty control for blockchain-based consensus systems. *Peer-to-Peer Networking and Applications*, *9*(2), 397–413. doi:10.100712083-015-0347-x

Krauth, O. (2018, February 9). *5 companies using blockchain to drive their supply chain*. TechRepublic. https://www.techrepublic.com/article/5-companies-using-blockchain-to-drive-their-supply-chain/

Kuhail, M. A., Mathew, S. S., Hammad, R., & Bahja, M. (2022). Blockchain Primer: Introduction to Blockchain Foundations and Implementation. In S. Khan, M. Syed, R. Hammad, & A. Bushager (Eds.), *Blockchain Technology and Computational Excellence for Society 5.0* (pp. 28–47). IGI Global. doi:10.4018/978-1-7998-8382-1.ch003

Kumar, R., & Sharma, V. (2013). *Auditing: Principles and Practice*. PHI Learning Private Limited.

Kundu, A. (2021). A Sound Framework for ICT Integration in Indian Teacher Education. *International Journal of Teacher Education and Professional Development*, *4*(1), 49–67. doi:10.4018/IJTEPD.2021010104

Kundu, A., & Bej, T. (2021). Experiencing e-assessment during COVID-19: An analysis of Indian students' perception. *Higher Education Evaluation and Development*, *15*(2), 114–134. doi:10.1108/HEED-03-2021-0032

Kundu, A., & Bej, T. (2021). Technology Adoption in Indian National Education Policy 2020: An Analysis of Pedagogical, Institutional and Human Aspects. *Journal of Social Sciences*, *17*(1), 145–157. doi:10.3844/jssp.2021.145.157

Kundu, A., & Bej, T. (2022). Reflective Teachers: A Probe into 21st Century Skills among Indian Trainee Teachers. *International Journal of Teacher Education and Professional Development*, *5*(1), 1–20. doi:10.4018/IJTEPD.2022010106

Kundu, A., Bej, T., & Rice, M. (2020). Time to Engage: Implementing Math and Literacy Blended Learning Routines in an Indian Elementary Classroom. *Education and Information Technologies*, *26*(1), 1201–1220. doi:10.100710639-020-10306-0

Kurubacak, G. (2007, July 1). Identify Research Priorities and Needs for Mobile Learning Technologies in Open and Distance Education: A Delphi Study. *Mobile Learning Technologies*, *19*(2), 1-31. Retrieved from https://files.eric.ed.gov/fulltext/ED495997.pdf

Kurubacak, G. (2011). E-Learning for Pluralism: The Culture of E-Learning in Building a Knowledge Society. *International Journal on E-Learning*, *10*(2), 145–167.

Kwok, A. O. J., & Koh, S. G. M. (2019). Is blockchain technology a watershed for tourism development? *Current Issues in Tourism*, *22*(20), 2447–2452. doi:10.1080/13683500.2018.1513460

Lam, T. Y., & Dongol, B. (2020). A blockchain-enabled e-learning platform. Interactive learning environments, 1–23. doi:10.1080/10494820.2020.1716022

Langard, V. (2019, 23 May). *B C Diploma raises €1.2 million to become the global standard for certifying diplomas and certificates online.* Retrieved from https://medium.com/bcdiploma/bcdiploma-raises-1-2m-58cdba9b3 b85

Laskowski, M., & Kim, H. M. (2016). Rapid Prototyping of a Text Mining Application for Cryptocurrency Market Intelligence. *2016 IEEE 17th International Conference on Information Reuse and Integration (IRI)*, 448–453. 10.1109/IRI.2016.66

Lee, Y. N. (2020, May 6). *5 charts show which travel sectors were worst hit by the coronavirus.* CNBC. https://www.cnbc.com/2020/05/06/coronavirus-pandemics-impact -on-travel-tourism-in-5-charts.html

Lemoie, K., & Soares, L. (2020). *Connected Impact: Unlocking Education and Workforce Opportunity Through Blockchain.* American Council on Education.

Li, D., Han, D., Zheng, Z., Weng, T.-H., Li, H., Castiglioane, A., & Li, K.-C. (2021). MOOCsChain: A Blockchain-Based Secure Storage and Sharing Scheme for MOOCs Learning. *Computer Standards & Interfaces*, *81*, 103597. doi:10.1016/j.csi.2021.103597

Li, J., & Kassem, M. (2021). Applications of distributed ledger technology (DLT) and Blockchain-enabled smart contracts in construction. *Automation in Construction*, *132*, 103955. doi:10.1016/j. autcon.2021.103955

Li, M., Shao, S., Ye, Q., Xu, G., & Huang, G. Q. (2020). Blockchain-enabled logistics finance execution platform for capital-constrained E-commerce retail. *Robotics and Computer-integrated Manufacturing*, *65*, 101962. doi:10.1016/j.rcim.2020.101962

Lin, R., Wang, L., Li, B., Lu, Y., Qi, Z., & Xie, L. (2022). Organizational governance in the smart era: the implications of blockchain. Nankai Business Review International. doi:10.1108/ NBRI-02-2021-0014

Linn, L. A., & Koo, M. B. (2016). *Blockchain For Health Data and Its Potential Use in Health IT and Health Care Related Research.* Academic Press.

Liu, M., Wu, K., & Xu, J. J. (2019). How will blockchain technology impact auditing and accounting: Permissionless versus permissioned blockchain. *Current Issues in Auditing*, *13*(2), A19–A29. doi:10.2308/ciia-52540

Liu, W., Zhang, J., & Wang, S. (2021). Factors influencing the smart supply chain innovation performance of commodity distribution enterprises: An investigation from China. *Industrial Management & Data Systems*, *121*(10), 2073–2099. doi:10.1108/IMDS-12-2020-0753

Li, Z., & Williams, M. (Eds.). (2006). *Environmental and geographical education for sustainability: cultural contexts*. Nova Publishers.

Lohmer, J., & Lasch, R. (2020). Blockchain in operations management and manufacturing: Potential and barriers. *Computers & Industrial Engineering*, *149*, 106789. doi:10.1016/j.cie.2020.106789

Lyons, T., Courcelas, L., Grandsenne, J., Carrel, E., & Timsit, K. (2018). *Blockchain Innovation in Europe.* European Union Blockchain Observatory and Forum.

Madden, D. (2021, January 14). *The Covid-19 Pandemic Has Cost the Global Tourism Industry $935 Billion.* Forbes. https://www.forbes.com/sites/duncanmadden/2021/01/14/the-covid-19-pandemic-has-cost-the-global-tourism-industry-935-billion/

Mahankali, S., & Chaudhary, S. (2020). Blockchain in Education: A Comprehensive Approach – Utility, Use Cases, and Implementation in a University. In R. Sharma, H. Yildirim, & G. Kurubacak (Eds.), *Blockchain Technology Applications in Education* (pp. 267–293). IGI Global. doi:10.4018/978-1-5225-9478-9.ch014

Mahroof, K. (2019). A human-centric perspective exploring the readiness towards smart warehousing: The case of a large retail distribution warehouse. *International Journal of Information Management, 45*, 176–190. doi:10.1016/j.ijinfomgt.2018.11.008

Maleh, Y., Baddi, Y., Alazab, M., Tawalbeh, L., & Romdhani, I. (2021). *Artificial Intelligence and Blockchain for Future Cybersecurity Applications.* Springer. doi:10.1007/978-3-030-74575-2

Mangina, E., Narasimhan, P. K., Saffari, M., & Vlachos, I. (2020). Data analytics for sustainable global supply chains. *Journal of Cleaner Production, 255*, 120300. doi:10.1016/j.jclepro.2020.120300

Mardamutu, K., Ponnusamy, V., & Zaman, N. (2016). Green energy in data centers. In Biologically-inspired energy harvesting through wireless sensor technologies (pp. 234–249). IGI Global. doi:10.4018/978-1-4666-9792-8.ch012

Marikyan, D., Papagiannidis, S., Rana, O. F., & Ranjan, R. (2022). Blockchain adoption: A study of cognitive factors underpinning decision making. *Computers in Human Behavior, 131*, 107207. doi:10.1016/j.chb.2022.107207

Marr, B. (2018, March 23). *How Blockchain Will Transform the Supply Chain and Logistics Industry.* Forbes. https://www.forbes.com/sites/bernardmarr/2018/03/23/how-blockchain-will-transform-the-supply-chain-and-logistics-industry/

Martin, R. (2018). *How blockchain will impact accounting.* https://igniteoutsourcing.com/blockchain/blockchain-accounting-applications/

Martínez-Ríos, F. O., Marmolejo-Saucedo, J. A., & Abascal-Olascoaga, G. (2020). A New Protocol Based on Blockchain Technology for Transparent Operation of Corporate Social Responsibility. In Strategy, Power and CSR: Practices and Challenges in Organizational Management. Emerald Publishing Limited. doi:10.1108/978-1-83867-973-620201012

Maryville University. (2021). *How Blockchain is used in education* [Blog]. Retrieved October 21, 2021 from https://online.maryville.edu/blog/blockchain-in-education/#:~:text=Blockchain%20gives%20students%20ownership%20of,what%20an%20employer%20can%20access

McArthur, D. (2018). Will blockchains revolutionize education. *Educause Review*. https://er.educause.edu/articles/2018/5/will-blockchains-revolutionize-education

McGreal, R. (2021, January 27). *What are the advantages of Blockchain as an educational resource? What is Blockchain and how could it improve access to Education?* World Economic Forum. Retrieved 27 December, 2021 from http://people.uis.edu/rschr1/?p=58137

McMurty, J., & Hiam, A. (2019). Marketing for Dummies. Pazarlama for Dummies. Nobel Yaşam.

Meit, Y. (2021). *National strategy on blockchain: Towards enabling trusted digital platforms.* Ministry of Electronics & Information Technology, Government of India.

Melkić, S., & Čavlek, N. (2020). The impact of blockchain technology on tourism intermediation. *Tourism (Zagreb)*, *68*(2), 130–143. doi:10.37741/t.68.2.2

Melko, M., & Uvarova, G. (2021). Blockchain Technology in the Tourist Sphere: Realities of Today. *International Conference on Economics, Accounting and Finance*. Retrieved from https://library.krok.edu.ua/media/library/category/materiali-konferentsij/melko_0034.pdf

Menon, M., & Mady, A. (2021). Blockchain: A non-technical primer for marketing academics. *Marketing Education Review*, 1–19.

Menon, M., & Mady, A. (2021). Blockchain: A Non-Technical Primer For Marketing Academics. *Marketing Education Review*, *32*(3), 1–19. doi:10.1080/10528008.2021.1943445

Meunier, S. (2018). Blockchain 101: What is Blockchain and how does this revolutionary technology work? In A. Mark (Ed.), *Edt.). Transforming Climate Finance and Green Investment with Blockchains* (pp. 23–34). Academic Press. doi:10.1016/B978-0-12-814447-3.00003-3

Mikroyannidis, A., Domingue, J., Bachler, M., & Quick, K. (2018). Smart blockchain badges for data science education. *2018 IEEE Frontiers in Education Conference (FIE)*, 1-5. 10.1109/FIE.2018.8659012

Mikroyannidis, A., Third, A., Domingue, J., Bachelor, M., & Quick, K. (2020). Blockchain applications in lifelong learning and the role of the semantic blockchain. In R. C. Sharma, H. Yildirim, & G. Kurubacak (Eds.), *Blockchain Technology Applications in Education*. IGI Global. doi:10.4018/978-1-5225-9478-9.ch002

Miraz, M. H., Hassan, M. G., & Sharif, K. I. M. (2020). Factors Affecting Implementation of Blockchain in Retail Market in Malaysia. *Inetranational Journal of Supply Chain Management*, *9*(1), 385–391.

Mogee, M. E., & Bean, A. S. (1978). The Role of the Purchasing Agent. In T. V. Bonoma & G. Zaltman (Eds.), *Industrial Innovation, "in Organizational Buying Behavior* (pp. 126–137). American Marketing Association.

Moore, S. (2019, October 16). *4 Ways Blockchain will transform higher education.* Retrieved from https://www.gartner.com/smarterwithgartner/4-ways-blockchain-will-transform-higher-education#:~:text=Improve

Morgan, B. (2020, May 7). *50 Stats That Show the Importance of Good Loyalty Programs, Even During A Crisis.* Forbes. https://www.forbes.com/sites/blakemorgan/2020/05/07/50-stats-that-show-the-importance-of-good-loyalty-programs-even-duri ng-a-crisis/

Morrison, O. (2019, August 6). *Anheuser-Busch InBev joins new IBM blockchain.* Foodnavigator.Com. https://www.foodnavigator.com/Article/2019/08/06/Anheuser-Busch-InBev-joins-new-IBM-blockchain

Mosinegutu, V. (2021, July 23). *What is Web 3.0 and how will it change the internet?* Retrieved from https://academy.coinzilla.com/what-is-web-3-0/

Mourtzis, D., Angelopoulos, J., & Panopoulos, N. (2020). *A Survey of Digital B2B Platforms and Marketplaces For Purchasing Industrial Product Service Systems: A Conceptual Framework.* Elsevier. doi:10.1016/j.procir.2020.05.246

Muharam, I. N., Tussyadiah, I. P., & Kimbu, A. N. (2023). A theoretical model of user acceptance of blockchain-based peer-to-peer accommodation. *Current Issues in Tourism.* doi:10.1080/13683500.2022.2164485

Murphy, J. (2021, March 8). *What Is an NFT? Inside the Next Billion-Dollar Crypto Sensation.* Entrepreneur. https://www.entrepreneur.com/article/366648

Nabil, A., Nafil, K., & Mounir, F. (2020). Blockchain Security and Privacy in Education: A Systematic Mapping Study. In Á. Rocha, H. Adeli, L. Reis, S. Costanzo, I. Orovic, & F. Moreira (Eds.), *Trends and Innovations in Information Systems and Technologies. WorldCIST 2020. Advances in Intelligent Systems and Computing* (Vol. 1161). Springer. doi:10.1007/978-3-030-45697-9_25

NAD. (n.d.). *Universities Grant Commission.* Ministry of Education. Govt. of India. https://nad.gov.in/

Nakamoto, S. (2008). Bitcoin: A peer-to-peer electronic cash system. *Decentralized Business Review*, 21260.

Nakamoto, S. (2008). Bitcoin: A peer-to-peer electronic cash system. *Decentralized Business Review*, 21260. https://bitcoin.org/bitcoin.pdf

Nakamoto, S. (2008, October 31). *Bitcoin: A Peer-to-Peer Electronic Cash System.* https://nakamotoinstitute.org/bitcoin/

Nandi, M. L., Nandi, S., Moya, H., & Kaynak, H. (2020). Blockchain technology-enabled supply chain systems and supply chain performance: A resource-based view. *Supply Chain Management, 25*(6), 841–862. doi:10.1108/SCM-12-2019-0444

Näslund, D., & Hulthen, H. (2012). Supply chain management integration: A critical analysis. *Benchmarking, 19*(4/5), 481–501. doi:10.1108/14635771211257963

NEP. (2020). *National Education Policy 2020*. Ministry of Education. New Delhi. India. https://www.education.gov.in/sites/upload_files/mhrd/files/NEP_Final_English_0.pdf

Newswire. (2017, October 8). *Melbourne first in Asia-Pacific to issue recipient-owned blockchain records*. Retrieved from https://www.newswire.com/news/university-of-melbourne-first-in-asia-pacific-to-issue-recipient-owned-19980513

NIC. (n.d.). *Drugs Logistics.* Government of Karnataka. Retrieved from https://blockchain.gov.in/drugslogistics.html

Niranjanamurthy, M., Nithya, B. N., & Jagannatha, S. J. C. C. (2019). Analysis of blockchain technology: Pros, cons, and SWOT. *Cluster Computing, 22*(6), 14743–14757. doi:10.100710586-018-2387-5

NISG. (2019). *National Strategy for Blockchain.* National Institute for Smart Government.

Novianti, S., Susanto, E., & Rafdinal, W. (2022). Predicting Tourists' Behaviour Towards Smart Tourism: The Case in Emerging Smart Destinations. *Journal of Tourism Sustainability, 2*(1), 19–30. doi:10.35313/jtospolban.v2i1.30

Nugent, T., Upton, D., & Cimpoesu, M. (2016). Improving data transparency in clinical trials using blockchain smart contracts. *F1000 Research, 5*, 2541. doi:10.12688/f1000research.9756.1 PMID:28357041

OECD. (2019). *The Policy Environment for Blockchain Innovation and Adoption.* OECD Blockchain Policy Series. Retrieved from https://www.oecd.org/finance/2019-OECD-Global-Blockchain-Policy-Forum-Summary-Report.pdf

Oh, J., & Shong, I. (2017). A case study on business model innovations using Blockchain: Focusing on financial institutions. *Asia Pacific Journal of Innovation and Entrepreneurship, 11*(3), 335–344. doi:10.1108/APJIE-12-2017-038

Oh, S.-C., Kim, M.-S., Park, Y., Roh, G.-T., & Lee, C.-W. (2017). Implementation of blockchain-based energy trading system. *Asia Pacific Journal of Innovation and Entrepreneurship, 11*(3), 322–334. doi:10.1108/APJIE-12-2017-037

Okay, A. (2013). *Kurum Kimliği.* Derin yayınları.

Olson, D. L., & Wu, D. D. (2015). *Enterprise Risk Management* (Vol. 3). World Scientific. doi:10.1142/9378

Onay, A. (2021). *Blok zincir teknolojisinin denetime etkisi ve literatüre iliskin bilimetrik bir arastirma.* In S. Karabulut (Ed.), *Muhasebe Konularinda Bilimsel Yaklasim Ve Arastirmalar* (pp. 189–220). Gazi Kitabevi.

Önder, I., & Treiblmaier, H. (2018). Blockchain and tourism: Three research propositions. *Annals of Tourism Research, 72*(C), 180–182. doi:10.1016/j.annals.2018.03.005

Önkan, Ö., & Arikan, Z. (2022). The Impact of Blockchain Technology on Tax and Accounting Practices. In P. Nijalingappa & M. Ghonge (Eds.), *Blockchain Technologies and Applications for Digital Governance* (pp. 1–36). IGI Global. doi:10.4018/978-1-7998-8493-4.ch001

Ontario Ministry of Education. (2010). *Green schools resource guide: a practical resource for planning and building green schools in Ontario.* ZAS Architechs Inc. http://www.edu.gov.on.ca/eng/policyfunding/GreenSchools_Guide.pdf

Özçelik, C., & Akyol, A. (2017). Endüstriyel alıcıların satın alma sürecinde yararlandıkları bilgi kaynakları ve KOBİ'ler üzerine bir araştırma. *Trakya Üniversitesi Sosyal Bilimler Dergisi, 19*(1), 167–182.

Özdemir, Ş. (2006). *Endüstriyel Ürün Pazarlaması.* Seçkin Yayıncılık.

Ozdemir, A. I., Ar, I. M., & Erol, I. (2020). Assessment of blockchain applications in travel and tourism industry. *Quality & Quantity, 54*(5), 1549–1563. doi:10.100711135-019-00901-w

Ozkul, F., & Alkan, B. S. (2020). Dijital çagda muhasebenin donusumu: "blockchain" teknolojisinde muhasebe ve mali kontroller. *World of Accounting Science, 22*(2), 218–236.

Öz, M. (2011). Lojistikte Yeni Yaklaşımlar. *Kahramanmaraş Sütçü İmam Üniversitesi İktisadi ve İdari Bilimler Fakültesi Dergi, 1*(1), 141–155.

Park, J. (2021). Promises and challenges of Blockchain in education. *Smart Learn. Environ., 8*(1), 33. doi:10.118640561-021-00179-2

Parlakkılıç, A. (2022). Blockchain Use Cases in Healthcare. In P. Nijalingappa & M. Ghonge (Eds.), *Blockchain Technologies and Applications for Digital Governance* (pp. 85–104). IGI Global. doi:10.4018/978-1-7998-8493-4.ch004

Patel, T., & Drozdiak, N. (2021, March 24). *Building a Covid Travel Passport Is a Serious Tech Challenge.* BloombergQuint. https://www.bloombergquint.com/businessweek/covid-19-vaccine-passports-face-security-concerns-technical-glitches

Pathak, A. (2021). *Blockchain Technology: A Guide for Beginners.* https://geekflare.com/finance/blockchain-technology-for-beginners/

Pathak, S. (2016, December 15). How blockchain might be useful in marketing and advertising. *Digiday.* https://digiday.com/marketing/blockchain-tech-might-useful-marketing/

Paton, E. (2021, April 20). LVMH, Richemont and Prada unite behind a blockchain consortium. *The New York Times*. https://www.nytimes.com/2021/04/20/business/lvmh-richemont-prada-blockchain.html

Patrizio, A. (2018, March 27). *Top 10 Blockchain as a Service Providers*. Datamation. https://www.datamation.com/data-center/top-10-blockchain-as-a-service-providers/

Paykamian, B. (2021). *Could ed tech and virtual learning help schools 'go green'?* https://www.govtech.com/education/k-12/could-ed-tech-and-virtual-learning-help-schools-go-green.html

Peltekoğlu, B. F. (2019). *Kavram ve Kuramlarıyla Reklam*. Beta Basım Yayım. İkinci Basım.

Penpece, D. (2013). *Dijital İçerik Pazarlaması*. Karahan Kitabevi.

Penuel, K. B., & Statler, M. (2013). *Encyclopedia of Crisis Management*. Sage. doi:10.4135/9781452275956

Philipson, G. (2011). A Comprehensive and Practical Green ICT Framework. In B. Unhelkar (Ed.), *Handbook of Research on Green ICT: Technology, Business and Social Perspectives* (pp. 131–145). IGI Global. doi:10.4018/978-1-61692-834-6.ch009

PIB. (n.d.). https://pib.gov.in/PressReleseDetailm.aspx?PRID=1777711

Pickett, K. H. (2011). *The Essential Guide to Internal Auditing*. British Library.

Pilkington, M. (2016). Blockchain technology: Principles and applications. *Research Handbook on Digital Transformations*. https://www.elgaronline.com/view/edcoll/9781784717759/9781784717759.00019.xml

Pina, P. (2022). Blockchain and Copyright: Challenges and Opportunities. In S. Khan, M. Syed, R. Hammad, & A. Bushager (Eds.), *Blockchain Technology and Computational Excellence for Society 5.0* (pp. 125–145). IGI Global. doi:10.4018/978-1-7998-8382-1.ch007

Prabawani, B., Hanika, I. M., Pradhanawati, A., & Budiatmo, A. (2017). Primary schools eco-friendly education in the frame of education for sustainable development. *International Journal of Environmental and Science Education*, *12*(4), 607–616.

Pradhan, D. (2018, July 18). *Blockchain This Week: Two Educational Institutions Adopt Blockchain, Oracle to Launch Blockchain Cloud Service in India, and more* [Blog]. Retrieved from https://inc42.com/buzz/blockchain-this-week-two-indian-education-institutions-adopt-blockchain-and-more/

Pranata, A. R., & Tehrani, P. M. (2022). The Legality of Smart Contracts in a Decentralized Autonomous Organization (DAO). In P. Tehrani (Ed.), *Regulatory Aspects of Artificial Intelligence on Blockchain* (pp. 112–131). IGI Global. doi:10.4018/978-1-7998-7927-5.ch006

Preindl, R., Nikolopoulos, K., & Litsiou, K. (2020). Transformation strategies for the supply chain: The impact of industry 4.0 and digital transformation. *Supply Chain Forum: An International Journal, 21*(1), 26–34. 10.1080/16258312.2020.1716633

Prescher, C. (2018, March 22). *Using Blockchain to Build Connections with Consumers.* Chief Marketer. https://www.chiefmarketer.com/using-blockchain-build-connections-consumers/

PricewaterhouseCoopers. (2017, February). *Blockchain: A new tool to cut costs.* PwC. https://www.pwc.com/m1/en/media-centre/articles/blockchain-new-tool-to-cut-costs.html

Progrebna, G., & Skilton, M. (2019). *Navigating New Cyber Risks: How Businesses Can Plan, Build and Manage Safe Spaces in the Digital Age.* Palgrave Macmillan. doi:10.1007/978-3-030-13527-0

Puri, V., Mondal, S., Das, S., & Vrana, V. G. (2023). Blockchain Propels Tourism Industry—An Attempt to Explore Topics and Information in Smart Tourism Management through Text Mining and Machine Learning. *Informatics (MDPI), 10*(1), 9. doi:10.3390/informatics10010009

PwC Global Power & Utilities. (2015). *PwC-Blockchain Ppportunity for Energy Producers and Consumers.pdf.* https://www.pwc.com/gx/en/industries/assets/pwc-blockchain-opportunity-for-energy-producers-and-consumers.pdf

Queiroz, M. M., Pereira, S. C. F., Telles, R., & Machado, M. C. (2019). Industry 4.0 and digital supply chain capabilities: A framework for understanding digitalisation challenges and opportunities. *Benchmarking, 28*(5), 1761–1782. doi:10.1108/BIJ-12-2018-0435

Raluca-Florentina, T. (2022). The Utility of Blockchain Technology in the Electronic Commerce of Tourism Services: An Exploratory Study on Romanian Consumers. *Sustainability (Basel), 14*(2), 943. doi:10.3390u14020943

Rao, S., Ellis, S. C., Goldsby, T. J., & Raju, D. (2019). On the "Invisible Inventory Conundrum" in RFID-Equipped Supply Chains: A Data Science Approach to Assessing Tag Performance. *Journal of Business Logistics, 40*(4), 339–358. doi:10.1111/jbl.12232

Rasyid, S., & Ardi, M. (2021). Development of environmentally friendly school model ın biak numfor regency of papua province. *Turkish Journal of Computer and Mathematics Education, 12*(13), 4096–4109.

Raut, R. D., Gotmare, A., Narkhede, B. E., Govindarajan, U. H., & Bokade, S. U. (2020). Enabling Technologies for Industry 4.0 Manufacturing and Supply Chain: Concepts, Current Status, and Adoption Challenges. *IEEE Engineering Management Review, 48*(2), 83–102. doi:10.1109/EMR.2020.2987884

Reiff, N. (2020, February 25). *Forget Bitcoin: Blockchain is the Future.* Investopedia. https://www.investopedia.com/tech/forget-bitcoin-blockchain-future/

Rejeb, A., Keogh, J. G., Simske, S. J., Stafford, T., & Treiblmaier, H. (2021). Potentials of blockchain technologies for supply chain collaboration: A conceptual framework. *International Journal of Logistics Management*, *32*(3), 973–994. doi:10.1108/IJLM-02-2020-0098

Reyna, A., Martín, C., Chen, J., Soler, E., & Díaz, M. (2018). On blockchain and its integration with IoT. Challenges and opportunities. *Future Generation Computer Systems*, *88*, 173–190. doi:10.1016/j.future.2018.05.046

Reynolds, P., & Irwin, A. S. (2017). Tracking digital footprints: Anonymity within the bitcoin system. *J. Money Laund. Control*, *20*(2), 172–189. doi:10.1108/JMLC-07-2016-0027

Roe, D. (2019, June 21). *What Blockchain Companies Need to Do to Succeed*. CMSWire.Com. https://www.cmswire.com/information-management/what-blockchain-companies-need-to-do-to-succeed/

Rosic, A. (2016, September 18). *What is Blockchain Technology? A Step-by-Step Guide for Beginners*. Blockgeeks. https://blockgeeks.com/guides/what-is-blockchain-technology/

Roy, V. (2021). Contrasting supply chain traceability and supply chain visibility: Are they interchangeable? *International Journal of Logistics Management*, *32*(3), 942–972. doi:10.1108/IJLM-05-2020-0214

Rustagi, N., & Chakraborti, A. (2021, Dec. 3). *Blockchain in schools and colleges*. Indian Express. https://indianexpress.com/article/opinion/blockchain-technology-education-nep-7696791/

Ryder, R. D., & Madhavan, A. (2019). *Cyber Crisis Management Overcoming the Challenges in Cyberspace*. Bloomsbury.

Saeidnia, H. R., & Lund, B. D. (2023). Non-fungible tokens (NFT): A safe and effective way to prevent plagiarism in scientific publishing. *Library Hi Tech News*, *40*(2), 18–19. doi:10.1108/LHTN-12-2022-0134

Saha, A. (2019). *UPS 2019 Industrial Buying Dynamics Study*. European Study. https://www.epressi.com/media/userfiles/136388/1562788200/whitepaper_ibd_v5_print.pdf

Salah, K., & Khan, M. (2017). IoT Security: Review, Blockchain Solutions, and Open Challenges. *Future Generation Computer Systems*, *82*, 395–411. doi:10.1016/j.future.2017.11.022

Sanders, N. R., & Premus, R. (2005). Modeling the Relationship Between Firm It Capability, Collaboration, and Performance. *Journal of Business Logistics*, *26*(1), 1–23. doi:10.1002/j.2158-1592.2005.tb00192.x

Sarkis, J., Kouhizadeh, M., & Zhu, Q. S. (2020). Digitalization and the greening of supply chains. *Industrial Management & Data Systems*, *121*(1), 65–85. doi:10.1108/IMDS-08-2020-0450

Saryatmo, M. A., & Sukhotu, V. (2021). The Influence of the Digital Supply Chain on Operational Performance: A Study of the Food and Beverage Industry in Indonesia. *Sustainability (Basel), 13*(9), 5109. doi:10.3390u13095109

Sathler, L. (2020). For an open innovation platform dedicated to education: a blockchain approach. In R. Sharma, H. Yildirim, & G. Kurubacak (Eds.), *Blockchain technology applications in education* (pp. 153–169). IGI Global. doi:10.4018/978-1-5225-9478-9.ch008

Schiller, B. (2018, February 5). *On This Blockchain-Based Version of Airbnb, There's No Middleman*. Fast Company. https://www.fastcompany.com/40524021/on-this-blockchain-based-version-of-airbnb-theres-no-middleman

Schmidt, C. G., & Wagner, S. M. (2019). Blockchain and supply chain relations: A transaction cost theory perspective. *Journal of Purchasing and Supply Management, 25*(4), 100552. doi:10.1016/j.pursup.2019.100552

Schreider, T. (2017). *Building Effective Cybersecurity Programs: A Security Manager's Handbook*. Rothstein Publishing.

Seebacher, S., & Schüritz, R. (2017). Blockchain Technology as an Enabler of Service Systems: A Structured Literature Review. *Proceedings of the International Conference on Exploring Services Science*, 12–23. 10.1007/978-3-319-56925-3_2

Selimoğlu, S., & Altunel, M. (2019). Siber Güvenlik Risklerinden Korunmada Köprü ve Katalizör Olarak İç Denetim [Internal Audit as a Bridge and Catalyst in the Protection of Cyber Security Risks]. *Denetişim,* (19), 5-16. Retrieved from https://dergipark.org.tr/en/download/article-file/750860

Shabani-Naeeni, F., & Ghasemy Yaghin, R. (2021). Incorporating data quality into a multi-product procurement planning under risk. *Journal of Business and Industrial Marketing, 36*(7), 1176–1190. doi:10.1108/JBIM-02-2020-0108

Shakir, Q. (2019). *Green schools (Eco-school).* https://www.researchgate.net/publication/331481308_Green_Sch
oolsEco-School

Sharma, M., & Joshi, S. (2020). Digital supplier selection reinforcing supply chain quality management systems to enhance firm's performance. *The TQM Journal*. doi:10.1108/TQM-07-2020-0160

Sharma, R. (2020, May 2). Toyota Set to Explore the Application of Blockchain Technology. *Business Blockchain HQ*. https://businessblockchainhq.com/business-blockchain-news/toyota-set-to-explore-the-application-of-blockchain-technology
/

Sharma, R. C., & Mishra, S. (2006). *Cases on Global E-Learning Practices: Successes and Pitfalls: Successes and Pitfalls*. IGI Global. https://www.researchgate.net/publication/307569541_Cases_on_Global_E-Learning_Practices_Successes_and_Pitfalls

Sharma, A., & Gupta, P. (2021). Blockchain Revolution: Adaptability in Business World and Challenges in Implementation. In S. Singh & A. Jurcut (Eds.), *Revolutionary Applications of Blockchain-Enabled Privacy and Access Control* (pp. 189–218). IGI Global. doi:10.4018/978-1-7998-7589-5.ch009

Sharma, M., Sehrawat, R., Daim, T., & Shaygan, A. (2021). Technology assessment: Enabling Blockchain in hospitality and tourism sectors. *Technological Forecasting and Social Change*, *169*, 120810. doi:10.1016/j.techfore.2021.120810

Sharma, R. C., Kawachi, P., & Bozkurt, A. (2019). Exploring Changing Perspectives in Distance Education. *Asian Journal of Distance Education*, *14*(1), 1–6. http://www.asianjde.com/ojs/index.php/AsianJDE/article/view/285

Sharma, R. C., Yildirim, H., & Kurubacak, G. (2020). Preface. In R. C. Sharma, H. Yildirim, & G. Kurubacak (Eds.), *Blockchain Technology Applications in Education*. IGI Global. doi:10.4018/978-1-5225-9478-9

Sharma, R. C., Yildirim, H., & Kurubacak, G. (Eds.). (2019). *Blockchain Technology Applications in Education*. IGI Global.

Sharples, M., & Domingue, J. (2016). The Blockchain and Kudos: A distributed system for educational record, reputation and reward. In K. Verbert, M. Sharples, & T. Klobučar (Eds.), *Adaptive and adaptable learning* (Vol. 9891, pp. 490–496). Springer. doi:10.1007/978-3-319-45153-4_48

Sheel, A., & Nath, V. (2020). Blockchain technology adoption in the supply chain (UTAUT2 with risk) – evidence from Indian supply chains. *International Journal of Applied Management Science*, *12*(4), 324–346. doi:10.1504/IJAMS.2020.110344

Shojafar, M., Cordeschi, N., & Baccarelli, E. (2016). Resource scheduling for energy-aware reconfigurable Internet data centers. In Q. Hassan (Ed.), *Innovative research and applications in next-generation high performance computing* (pp. 21–46). IGI Global. doi:10.4018/978-1-5225-0287-6.ch002

Singh, S. (2021, July 16). *Potential Use Cases of Blockchain Technology for Cybersecurity*. Retrieved from https://www.itbusinessedge.com/security/potential-use-cases-of-blockchain-technology-for-cybersecurity/

Singh, S., & Gond, S. (2017). Green computing and its impact. In *Nature-inspired computing: Concepts, methodologies, tools, and applications* (pp. 1628–1642). IGI Global. doi:10.4018/978-1-5225-0788-8.ch062

Single Grain Blog. (2018, March 13). *Case Studies—Companies Doing Blockchain Marketing Well*. Single Grain. https://www.singlegrain.com/blockchain/blockchain-marketing-case-studies/

Skiba, D. J. (2017). The potential of Blockchain in education and health care. *Nursing Education Perspectives*, *38*(4), 220–221. doi:10.1097/01.NEP.0000000000000190 PMID:28622267

Skinner, B. F. (1938). The Behavior of Organisms. Academic Press.

Smits, D., & Hillegersberg, J. V. (2015). IT Governance Maturity: Developing a Maturity Model Using the Delphi Method. In *48th Hawaii International Conference on System Sciences* (pp. 4534-4543). Institute of Electrical and Electronics Engineers (IEEE). 10.1109/HICSS.2015.541

Sonmez, Y., & Demir, V. (2021). muhasebe ve finansmanda dijital gelecek. *Mali Cozum*, *31*(167), 39–53.

Soohoo, S. (2019, April). *Worldwide Blockchain 2019-2022 Forecast: Market Opportunity by Use Case—1H18 Update*. https://www.reportlinker.com/p05770124/Worldwide-Blockchain-Forecast-Market-Opportunity-by-Use-Case-1H18-Update.html

Spafford, M. L., Stanaway, D. F., & Chung, S. (2019). Blockchain and cryptocurrencies: A cross-border conundrum. *Journal of Investment Compliance*, *20*(3), 10–19. doi:10.1108/JOIC-05-2019-0027

Srai, J. S., & Lorentz, H. (2019). Developing design principles for the digitalisation of purchasing and supply management. *Journal of Purchasing and Supply Management*, *25*(1), 78–98. doi:10.1016/j.pursup.2018.07.001

Srivastava, A., Bhattacharya, P., Singh, A., & Mathur, A. (2018). A systematic review on evolution of blockchain generations. *International Journal of Information Technology and Electrical Engineering*, *7*(6), 1–8.

Steiu, M.-F. (2020). Blockchain in education: Opportunities, applications, and challenges. *First Monday*, *25*(9). Retrieved from https://journals.uic.edu/ojs/index.php/fm/article/download/10654/9726 doi:10.5210/fm.v25i9.10654

Suekinci, C., & Catikkas, D. Ö. (2020). Blok zinciri teknolojisinin muhasebe ve vergilendirme uzerine etkileri. *Mali Cozum Dergisi*, *30*(162), 51–65.

Susanto, E., Novianti, S., Rafdinal, W., Prawira, M. F. A., & Septyandi, C. B. (2020). Visiting Tourism Destination: Is It Influenced by Smart Tourism Technology? *Journal of Indonesian Tourism and Development Studies*, *8*(3), 145–155. doi:10.21776/ub.jitode.2020.008.03.04

Sutanto, J. E., & Japutra, A. (2021). The Impact of Supply Chain Integration and Trust on Supply Chain Performance: Evidence from Indonesia Retail Sector. *International Journal of Economics & Business Administration*, *IX*(1), 211–224. doi:10.35808/ijeba/668

Swam, M. (2015a). Blockchain thinking: The brain as a decentralised autonomous corporation. *IEEE Technology and Society Magazine*, *34*(4), 41–52. doi:10.1109/MTS.2015.2494358

Swam, M. (2015b). *Blockchain: Blueprint for a new economy*. O'Reilly.

Swan, M. (2015). *Blockchain: Blueprint for a new economy*. O'Reilly Media, Inc.

Szabo, N. (1997). *The Idea of Smart Contracts*. Retrieved from https://nakamotoinstitute.org/literature/the-idea-of-smart-c ontracts

Ta, N. B. D., & Yang, T. J. J. (2021). Ether Learn: Decentralizing learning via blockchain. *Proceedings of the IEEE TALE 2021*, 1-6. https://ink.library.smu.edu.sg/sis_research/6838

Tait, A. (2003). Management of services to students. In S. Panda (Ed.), *Planning and management in open and distance education*. Kogan Page.

Tama, B. A., Kweka, B. J., Park, Y., & Rhee, K. H. (2017). A critical review of blockchain and its current applications. *2017 International Conference on Electrical Engineering and Computer Science (ICECOS)*, 109-113. 10.1109/ICECOS.2017.8167115

Tan, B., Yan, J., Chen, S., & Liu, X. (2018). The Impact of Blockchain on Food Supply Chain: The Case of Walmart. In M. Qiu (Ed.), *Smart Blockchain* (pp. 167–177). Springer International Publishing. doi:10.1007/978-3-030-05764-0_18

Tang, Y., Xiong, J., Becerril-Arreola, R., & Iyer, L. (2020). Ethics of blockchain: A framework of technology, applications, impacts, and research directions. *Information Technology & People*, *33*(2), 602–632. doi:10.1108/ITP-10-2018-0491

Tapscott, D., & Kaplan, A. (2019). *Blockchain Revolution in Education and Lifelong Learning*. Blockchain Research Institute and IBM Institute for Business. https://www.ibm.com/downloads/cas/93DDVAKE

Tapscott, D., & Tapscott, A. (2018). *Blockchain Revolution*. Portfolio/Penguin. https://blockchain-revolution.com/

Tapscott, D., & Kaplan, A. (2019). *Blockchain revolution in education and lifelong learning: Preparing for disruption, leading the transformation*. Blockchain Research Institute and IBM Institute for Business Value. Retrieved from https://www.ibm.com/downloads/cas/93DDVAKE

Tapscott, D., & Tapscott, A. (2017). The blockchain revolution and her education. *EDUCAUSE Review*, *52*(2), 11–24.

Tapscott, D., & Tapscott, A. (2018). *Blockchain Revolution*. Portfolio Penguin. doi:10.1108/JHTT-05-2019-0069

Tekin, V. N. & Şenol, L. (2011). *İşletme, Kavramlar. İlkeler. İşlevler.* Seçkin Yayıncılık.

Tektaş, Ö. Ö. (2008). Endüstriyel Pazarlarda Bilgi Kaynaklarına Verilen Önemin Kültürel Farklılıklar Açısından Kavramsal Olarak İncelenmesi. *H. Ü. İktisadi ve İdari Bilimler Fakültesi*, *26*(2), 239-255.

Tengilimoğlu, D. & Öztürk, Y. (2011). *İşletmelerde Halkla İlişkiler.* Seçkin Yayıncılık.

Tenorio, E. M. (2021). *Advantages and disadvantages of Blockchain*. https://www.bbva.ch/en/news/advantages-and-disadvantages-of-blockchain/

Thake, M. (2018). *What's the difference between blockchain and DLT?* https://medium.com/nakamo-to/whats-the-difference-between-bl ockchain-and-dlt-e4b9312c75dd

Thayer, T. L. (2018). *4 Promising and ambitious blockchain initiatives for higher education*. Gartner Research. Retrieved from https://www.gartner.com/en/documents/3893582/4-promising-and -ambitious-blockchain-initiatives-for-hig

Thayer, T. L., & Yanckello, R. (2019). *Education Administrative Systems Innovation Primer for 2019*. Gartner, Inc. https://www.gartner.com/en/documents/3901065/education-admin istrative-systemsinnovation-primer-for-2

The Organisation for Economic Co-operation and Development. (2019). *Financing climate futures: blockchain technologies as a digital enabler for sustainable infrastructure. in financing climate futures: rethinking infrastructure initiative*. https://www.oecd.org/finance/Blockchain-technologies-as-a-di gital-enabler-for-sustainable-infrastructure-key-findings.pd f

Thejaswini, S., & Ranjitha, K. R. (2021). Blockchain for management of natural resources using energy trading as a platform. In *International Conference on Communication, Computing and Electronics Systems: Proceedings of ICCCES 2020* (*Vol. 733*, p. 475). Springer Nature. 10.1007/978-981-33-4909-4_36

Thomas, S. (2020, June 29). *Fundamentals of Blockchain, Data Driven Investor*. Retrieved 11 April 2022 from https://medium.datadriveninvestor.com/fundamentals-of-blockc hain-180de46b472f

Thomason, J. (2022). Fintech and Blockchain: Maximizing Benefit and Minimizing Harm. In I. Vasiliu-Feltes & J. Thomason (Eds.), *Applied Ethics in a Digital World* (pp. 185–198). IGI Global. doi:10.4018/978-1-7998-8467-5.ch013

Tian, F. (2017). A supply chain traceability system for food safety based on HACCP, blockchain amp; Internet of things. *2017 International Conference on Service Systems and Service Management*, 1–6. 10.1109/ICSSSM.2017.7996119

Tipton, H. F., & Krause, M. (2007). *Information Security Management Handbook*. Taylor&Francis Group. doi:10.1201/9781439833032

Tiwari, C.K., & Pal, A. (2022). Using blockchain for global governance: past, present and future. *South Asian Journal of Business Studies*. doi:10.1108/SAJBS-07-2022-0252

Tiwari, T. (2016, December 6). *Profit Alert: Walmart is adopting the blockchain right now...* Palm Beach Research Group. https://www.palmbeachgroup.com/palm-beach-daily/profit-alert -walmart-is-adopting-the-blockchain-right-now/

Tönnissen, S., & Teuteberg, F. (2020). Analysing the impact of blockchain-technology for operations and supply chain management: An explanatory model drawn from multiple case studies. *International Journal of Information Management*, *52*, 101953. doi:10.1016/j.ijinfomgt.2019.05.009

Tranfield, D., Denyer, D., & Smart, P. (2003). Towards a Methodology for Developing Evidence-Informed Management Knowledge by Means of Systematic Review. *British Journal of Management*, *14*(3), 207–222. doi:10.1111/1467-8551.00375

Treiblmaier, H. (2020). Blockchain and Tourism. In Z. Xiang, M. Fuchs, U. Gretzel, & W. Höpken (Eds.), Handbook of e-Tourism (pp. 1–21). Springer International Publishing. doi:10.1007/978-3-030-05324-6_28-1

Treiblmaier, H. (2018). The impact of the blockchain on the supply chain: A theory-based research framework and a call for action. *Supply Chain Management*, *23*(6), 545–559. doi:10.1108/SCM-01-2018-0029

Treiblmaier, H. (2021). *Need some inspiration for blockchain research? Try tokenization!* Academia Letters. doi:10.20935/AL2937

Treiblmaier, H., & Onder, I. (2019). The impact of blockchain on the tourism industry: A theory-based research framework. In H. Treiblmaier & R. Beck (Eds.), *Business Transformation through blockchain* (pp. 3–21). Palgrave Macmillan. doi:10.1007/978-3-319-99058-3_1

TreiblmaierH.ŠpanŽ. (2022). Will Blockchain Really Impact Your Business Model? Empirical Evidence from Slovenian SMEs. *Economic and Business Review, 24*(2), 132–140. doi:10.15458/2335-4216.1302

Tripathi, S., & Gupta, M. (2020). Transforming towards a smarter supply chain. *International Journal of Logistics Systems and Management*, *36*(3), 319–342. doi:10.1504/IJLSM.2020.108694

Ts, D., & v, R.TS. (2020). An integrated ANP–QFD approach for prioritization of customer and design requirements for digitalization in an electronic supply chain. *Benchmarking*, *28*(4), 1213–1246. doi:10.1108/BIJ-06-2020-0298

UGC. (2021, July 29). *The University Grants Commission (Establishment and Operation of Academic Bank of Credits in Higher Education) Regulations' 2021.* Retrieved from https://www.ugc.ac.in/pdfnews/1767992_Notice-Academic-Bank-of-Credits.pdf

UGC. (2022, January 3). *Acceptance of Degree, Mark-sheet and other educational documents presented through DigiLocker platform – reg.* New Delhi: University Grants Commission. Retrieved from https://www.ugc.ac.in/pdfnews/0790412_DigiLocker-Platform_001.pdf

UIAI. (n.d.). *What is Aadhar?* New Delhi: Unique Identification Authority of India, Government of India. Retrieved from https://uidai.gov.in/my-aadhaar/about-your-aadhaar.html

United Nations Framework Convention on Climate Change. (2016). *ICT Sector Helping to Tackle Climate Change*, https://www.unfccc.int/news/ict-sector-helping-to-tackle-climate-change

United Nations Population Fund. (2017). *World population trends*. https://www.unfpa.org/world-population-trends

University of North Dakota. (2021). *5 Blockchain Security Issues & Tips for Solving Them*. Retrieved from https://onlinedegrees.und.edu/blog/5-blockchain-security-issues/

Upah, G. D. (1980). *Applying the Concept of Perceived Risk to Buying Influence in Industrial Firms*. Assocation for Consumer Research.

Uysal, T. U., & Kurt, G. (2018). Muhasebede ve denetimde blok zinciri teknolojisi. *Suleyman Demirel Universitesi İktisadi ve İdari Bilimler Fakultesi Dergisi*, *23*(2), 467–481.

Vaccaro, S. (2019, April 2). *The Role of Blockchain in Fighting Chargeback and Ecommerce Fraud*. Chargeback. https://chargeback.com/the-role-of-blockchain-in-fighting-chargeback-and-ecommerce-fraud/

Vacchio, E. D., & Bifulco, F. (2022). Blockchain in Cultural Heritage: Insights from Literature Review. *Sustainability (Basel)*, *14*(4), 2324. doi:10.3390u14042324

Valeri, M., & Baggio, R. (2021). A critical reflection on the adoption of blockchain in tourism. *Information Technology & Tourism*, *23*(2), 121–132. doi:10.100740558-020-00183-1

van Hoek, R. (2019). Exploring blockchain implementation in the supply chain: Learning from pioneers and RFID research. *International Journal of Operations & Production Management*, *39*(6/7/8), 829–859. doi:10.1108/IJOPM-01-2019-0022

Varelas, S., Georgitseas, P., Nechita, F., & Sahinidis, A. (2019). Strategic Innovations in Tourism Enterprises Through Blockchain Technology. In A. Kavoura, E. Kefallonitis, & A. Giovanis (Eds.), *Strategic Innovative Marketing and Tourism* (pp. 885–891). Springer International Publishing. doi:10.1007/978-3-030-12453-3_102

Varghese, B., Villari, M., Rana, O., James, P., Shah, T., Fazio, M., & Ranjan, R. (2018). Realizing Edge Marketplaces: Challenges and Opportunities. *IEEE Cloud Computing.*, *5*(6), 9–20. doi:10.1109/MCC.2018.064181115

Varriale, V., Cammarano, A., Michelino, F., & Caputo, M. (2021). Sustainable Supply Chains with Blockchain, IoT and RFID: A Simulation on Order Management. *Sustainability (Basel)*, *13*(11), 6372. doi:10.3390u13116372

Vasarhelyi, M. A., Lombardi, D., & Bloch, R. (2014). The Future of Audit: A Modified Delphi Approach. In *Canadian Academic Accounting Association (CAAA) Annual Conference 2011* (pp. 1-31). Social Science Research Network (SSRN).

K. Verbert, M. Sharples, & T. Klobucar (Eds.). (2016). Adaptive and adaptable learning. In *Proceedings of the Conference on Technology Enhanced Learning*. Springer. 10.1007/978-3-319-45153-4_7

Verma, S., & Sheel, A. (2022). Blockchain for government organizations: Past, present and future. *Journal of Global Operations and Strategic Sourcing*, *15*(3), 406–430. doi:10.1108/JGOSS-08-2021-0063

Vodenicharova, M. S. (2020). Supply chain study in food industry in Bulgaria. *International Journal of Retail & Distribution Management*, *48*(9), 921–938. doi:10.1108/IJRDM-03-2019-0080

Vukolić, M. (2016). The Quest for Scalable Blockchain Fabric: Proof-of-Work vs. BFT Replication. In J. Camenisch & D. Kesdoğan (Eds.), Lecture Notes in Computer Science: Vol. 9591. *Open Problems in Network Security. iNetSec 2015*. Springer. doi:10.1007/978-3-319-39028-4_9

Waldo, J. (2019). A Hitchhiker's Guide to the Blockchain Universe? Blockchain remains a mystery, despite its growing acceptance. *ACM*, *16*(6), 38-42. Retrieved from https://queue.acm.org/detail.cfm?id=3305265

Wamba, S. F., Gunasekaran, A., Akter, S., Ren, S. J., Dubey, R., & Childe, S. J. (2017). Big data analytics and firm performance: Effects of dynamic capabilities. *Journal of Business Research*, *70*, 356–365. doi:10.1016/j.jbusres.2016.08.009

Wamba, S. F., & Queiroz, M. M. (2020). Industry 4.0 and the supply chain digitalisation: A blockchain diffusion perspective. *Production Planning and Control*, *0*(0), 1–18. doi:10.1080/09537287.2020.1810756

Wan, H., Li, K., & Huang, Y. (2021). Blockchain: A Review from The Perspective of Operations Researchers. 2020 Winter Simulation Conference (WSC), 75-89. doi:10.1109/WSC48552.2020.9383924

Wang, H., Chen, K., & Xu, D. (2016, December). A maturity model for blockchain adoption. *Financial Innovation*, *2*(1), 12. doi:10.118640854-016-0031-z

Wang, Y., & Kogan, A. (2018). Designing confidentiality-preserving Blockchain-based transaction processing systems. *International Journal of Accounting Information Systems*, *30*, 1–18. doi:10.1016/j.accinf.2018.06.001

Wang, Y., Singgih, M., Wang, J., & Rit, M. (2019). Making sense of blockchain technology: How will it transform supply chains? *International Journal of Production Economics*, *211*, 221–236. doi:10.1016/j.ijpe.2019.02.002

Webster, F. E. Jr. (1965). Modelling the Industrial Buying Process. *JMR, Journal of Marketing Research*, *II*(4), 370–376. doi:10.1177/002224376500200404

Westby, J. R. (2004). International Guide to Cyber Security. American Bar Association (ABA).

Wharton School of the University of Pennsylvania. (2018, November 28). *How the Blockchain Brings Social Benefits to Emerging Economies.* Wharton School of the University of Pennsylvania. https://knowledge.wharton.upenn.edu/article/blockchain-brings-social-benefits-emerging-economies/

Williams, P. (2019). Does competency-based education with blockchain signal a new mission for universities? *Journal of Higher Education Policy and Management*, *41*(1), 104–117. doi:10.1080/1360080X.2018.1520491

Winstone, N. E., Ajjawi, R., Dirkx, K., & Boud, D. (2021). Measuring what matters: The positioning of students in feedback processes within national student satisfaction surveys. *Studies in Higher Education*, *47*(7), 1524–1536. doi:10.1080/03075079.2021.1916909

Wong, L.-W., Tan, G. W.-H., Lee, V.-H., Ooi, K.-B., & Sohal, A. (2020). Unearthing the determinants of Blockchain adoption in supply chain management. *International Journal of Production Research*, *58*(7), 2100–2123. doi:10.1080/00207543.2020.1730463

Woodard, H. C., & Orr, R. L. (2015). Green technology for green schools. In *Marketing the green School: Form, function, and the future* (pp. 106–117). IGI Global. doi:10.4018/978-1-4666-6312-1.ch007

Woodside, J. M., Augustine, F. K. Jr, & Giberson, W. (2017). Blockchain technology adoption status and strategies. *Journal of International Technology and Information Management*, *26*(2), 65–93. doi:10.58729/1941-6679.1300

World Economic Forum. (2018). *Building Block(chain)s for a better planet.* The World Economic Forum Report 2018. Retrieved from https://www.weforum.org/reports?year=2017#filter

Xiao, Y., & Watson, M. (2019). Guidance on Conducting a Systematic Literature Review. *Journal of Planning Education and Research*, *39*(1), 93–112. doi:10.1177/0739456X17723971

Xu, X., Pautasso, C., Zhu, L., Gramoli, V., Ponomarev, A., Tran, A. B., & Chen, S. (2016). The Blockchain as a Software Connector. *2016 13th Working IEEE/IFIP Conference on Software Architecture (WICSA)*, 182–191. 10.1109/WICSA.2016.21

Yamada, Y., Nakajima, T., & Sakamoto, M. (2016). Blockchain-LI: a study on implementing activity-based micro-pricing using cryptocurrency technologies. In *Proceedings of the 14th International Conference on Advances in Mobile Computing and Multi Media* (pp. 203-207). https://doi.org/10.1145/3007120.3007151

Yan, C., Zhu, J., Ouyang, Y., & Zeng, X. (2021). Marketing Method and System Optimization Based on the Financial Blockchain of the Internet of Things. *Wireless Communications and Mobile Computing*, *2021*, 1–11. doi:10.1155/2021/9354569

Yang, M., Kuo, C., & Yeh, Y. (2011). Dynamic rightsizing with quality-controlled algorithms in virtualization environments. *International Journal of Grid and High Performance Computing*, *3*(2), 29–43. doi:10.4018/jghpc.2011040103

Yermack, D. (2019). Blockchain technology's potential in the financial system. *Blockchain Technology*, 20.

Yıldırım, H. (2018). Açık ve uzaktan öğrenmede blokzincir teknolojisinin kullanımı. *AUAd*, *4*(3), 142–153.

Yılmaz, K. M., & Erçiş, A. (2012). Endüstriyel Pazarlarda Kişisel ve Kişisel Olmayan Bilgi Kaynaklarının Marka Değeri Üzerindeki Etkilerinin İncelenmesi. *Uluslararası Alanya İşletme Fakültesi Dergisi*, *4*, 27–44.

Yuan, Y. & Wang, F.–Y. (2018). Blockchain and cryptocurrencies: Model, techniques, and applications, *IEEE Transactions on Systems, Man, and Cybernetics: Systems*, *48*(9), 1,421–1,428. doi:10.1109/TSMC.2018.2854904

Zaidi, T. (2022). Blockchain: Emerging Trends, Applications, and Challenges. In S. Khan, M. Syed, R. Hammad, & A. Bushager (Eds.), *Blockchain Technology and Computational Excellence for Society 5.0* (pp. 189–203). IGI Global. doi:10.4018/978-1-7998-8382-1.ch009

Zantout, H., & Seyff, N. (2018). *Green ict for sustainable development.* https://www.goodfellowpublishers.com/free_files/Chapter%206-600bcc6f6d9ccd4d6521451487b23bfe.pdf

Zeng, F., Chan, H. K., & Pawar, K. (2021). The effects of inter- and intraorganizational factors on the adoption of electronic booking systems in the maritime supply chain. *International Journal of Production Economics*, *236*, 108119. doi:10.1016/j.ijpe.2021.108119

Zet, C., & Dumitriu, G. C. (2021). Using blockchain technology for ensuring students results traceability for instrumentation classes. *Measurement: Sensors*, *18*, 100315. doi:10.1016/j.measen.2021.100315

Zhang, Y., Xiong, F., Xie, Y., Fan, X., & Gu, H. (2020). The impact of artificial intelligence and blockchain on the accounting profession. *IEEE Access : Practical Innovations, Open Solutions*, *8*, 110461–110477. doi:10.1109/ACCESS.2020.3000505

Zhang, Z., Ma, M., Popkowski Leszczyc, P. T. L., & Zhuang, H. (2020). The influence of coupon duration on consumers' redemption behavior and brand profitability. *European Journal of Operational Research*, *281*(1), 114–128. doi:10.1016/j.ejor.2019.08.029

Zhao, J.L., Fan, S., & Yan, J. (2016). Overview of business innovations and research opportunities in blockchain and introduction to the special issue. *Financ Innov*, *2*(28). doi:10.1186/s40854-016-0049-2

Zhao, W. (2018, March 15). *Anheuser-Busch Owner Pilots Blockchain for Shipping.* CoinDesk. https://www.coindesk.com/anheuser-busch-owner-pilots-blockchain-shipping

Zheng, Z., Xie, S., Dai, H.-N., Chen, X., & Wang, H. (2017, June 25). An Overview of Blockchain Technology: Architecture, Consensus, and Future Trends. *IEEE International Congress on Big Data (BigData Congress)*. 10.1109/BigDataCongress.2017.85

Zheng, Z., Xie, S., Dai, H. N., Chen, X., & Wang, H. (2018). Blockchain challenges and opportunities: A survey. *International Journal of Web and Grid Services, 14*(4), 352–375. doi:10.1504/IJWGS.2018.095647

Zhong, J., Xie, H., Zou, D., & Chui, D. K. (2018). *A blockchain model for word-learning systems.* Paper presented at the 2018 5th International Conference on Behavioral, Economic, and Socio-Cultural Computing (BESC), Kaohsiung City, Taiwan. 10.1109/BESC.2018.8697299

Zhong, H., & Nof, S. Y. (2015). The dynamic lines of collaboration model: Collaborative disruption response in cyber–physical systems. *Computers & Industrial Engineering, 87*, 370–382. doi:10.1016/j.cie.2015.05.019

Zhou, H., de Laat, C., & Zhao, Z. (2018). Trustworthy cloud service level agreement enforcement with blockchain based smart contract. *2018 IEEE International Conference on Cloud Computing Technology and Science (cloudcom),* 255–260. 10.1109/CloudCom2018.2018.00057

About the Contributors

Gulsun Kurubacak is a professor in Distance Education at the College of Open Education of Anadolu University. She undertook graduate studies at Anadolu University, Turkey (MA. Educational Technology) and the University of Cincinnati, USA (Ed.D. Curriculum & Instruction), and also has worked a post-doctoral fellow at the College of Education at New Mexico State University, USA (2001-2002). Dr. Kurubacak earned her B.S. degree in Computer Engineering from the College of Informatics Technologies and Engineering of Ahmet Yesevi University in the year 2012-2013. She has over thirty-five years of experience in focusing on the egalitarian and ecological aspects of open and distance learning through new communication technologies. She had been the Group Coordinator of R&D and International Relations at the College of Open Education of Anadolu University between 2014-2018. She is the founder and editor-in-chief of AUAd as well as associated editor of TOJDE. Dr. Kurubacak is also the founding member of Open and Distance Education Programs Accreditation Agency (AUDAK), and has the head of the board of management member at the same accreditation agency since 2017.

Ramesh Chander Sharma, Ph.D., is currently working as an Associate Professor of Instructional Design at Ambedkar University Delhi, India. Previously he has served as an Associate Professor of Educational Technology and Learning Resources in the Educational Technology Unit at Wawasan Open University, Malaysia; Visiting Professor at University of Fiji, Fiji; Commonwealth of Learning as Director of the Commonwealth Educational Media Centre for Asia, New Delhi; Indira Gandhi National Open University, India; and University of Guyana, Guyana, South America. He is the co-Editor of the 'Asian Journal of Distance Education'. In addition, he has been associated with peer reviewed journals as Reviewer, Editor and Editorial Advisory Board member in the field of Open and Distance Learning such as "Distance Education", " International Review of Research in Open and Distributed Learning (IRRODL)", "International Journal of Distance Education Technologies (IJDET)", and "Indian Journal of Open Learning (IJOL)". He is also on the Editorial Advisory Board and an author for the "Encyclopedia of Distance Learning (4 volumes), 2005" ().

Hakan Yıldırım is a lecturer in the Department of Computer Programming, at Eskisehir Osmangazi University, Turkey. He was a member of the Center for Distance Education at the same university. He was a visiting Ph.D. scholar in the School of Engineering Education, Purdue University, the USA in 2021-2022 as a TUBITAK fellow. He received his B.A. in Computer Education and Instructional Technology from Anadolu University, Eskisehir, Turkey in 2007. He received his M.A. degree in the Department of Distance Education in 2017 at Anadolu University and is a Ph.D. candidate in the same department. He has more than 10 articles published in refereed journals and conferences. He is a co-founder and the CEO of Asystee EdTech Ltd., an educational technology start-up that aims to apply emerging technologies in open and distance learning, located at the Anadolu University Technopark funded by TUBITAK in 2020. The company's main operations include designing, developing, and integrating sustainable e-learning environments in the cloud or on-premises.

* * *

Syed Imran Ali is a Senior Lecturer of Logistics and Supply Chain Management at Huddersfield Business School, University of Huddersfield. Imran engages widely with the industry and has research interests in digitalisation of supply chains, supply chain data analytics, industry 4.0 applications (AI, IoT, Blockchain, Digital Twin) and innovation. He is currently conducting research in the usage of technologies broadly in the logistics and supply chain to make the data-driven decisions with more accuracy and efficiency. He has published in academic and managerial journals including International Journal of Production Economics, International Journal of Production Research, International Business Review, Computers and Operations Research.

Mehtap Altunel was born in Artvin in 1989. After completing her primary and secondary education in Artvin, she graduated from Akdeniz University, the School of Economics and Administrative Sciences, Faculty of Business in 2013. She completed her master's degree in Recep Tayyip Erdoğan University, Institute of Social Sciences, Department of Business Administration in 2018. She was accepted doctoral program in Anadolu University, Institute of Social Sciences, Department of Accounting in 2018. Currently, she continues doctoral program in Anadolu University. She is currently at the dissertation stage and the doctoral process is still ongoing. She has published papers in national and international symposiums, articles and book chapters auditing in the fields of independent, internal auditing and accounting.

Anjana working as Assistant Regional Director at Indira Gandhi National Open University, India has the professional experience of more than a decade in Open and Distance Education System. Her academic qualifications include Master of Arts in Distance Education, Doctorate in Botany (Plant Physiology) and Masters of Science in Environmental Botany. She is the Gold Medalist at Masters' Degree level and Merit holder at Bachelors' Degree level and also an awardee of Postgraduate merit scholarship. She has been the Junior Research Fellow, Senior Research Fellow as well as Research Associate of Council of Scientific and Industrial Research (India), a premier national Research and Development organization. With more than 20 research papers and review articles published in journals of national and international repute as well as many book chapters, she has 3 books to her credit. She has participated and presented research papers in various national and international conferences, symposia and webinars. She has been the resource person for various workshops and professional development programmes organized by Teaching Learning Centres for teachers and academics. She has also translated various lectures of Cell Culture Technologies for NPTEL (National Programme on Technology Enabled Learning) and completed various MOOCs, Professional Development Programmes and Online Courses. Her areas of interest include Technology-Enabled Learning, Open and Distance Learning, Inclusive Education, Technology integration in distance learning, etc.

Ayla Avci graduated from Anadolu University, Department of Business Administration. She completed her master's thesis with her thesis on nepotism at Istanbul Sabahattin Zaim University. In 2021, she completed his doctoral thesis with her research on the workload and job boredom of employees. There are articles and book chapters in many national and international journals.

Selin Göçen is a research assistant at Dicle University, Faculty of Education. She completed her undergraduate education at Hacettepe University (Department of Primary Education) and master's degree at Dicle University (Department of Educational Administration). Now she is a Ph.D. student at Anadolu University, Department of Distance Education. She works with children and prospective teachers.

Stelios A. K. Ioannidis is a PhD candidate and Researcher at the University of the Aegean in Greece. He specializes in the field of Metaverse Tourism and Human Behaviour in Metaverse, and has a proven track record of conducting cutting-edge research in this emerging field. Stelios holds a Bachelor's degree in Medical Laboratories from the University of Thessaly, as well as two Master's Degrees in Tourism Management from the University of the Aegean and the University of Thessaly, which uniquely position him to study the intersection of tourism marketing and biometrics research methods in Metaverse environments. Within his PhD research, he describes the concept of phygital (physical + digital) tourism destinations from a tourism marketing perspective. Stelios has also gained valuable practical experience in the tourism industry, having worked as a reservations manager and marketing manager for several hotels and yacht charter companies. He brings this real-world expertise to his academic work, which focuses on using marketing research methods to better understand human behaviour within Metaverse spaces. In recognition of his expertise in the field, Stelios has been appointed as a core group leader at the World Metaverse Council, where he designs and conducts cutting-edge research on Metaverse adoption. He is deeply committed to bridging the gap between academia and industry, and is passionate about exploring new and innovative ways to approach Tourism in the Metaverse.

Arnab Kundu, an independent researcher based in West Bengal, India, has received Master's degrees in English and Education, M.Phil. & Ph. D. in Education. He stood all India topper in CUCET-2018 for Ph.D. Entrance in Education. His research concentrates on different issues of teachers' work especially in promoting technology-integrated pedagogy at the school level. So far his 30 research papers published in SCOPUS-indexed international journals.

Ashraf Mady is an associate professor in the Computer Science and Information Systems Department at the University of North Georgia. He is also the Information Systems Curriculum Chair at the Mike Cottrell College of Business (MCCB). Dr. Mady also served as the head of the Computer Science and Information Systems Department in the MCCB at the University of North Georgia from 2017 until 2021. During his term as a department head, he successfully led the department to acquire ABET program accreditation, launch several new academic programs including the undergraduate degree of Cyber security, the Graduate Certificate of Cyber Security, Minor in Machine Learning, and many more significant milestone achievements for the department to support students' success. His research interest and active projects are in the areas of Cybersecurity, Artificial Intelligence, and Financial Technology or Fintech. He has led several projects and awarded grants in work related to cybersecurity, blockchain, technology adoption, and artificial intelligence. He is a member of the Special Interest Group on Information Security and Privacy. He has over twenty-five years of professional experience in academia, management, and software engineering. Mady received his Ph.D. degree in business administration with a research focus in information security from Kennesaw State University. Dr.

Mady also earned a Master of Science in Information Systems, Bachelor of Science in Computer Science, and a Bachelor of Science in Chemistry and Physics.

Mohan Menon is a Professor of Marketing and the Department Head of the Management and Marketing department in the Mike Cottrell College of Business at the University of North Georgia. Prior to this, Dr. Menon was the Chair of the Department of Marketing at University of South Alabama. Dr. Menon received his DBA from University of Memphis in the areas of Marketing and International Business. His teaching interests include marketing, consumer behavior, advertising, and digital marketing. While his scholarship interest dovetails areas of teaching, he has explored various technology-based application research in the marketing space. His recent co-authored publication in Blockchain is one of them. Currently, he is involved research the use of TikTok as a learning engagement platform and the use of ChatGPT in student learning.

Siran Mukerji, a Jawahar Lal Nehru scholar for Doctoral Studies in Human Resource Development, has Masters in Distance Education and Public Administration. She has been International Research Fellow of Open University Business School (2009), Open University (UK). At Arab Open University Saudi Arabia, she was a faculty member in Business Administration. She is one of the Editors-in-Chief of International Journal of Technology and Educational Marketing and Author/Editor of Teaching Case books on Innovations in Educational Marketing, Interactive Technology Environments, Technology Enhanced Learning, Transnational Learning & Technologically enabled Environments, and Technological Adaptability and Transnational Learning. She has contributed articles in standard national and international journals and presented papers in national and international conferences. Dr. Mukerji is a member of review committees for numerous international conferences and journals. Her research interests include Performance Management and HRM in Open and Distance Learning institutions. As Additional Director in IGNOU (India), she is responsible for student recruitment and related support services management .

Siran Mukerji has been performing multifaceted roles of an educator, trainer, and facilitator, responsible for key roles in learner recruitment management, student performance assessment process, and training, skill development and competency enhancement of higher education professionals. She is presently Additional Director, responsible for student performance assessment and related support services management in Indira Gandhi National Open University, New Delhi (India). Actively engaged in academic research and publication, she has more than 12 books to her credit on Interactive Technology Environments in Higher Education, and Transnational Collaborations and Learning. Her current research interests include

Dimensions of Human Resource Development and Performance Management, and Higher Education Management.

Kalpana Muthireddi is a dual licensed business lawyer, licensed to practice in India and US (New York Bar) She is a technology, commercial and outsourcing transactions attorney, with experience handling legal matters in emerging and cutting edge technologies viz., Blockchain, cloud computing (XaaS agreements) & other IT space enterprises and Biotechnology (adult stem cell based therapies). She is also the author of – "LEGALESE Decoded – Tips and Strategies To Make Sense of Legal Concepts in The Business World". She advises start-ups, small and medium industries on legal, contract and compliance matters.

S. K. Pulist got his Master's Degree in the disciplines of Sociology, English Literature, Management Studies and Distance Education. He acquired his Ph D Degree in Education Technology from Jamia Millia Islamia, New Delhi. He is a Certified Trainer (a member of India Society for Training and Development). He possesses the PG Diploma in eLearning from IGNOU. He has completed the Study "eLearning in Commonwealth Asia 2013"and "OER Policy Implementation and Use in Open and Distance Learning" published by Commonwealth Educational Media Centre for Asia (Commonwealth of Learning). He has co-edited a book entitled "Education in the Digital World" and has published a number of articles in national and international journals. He has been associated with Indian Journal of Open Learning as Co-Editor during the period 2011-2018 and 2021 onwards. He is involved with the ODL activities for the last over 28 years. Currently, he is posted as Deputy Director in Student Registration Division, IGNOU, New Delhi.

Seval Selimoğlu is a Professor in the Faculty of Economics and Administrative Sciences, Department of Business at Anadolu University, Turkey. She completed her master's degree in Anadolu University, Institute of Social Sciences, Department of Accounting in 1987. She obtained International MBA degree in the United States of America in Baldwin-Wallace University in 1990. She completed her doctorate degree in Anadolu University in 1995. She has received the title of Assistant Professor Doctor in the field of Accounting-Finance in 1995, Associate Professor Doctor in 2002 and Professor in 2007. Her main academic studies are in the fields of Independent Audit, Ethics, Accounting Fraud, Corruption, Internal audit, Financial Statements Analysis and Accounting. She has lots of books, articles, assertions and studies on EU projects published nationally and internationally. She is serving as member of The Governmental Accounting Standards Board, the Certified Public Accountant and Independent Auditor and serves as expert in courts relate to her area of specialization.

Gül Yeşilçelebi is an Associate Professor in the Academy of Applied Sciences, Department of Aviation Management at Gumushane University, Turkey. Her research activities lie in the areas of independent audit, internal audit, financial accounting, corporate sustainability, and integrated reporting. She graduated in Business Administration in English within the Faculty of Economics and Administrative Sciences, and, afterward, she obtained a master's degree in Accounting. She received her Ph.D. in Department of Accounting from the Anadolu University, Turkey. She wrote her thesis on integrated reporting and combined assurance. She has published a series of papers on her research interests, and she has presented papers in various conferences. She currently teaches financial accounting, corporate accounting, cost accounting, managerial accounting, and financial statement analysis, in undergraduate degrees at Gumushane University. Also, she teaches internal audit and risk management and independent auditing standards in master's degrees, sustainability, and integrated reporting doctoral degrees at Gumushane University.

Xinyu Zheng received the B.S. degree in Logistics Management from Shandong University, China, in 2020 and the M.S. degree in Logistics and Supply Chain Management from Cranfield University, UK, in 2021. Her research interests include IoT, Blockchain and Emergency logistics.

Index

E

F

G

I

J

L

M

N

O

P

R

S

T

Recommended Reference Books

IGI Global's reference books are available in three unique pricing formats:
Print Only, E-Book Only, or Print + E-Book.

Order direct through IGI Global's Online Bookstore at **www.igi-global.com** or through your preferred provider.

ISBN: 9781799884552
EISBN: 9781799884576
© 2022; 394 pp.
List Price: US$ **270**

ISBN: 9781799897101
EISBN: 9781799897125
© 2022; 335 pp.
List Price: US$ **250**

ISBN: 9781668441398
EISBN: 9781668441411
© 2022; 291 pp.
List Price: US$ **270**

ISBN: 9781668441534
EISBN: 9781668441558
© 2023; 335 pp.
List Price: US$ **270**

ISBN: 9781799885160
EISBN: 9781799885184
© 2022; 240 pp.
List Price: US$ **270**

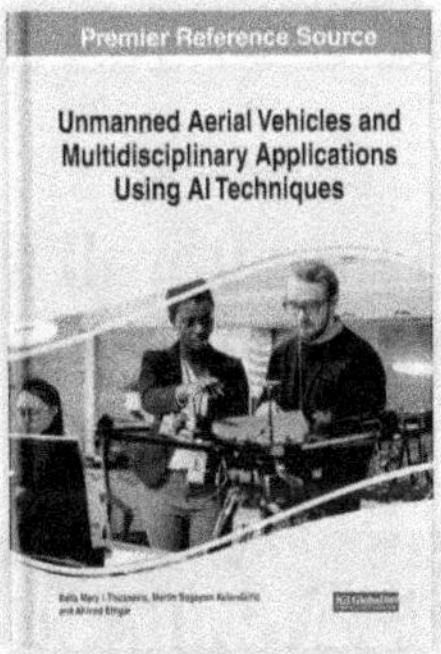

ISBN: 9781799887638
EISBN: 9781799887652
© 2022; 306 pp.
List Price: US$ **270**

Do you want to stay current on the latest research trends, product announcements, news, and special offers?
Join IGI Global's mailing list to receive customized recommendations, exclusive discounts, and more.
Sign up at: **www.igi-global.com/newsletters.**

Publisher of Timely, Peer-Reviewed Inclusive Research Since 1988

IGI Global
PUBLISHER of TIMELY KNOWLEDGE

www.igi-global.com Sign up at www.igi-global.com/newsletters facebook.com/igiglobal twitter.com/igiglobal

Ensure Quality Research is Introduced to the Academic Community

Become an Evaluator for IGI Global Authored Book Projects

The overall success of an authored book project is dependent on quality and timely manuscript evaluations.

Applications and Inquiries may be sent to:
development@igi-global.com

Applicants must have a doctorate (or equivalent degree) as well as publishing, research, and reviewing experience. Authored Book Evaluators are appointed for one-year terms and are expected to complete at least three evaluations per term. Upon successful completion of this term, evaluators can be considered for an additional term.

If you have a colleague that may be interested in this opportunity, we encourage you to share this information with them.

Easily Identify, Acquire, and Utilize Published Peer-Reviewed Findings in Support of Your Current Research

IGI Global OnDemand

Purchase Individual IGI Global OnDemand Book Chapters and Journal Articles

For More Information:
www.igi-global.com/e-resources/ondemand/

Browse through 150,000+ Articles and Chapters!

Find specific research related to your current studies and projects that have been contributed by international researchers from prestigious institutions, including:

Massachusetts Institute of Technology
HARVARD UNIVERSITY
COLUMBIA UNIVERSITY IN THE CITY OF NEW YORK
Australian National University

- Accurate and Advanced Search
- Affordably Acquire Research
- Instantly Access Your Content
- Benefit from the InfoSci Platform Features

“*It really provides* **an excellent entry into the research literature of the field**. *It presents a manageable number of* **highly relevant sources** *on topics of interest to a wide range of researchers. The sources are* **scholarly, but also accessible** *to ‘practitioners’.*”

- Ms. Lisa Stimatz, MLS, University of North Carolina at Chapel Hill, USA

Interested in Additional Savings?

Subscribe to
IGI Global OnDemand *Plus*

Learn More

Acquire content from over 128,000+ research-focused book chapters and 33,000+ scholarly journal articles for as low as US$ 5 per article/chapter (original retail price for an article/chapter: US$ 37.50).

7,300+ E-BOOKS. ADVANCED RESEARCH. INCLUSIVE & AFFORDABLE.

IGI Global
PUBLISHER of TIMELY KNOWLEDGE

IGI Global e-Book Collection

- **Flexible Purchasing Options** (Perpetual, Subscription, EBA, etc.)
- Multi-Year Agreements with **No Price Increases** Guaranteed
- **No Additional Charge** for Multi-User Licensing
- No Maintenance, Hosting, or Archiving Fees
- Continually Enhanced & Innovated **Accessibility Compliance Features** (WCAG)

Handbook of Research on Digital Transformation, Industry Use Cases, and the Impact of Disruptive Technologies
ISBN: 9781799877127
EISBN: 9781799877141

Handbook of Research on New Investigations in Artificial Life, AI, and Machine Learning
ISBN: 9781799886860
EISBN: 9781799886877

Handbook of Research on Future of Work and Education
ISBN: 9781799882756
EISBN: 9781799882770

Research Anthology on Physical and Intellectual Disabilities in an Inclusive Society (4 Vols.)
ISBN: 9781668435427
EISBN: 9781668435434

Innovative Economic, Social, and Environmental Practices for Progressing Future Sustainability
ISBN: 9781799895909
EISBN: 9781799895923

Applied Guide for Event Study Research in Supply Chain Management
ISBN: 9781799889694
EISBN: 9781799889717

Mental Health and Wellness in Healthcare Workers
ISBN: 9781799888130
EISBN: 9781799888147

Clean Technologies and Sustainable Development in Civil Engineering
ISBN: 9781799898108
EISBN: 9781799898122

Request More Information, or Recommend the IGI Global e-Book Collection to Your Institution's Librarian

For More Information or to Request a Free Trial, Contact IGI Global's e-Collections Team: eresources@igi-global.com | 1-866-342-6657 ext. 100 | 717-533-8845 ext. 100

Are You Ready to **Publish Your Research ?**

IGI Global
PUBLISHER of TIMELY KNOWLEDGE

IGI Global offers book authorship and editorship opportunities across 11 subject areas, including business, computer science, education, science and engineering, social sciences, and more!

Benefits of Publishing with IGI Global:

- Free one-on-one editorial and promotional support.
- Expedited publishing timelines that can take your book from start to finish in less than one (1) year.
- Choose from a variety of formats, including Edited and Authored References, Handbooks of Research, Encyclopedias, and Research Insights.
- Utilize IGI Global's eEditorial Discovery® submission system in support of conducting the submission and double-blind peer review process.
- IGI Global maintains a strict adherence to ethical practices due in part to our full membership with the Committee on Publication Ethics (COPE).
- Indexing potential in prestigious indices such as Scopus®, Web of Science™, PsycINFO®, and ERIC – Education Resources Information Center.
- Ability to connect your ORCID iD to your IGI Global publications.
- Earn honorariums and royalties on your full book publications as well as complimentary content and exclusive discounts.

Join Your Colleagues from Prestigious Institutions, Including:

Australian National University
Massachusetts Institute of Technology
Johns Hopkins University
Harvard University
Tsinghua University
Columbia University in the City of New York

Learn More at: www.igi-global.com/publish

or by Contacting the Acquisitions Department at: acquisition@igi-global.com

Individual Article & Chapter Downloads

US$ 29.50/each

Easily Identify, Acquire, and Utilize Published Peer-Reviewed Findings in Support of Your Current Research

- Browse Over ***170,000+ Articles & Chapters***
- ***Accurate & Advanced*** Search
- Affordably Acquire ***International Research***
- ***Instantly Access*** Your Content
- Benefit from the ***InfoSci® Platform Features***

THE UNIVERSITY *of* NORTH CAROLINA *at* CHAPEL HILL

"*It really provides **an excellent entry into the research literature of the field**. It presents a manageable number of **highly relevant sources** on topics of interest to a wide range of researchers. The sources are **scholarly, but also accessible** to 'practitioners'.*"

- Ms. Lisa Stimatz, MLS, University of North Carolina at Chapel Hill, USA

Interested in Additional Savings?

Subscribe to

IGI Global OnDemand Plus

Learn More

Acquire content from over 128,000+ research-focused book chapters and 33,000+ scholarly journal articles for as low as US$ 5 per article/chapter (original retail price for an article/chapter: US$ 37.50).

IGI Global
PUBLISHER of TIMELY KNOWLEDGE

IGI Global Proudly Partners with

eContent Pro®
International

Editorial Services

Providing you with High-Quality, Affordable, and Expeditious Editorial Support from Manuscript Development to Publication

Copy Editing & Proofreading

Perfect your research paper before publication. Our expert editors will correct faulty spelling, grammar, punctuation, and word usage.

Scientific & Scholarly Editing

Increase your chances of being published. Our expert editors will aid in strengthening the quality of your research before submission.

Figure, Table, Chart & Equation Conversions

Enhance the visual elements of your research. Let our professional designers produce or correct your figures before final submission.

Journal Recommendation

Save time and money when you rely on our expert journal selectors to provide you with a comprehensive journal recommendation report.

Order now to receive an automatic **10% Academic Discount** on all your editorial needs.

Scan the QR Code to Learn More

Upload Your Manuscript, Select Your Desired Editorial Service, and Receive a Free Instant Quote

Email: customerservice@econtentpro.com

econtentpro.com